An Annotated Checklist of Southeast Asian Begonia

MARK HUGHES

ROYAL BOTANIC GARDEN EDINBURGH
2008

ISBN 978-1-906129-14-9

Royal Botanic Garden Edinburgh
20A Inverleith Row
Edinburgh, EH3 5LR
United Kingdom

Designed and typeset by Paul Barrett Book Production, Cambridge
Printed at the University Press, Cambridge

Contents

Foreword

It might be thought, perhaps by those for whom the name *Begonia* conjures up images of garden centre displays packed with those most familiar and popular plants for home and garden, that no great fanfare should announce the publication of this checklist of southeast Asian *Begonia*. Such a view would be entirely wrong. This is a publication of great significance. *Begonia* is one of the largest genera of flowering plants. I can think of no other popular horticultural genus in which, despite the seemingly endless assortment of cultivars available, the number of species is such that the diversity in the wild so greatly exceeds that in cultivation. Documenting this diversity is therefore an important task and one that is made more urgent by the destruction, around the world, of habitats favoured by begonias, placing many species at risk in the wild. This new checklist resolves for each species its correct name, following the International Code of Botanical Nomenclature, and lists names known to be synonyms. Of primary importance is the fact that the list also cites types and other specimens, and thus provides much needed clarification on species delimitation and distribution. It also provides a preliminary assessment of the status of each species in the wild according to the internationally agreed Red List categories of the World Conservation Union (IUCN). This shows that of the 521 species listed here, one species is extinct in the wild and a further 119 are considered vulnerable, endangered or critically endangered. It is reassuring that a further 122 species are classified as of 'least concern', with only 13 listed as 'near threatened', but it should be noted that for 266 species there is currently insufficient information even to make a preliminary assessment of their status in the wild. *The Global Strategy for Plant Conservation* formally adopted under the *Convention on Biological Diversity* (CBD) has, as its second target, 'A preliminary assessment of the conservation status of all known plant species, at national, regional and international levels'. It is obvious why knowing the conservation status of plants is so fundamental to their conservation but it is a sad fact that for far too many plant species we know very little about how they are doing in the wild. Such is the extent of our ignorance that even though roughly half of the *Begonia* species listed here are 'data deficient' and therefore cannot be assessed, this is one of the best examples of conservation status assessment in recent years. Highlighting how little we know is one way of drawing attention to the opportunities for important and interesting research that remains to be done. It is the hope of the author that the publication of the checklist will serve to stimulate new research on southeast Asian *Begonia*. I have no doubt that it will have the desired effect and congratulate Mark Hughes on a timely and important publication. I would also like to take the opportunity to express my appreciation to the M.L. MacIntyre *Begonia* Trust who have done so much to support this study and other projects that the Royal Botanic Garden Edinburgh has been delighted to be part of.

Stephen Blackmore FRSE
Regius Keeper

Acknowledgements

This work was supported by the M.L. MacIntyre *Begonia* Trust, the Sibbald Trust, SYNTHESYS (awards NL-TAF 1608, FR-TAF 1416 and DE-TAF 2181) and the Rural and Environment Research and Analysis Directorate. The list would not have been possible to complete in the time available were it not for the groundwork carried out by J. Golding, D. C. Wasshausen, L. B. Smith and C. E. Karegeannes on the global species list for *Begonia*, and the work on the sectional placement of *Begonia* species by J. Doorenbos, M. S. M. Sosef and J. J. F. E. de Wilde. I am also grateful to: the curators of the herbaria A, B, BM, BO, E, FI, K, L, P, SING, U and WAG for letting me consult their specimens; the staff of the RBGE library and J. Golding for helping me locate protologues; R. Kiew for information on the Vietnamese species and reviewing the manuscript; D. Middleton, M. Tebbitt and M. Sands for reviewing the manuscript; R. Rubite for assistance with conservation assessments on the Philippine species; V. Lisewski and S. Bachman for assisting with conservation assessments; J. Gagul for information on sect. *Symbegonia*; A. Dorward for the plant illustrations; E. Schwarz for proofreading the manuscript; H. Adamson and C. Mouat for assistance during publication, and M. Pullan for tireless technical support.

Introduction

Begonia is one of the largest angiosperm genera, containing around 1500 species with many more yet to be described. Approximately 150 species occur in Africa, with the remaining species being split almost equally between the American and Asian tropics. The *Begonia* of southeast Asia are amongst the most taxonomically neglected, and the purpose of this checklist is to stimulate the production of much needed floras through the provision of regional species lists. The lists are based almost entirely on specimen data, with the aim of making them verifiable and updateable. The vast majority of the 5827 specimens cited, including types, are available on-line as digital photographs. Images of protologues for each species have also been made accessible on-line, copyright permitting, and all the information is available via the website of the Royal Botanic Garden Edinburgh (www.rbge.org.uk).

The checklist is in two parts. The first is an annotated alphabetical list of all species, and the second is a collection of regional lists of species names. The geographical scope of the checklist runs from Burma to Fiji, and is detailed in Fig. 1 and Table 1. The highest geographic level reported (in bold in Table 1) corresponds to the Botanical Country concept (Level 3) of Brummitt (2001). These are also the basis for the geographic species lists on pages 142–152. The next geographic reporting level (indented in Table 1) has been broken down as far as geographically stable

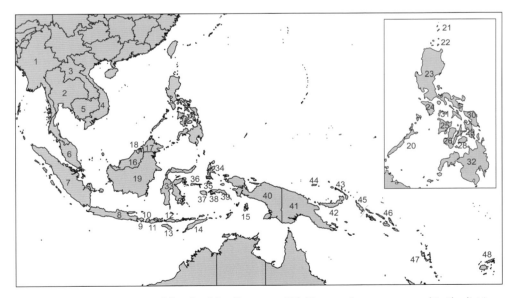

FIG. 1. Geographical extent of the checklist (Burma to Fiji). The numbers correspond to the list in Table 1. Inset: Philippine islands.

TABLE 1. Reporting order for geographic units. Those in bold correspond to the Botanical Countries of the TDWG World Geographical Scheme for Recording Plant Distributions (Brummitt, 2001). Numbers correspond to those on the map in Fig. 1. Only localities where *Begonia* specimens have been collected are listed.

Burma	1	Cebu	27
Thailand	2	Bohol	28
Laos	3	Leyte	29
Vietnam	4	Samar	30
Cambodia	5	Sibuyan	31
Peninsular Malaysia	6	Mindanao	32
Sumatra	7	**Sulawesi**	33
Java	8	**Moluccas**	
Lesser Sunda Islands		Halmahera	34
Bali	9	Obi	35
Lombok	10	Sula Islands	36
Sumbawa	11	Buru	37
Flores	12	Ambon	38
Sumba	13	Seram	39
Timor	14	**New Guinea**	
Aru Islands	15	Papua	40
Borneo		Papua New Guinea	41
Sarawak	16	**Bismarck Archipelago**	
Sabah	17	New Britain	42
Brunei	18	New Ireland	43
Kalimantan	19	Manus	44
Philippines		**Solomon Islands**	
Palawan	20	North Solomons	45
Batan Islands	21	South Solomons	46
Babuyan Islands	22	**Vanuatu**	
Luzon	23	Santo Island	47
Mindoro	24	**Fiji**	
Panay	25	Vanua Levu	48
Negros	26		

borders (usually coastlines) allow. Collection localities have been georeferenced where possible to allow distribution maps to be produced via the website; all are shown in Fig. 2.

There are 521 species in the study area, with the three most species rich floras being the Philippines (104 spp.), Borneo (95 spp.) and New Guinea (79 spp.) (Fig. 3). Only Peninsular Malaysia, Java and perhaps the Philippines are reasonably well collected, with the remaining parts of southeast Asia harbouring a considerable proportion of undescribed taxa. Not surprisingly, the level of endemism in the continental regions is generally lower than on the large Malesian islands (Fig. 3). The most species rich *Begonia* section in southeast Asia is § *Petermannia*, at 247 species accounting for nearly half of the total (Fig. 4). The centres of diversity for this section are Borneo, Sulawesi, the Philippines and New Guinea. The next largest sections are § *Diploclinium* (82 spp.) and § *Platycentrum* (60 spp.), which have their centres of diversity in the Philippines and Indo-China, respectively.

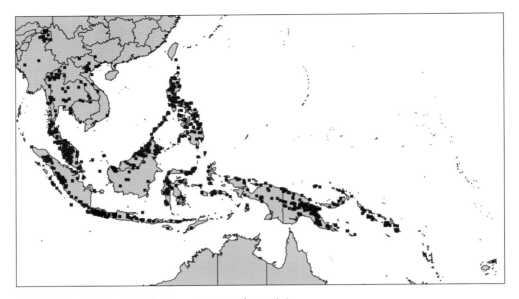

FIG. 2. Collection localities for *Begonia* in southeast Asia.

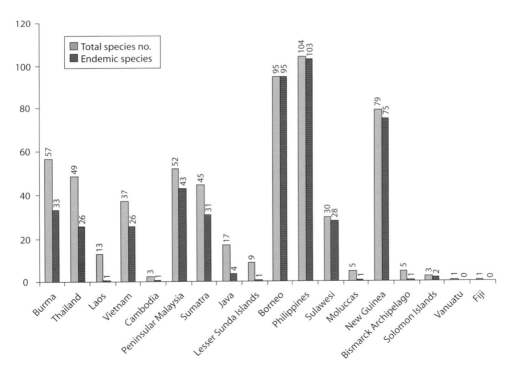

FIG. 3. The number of *Begonia* species (total and endemic) per Botanical Country. Statistics are taken from this checklist only; some of the species noted as endemic to countries in the Indo-China region may also be represented in India and China.

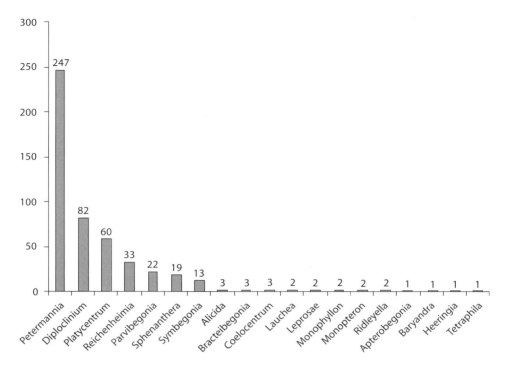

FIG. 4. The number of *Begonia* species per section in the geographic area covered by the checklist. There are a further 22 species which are unplaced to section.

Explanatory Notes

Only species I consider as native are reported. The main alien species one comes across in southeast Asia is *Begonia hirtella* Link.

Author abbreviations follow Brummitt and Powell (1992). Titles for books and periodicals have been abbreviated following *Botanico-Periodicum-Huntianum* and supplement (BPH/S; Bridson & Smith, 1991).

Beccari's catalogue numbers, with prefixes PB (*Plante Bornensi*), PP (*Plante Papuane*) or PS (*Plante Sumatrane*), have been used to refer to his specimens when available. When no catalogue number has been assigned by Beccari, I have used the Florence Herbarium '*Collezioni Beccari*' reference numbers which are to be found on each sheet; these can be identified here by the prefix CB. Fragments of Beccari's collections taken by Irmscher and kept in Berlin are referred to as merotypes (Fuchs, 1958) when they come from a type collection.

Citations are listed after a name when it has either (i) been cited as an accepted name for whatever reason or (ii) undergone a nomenclatural change. This inevitably involves a small amount of repetition, but it should allow nomenclatural trails to be followed more easily.

Paratypes have been highlighted in the *exsiccatae* lists in order to communicate the taxon concept intended by the author of the name. Also, specimen determinations have been qualified as '*cf.*' (*confer*), '*aff.*' (*affinis*) or '?' where the information was available on the label. Although not all taxonomists will have used these terms in an exactly equivalent manner, they do give

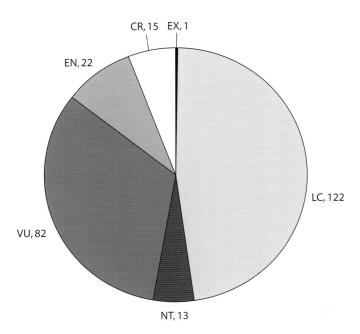

FIG. 5. The number of *Begonia* species in each IUCN Red List Category (LC, Least Concern; NT, Near Threatened; VU, Vulnerable; EN, Endangered; CR, Critically Endangered; EX, Extinct). A further 266 species are classed as DD (Data Deficient).

some indication as to the certainty of the name applied. As far as possible, I have checked each determination and have endeavoured to weed out the incorrect and ridiculous; however, there may be one or two that have slipped through. All specimens have been seen by the author unless annotated *n.v.* (*non visus*).

The conservation status of each species has been assessed using IUCN guidelines (Standards and Petitions Working Group, 2006). Each species has been assigned to one of seven categories, namely Data Deficient (DD), Least Concern (LC), Near Threatened (NT), Vulnerable (VU), Endangered (EN), Critically Endangered (CR) or Extinct (EX). Preliminary assessments were carried out using an algorithm to calculate the area of occupancy and extent of occurrence for each species which had a sufficient number of georeferenced specimens (Willis *et al.*, 2003; Moat, 2007). The preliminary assessments were then checked and expanded by using information (when available) from (i) floras and monographs, (ii) specimen labels, (iii) satellite images of habitat, (iv) maps of protected areas, (v) World Wildlife Fund Ecoregion Assessments (Wikramanayake *et al.*, 2001) and (vi) regional expert opinion. The assessments are all global in scope (i.e., cover the entire range of a species), and sub-specific taxa were not assessed separately. A synopsis of the assessments is presented in Fig. 5. For a large proportion of the species there was insufficient information available to perform an assessment (266; 51% of the total). The information used to inform the assessments was largely geographic in nature, hence criteria B (geographic range) and D (very small or restricted population) appear more frequently than the other criteria available (A, population reduction; C, small population size and decline; E, quantitative analysis). It is hoped that botanists in southeast Asia will be able to tackle the species listed here as Data Deficient, and also expand upon the other assessments given here with more detailed information on population size and specific threats.

Taxonomic Changes

There are a small number of taxonomic changes made in the list, carried out where (i) the case was obvious or (ii) I came across unpublished information which I was able to quickly confirm. There was insufficient time for the project to be a critical revision of all 521 species, and there is of course much work still to be done. However, I have highlighted where relevant in the species notes if I consider a name likely to be a synonym, or if one name appears to be covering more than one species.

The taxonomic changes made in the list are:

Begonia altissima Ridl. has been transferred from § *Petermannia* to § *Platycentrum*
Begonia beccariana Ridl. has been confirmed as belonging to § *Platycentrum*
Begonia bifolia has been moved from § *Petermannia* to § *Platycentrum*
Begonia cristata Warb. ex L.B. Sm. & Wassh. is synonymised under *B. aptera* Blume
Begonia divaricata is assigned to § *Petermannia*
Begonia guertiziana Gibbs has been moved from § *Platycentrum* to § *Diploclinium*
Begonia heteroclinis Miq. ex Koord. has been transferred from § *Sphenanthera* to § *Petermannia*
Begonia inversa Irmsch. has been transferred from § *Diploclinium* to § *Reichenheimia*
Begonia lecomtei Gagnep. has been lectotypified and is synonymised under *B. handelii* Irmsch.
Begonia littleri Merr. has been transferred from § *Platycentrum* to § *Petermannia*
Begonia moulmeinensis C.B. Clarke is considered a superfluous name, and a synonym of *B. parvuliflora* var. *pubescens* A. DC.
Begonia sootepensis Craib is synonymised under *B. modestiflora* Kurz as *B. modestiflora* var. *sootepensis* (Craib) Z. Badcock ex M. Hughes, *comb. nov.*
Begonia yunnanensis H. Lév., *B. sootepensis* var. *thorelii* Gagnep. and *B. lushaiensis* C.E.C. Fisch. have been synonymised under *B. modestiflora* Kurz
Casparya robusta var. *glabriuscula* A. DC. has been lectotypified and is considered a synonym of *Begonia multangula* Blume

The bulk of the species published by Sands in Coode *et al.* (1997) (*Begonia awongii*, *B. bahakensis*, *B. bruneiana*, *B. chlorandra*, *B. cyanescens*, *B. eutricha*, *B. fuscisetosa*, *B. hexaptera*, *B. leucochlora*, *B. leucotricha*, *B. papyraptera*, *B. sibutensis*, *B. stenogyna*, *B. temburongensis*) are assigned to § *Petermannia*. One of the species, *Begonia laccophora*, remains unplaced to section.

Annotated Species List

Begonia abdullahpieei Kiew [§ Platycentrum], Begonias Penins. Malaysia 252 (2005). – Type: Peninsular Malaysia, Perak, Kelian Gunung, 17 ii 2000, *R. Kiew* 4907 (holo SING *n.v.*; iso K *n.v.*, KEP *n.v.*, L *n.v.*).

■ **PENINSULAR MALAYSIA:** Perak, Kelian Gunung, 17 ii 2000, *R. Kiew* 4907 (holo SING *n.v.*; iso K *n.v.*, KEP *n.v.*, L *n.v.*) – Type of *Begonia abdullahpieei* Kiew.

■ **IUCN category: VU D2.** The type (and only) locality for this species is in the vicinity of the proposed Selama Wildlife Reserve.

Begonia aberrans Irmsch. [§ Bracteibegonia], Webbia 9: 483 (1953). – Type: Sumatra, Padang, Ajer Mantjoer, 1878, *O. Beccari* CB4514 (syn B[2], FI); Sumatra, Padang, Ajer Mantjoer, 1878, *O. Beccari* CB4514A (syn B, FI).

■ **SUMATRA:** Padang, Ajer Mantjoer, 1878, *O. Beccari* CB4514 (syn B[2], FI) – Type of *Begonia aberrans* Irmsch.; Padang, Ajer Mantjoer, 1878, *O. Beccari* CB4514A (syn B, FI) – Type of *Begonia aberrans* Irmsch.

■ **IUCN category: VU D2.** Known only from the type locality, which is near to the main road into Padang. However, the forests immediately to the north on the footslopes of Mt. Tandikat are relatively intact.

Begonia acaulis Merr. & L.M. Perry [§ Diploclinium], J. Arnold Arbor. 24: 43 (1943); Tebbitt, Begonias 72 (2005). – Type: New Guinea, Papua New Guinea, Rona, Laloki River, iii 1933, *L.J. Brass* 3599 (holo A *n.v.*; iso B).

■ **NEW GUINEA: Papua New Guinea:** Rona, Laloki River, iii 1933, *L.J. Brass* 3599 (holo A *n.v.*; iso B) – Type of *Begonia acaulis* Merr. & L.M. Perry.

■ **IUCN category: NT.** Merrill states this is a common plant in light rain forests. The forest at the altitude of the collection in the vicinity of the Laloki River (450 m) seems to be in good condition, although the proximity of Port Moresby (c. 40 km away) and the lack of formal protection means this species must be considered NT.

Begonia aceroides Irmsch. [§ Diploclinium], Bot. Jahrb. Syst. 76: 100 (1953). – *Begonia burkillii* Irmsch., Mitt. Inst. Allg. Bot. Hamburg 8: 116 (1929); Irmscher, Bot. Jahrb. Syst. 76: 100 (1953). – Type: Thailand, Pungali, 6 xii 1918, *M. Haniff, Md. Nur* 3898 (holo K ex SING; iso BM).

■ **THAILAND:** Pungali, 6 xii 1918, *M. Haniff, Md. Nur* 3898 (holo K ex SING; iso BM) – Type of *Begonia burkillii* Irmsch.

■ **IUCN category: DD.** Insufficient specimens could be georeferenced with certainty.

Begonia acetosella Craib [§ Sphenanthera], Bull. Misc. Inform. Kew 1912: 153 (1912); Gagnepain, Bull. Mus. Hist. Nat. (Paris) 25: 282 (1919); Irmscher, Mitt. Inst. Allg. Bot. Hamburg 6: 347 (1926); Craib, Fl. Siam. 1: 770 (1931); Tebbitt, Brittonia 55(1): 22 (2003). – Type: Thailand,

Chiangmai, Doi Sootep, 21 iii 1909, *A.F.G. Kerr* 557 (lecto K; isolecto B, K); Thailand, Chiangmai, 1 iv 1911, *A.F.G. Kerr* 1744 (syn B, BM, E *n.v.*, K[2], L, P).

Begonia tetragona Irmsch., Mitt. Inst. Allg. Bot. Hamburg 10: 515 (1939); Huang & Shui, Acta Bot. Yunnan. 21(1): 12 (1999); Tebbitt, Brittonia 55(1): 22 (2003). – Type: China, Yunnan, Mengtze, *Henry* 10737A (holo B; iso E *n.v.*).

Begonia aptera auct. non Blume: Gagnepain, Bull. Mus. Hist. Nat. (Paris) 25: 281 (1919); Gagnepain, in Lecomte (ed.), Fl. Indo-Chine 2: 1110 (1921); Hô, Ill. Fl. Vietnam 1: 729 (1991).

Begonia acetosella var. acetosella
- **BURMA:** Southern Shan States, Mong Nai, 17 iii 1911, *Robertson* 273 (B, K); Putao Distr., 12 ix 1920, *Farrer* 1503 (E *n.v.*); Keng Tung, 27 i 1922 – 28 i 1922, *J.F.C. Rock* 2124 (B, NY *n.v.*, US *n.v.*); Kachin, Sumprabum sub-division, Ning W'Krok – Kanang, iii 1962, *J. Keenan et al.* 3962 (E) [as *Begonia acetosella* ?]; Kachin, Sumprabum sub-division, Ning W'Krok – Kanang, iii 1962, *J. Keenan et al.* 3963 (E) [as *Begonia acetosella* ?]; Bhamo District, Bhamo, Lapyeka Wdweje, 7 iv 1912, *J.H. Lace* 5759 (E, K); Senighku-Adung confluence, 13 viii 1926, *F.K. Ward* 7288 (K); Valley of the Taping, *G. Forrest* 12155 (E).
  THAILAND: Khao Yai National Park, 26 v 2000, *S. Chongko* 70 (L); Chiangmai, Shristian Hill, *Maxwell* 90-290 (E *n.v.*, L); Chiangmai, Chiang Dao, Pa Blawag Cave, 11 iii 1989, *Maxwell* 89-323 (L); Chiangmai, Doi Chiang Dao Animal Sanctuary, 14 iii 1996, *Maxwell* 96-349 (A, L); Chiangmai, Doi Sootep, 21 iii 1909, *A.F.G. Kerr* 557 (lecto K; isolecto B, K) – Type of *Begonia acetosella* Craib; Chiangmai, Doi Sootep, 10 ii 1926, *D.J. Collins* 1221 (K); Payap, Doi Chiengdao, 20 iii 1950, *H.B.G. Garrett* 1288 (K, L, P *n.v.*); Payap, Doi Chiengdao, 4 iv 1955, *H.B.G. Garrett* 1450 (A, K, L); Chiangmai, Doi Sootep, 24 iii 1937, *H.G. Deignan* 1550 (A); Loei Province, Phu Luang, Loei, 8 ii 1968, *K. Bunchuai* 1607 (L); Chiangmai, 1 iv 1911, *A.F.G. Kerr* 1744 (syn B, BM, E *n.v.*, K[2], L, P) – Type of *Begonia acetosella* Craib; Doi Wao, 24 ii 1912, *A.F.G. Kerr* 2440 (BM, K); Doi Wao, 24 ii 1912, *A.F.G. Kerr* 2440a (BM); Teen Tok, *Larson et al.* 3098 (E *n.v.*, L); Doi Pu Ka, 20 iii 1950, *A.F.G. Kerr* 4940 (ABD, BM, K); Hue Sala, 10 iii 1921, *A.F.G. Kerr* 5075 (ABD, BM, K); Kao Paga Paw, 4 iii 1931, *A.F.G. Kerr* 20325 (BM); Chiangmai, Fang, Ban Bu Meam, 22 iii 1965, *C. Pengkali, B. Sangkhachand* 32329 (P); Chiang Rai Prov., Doi Tung, 14 ii 1983, *H. Koyama* 33520 (A).
  LAOS: Laos, 4 iv 1932, *E. Poilane* 20721 (P); Tawieng Chieng Kwang, 6 iv 1932, *A.F.G. Kerr* 20929 (BM, K, L, P).
  VIETNAM: Vinh Yen Province, Tam Dao, ix 1908, *D'Alleizette s.n.* (L); Lao Cai Prov., Sapa, 11 xii 1964, *C.Y. Wu, Sino-Vietnam expedition* 379 (KUN *n.v.*); Tonkin, Mt. Bavi, 24 ii 1941, *P.A. Petelot* 7084 (B[2]).

Begonia acetosella var. hirtifolia Irmsch., Mitt. Inst. Allg. Bot. Hamburg 10: 515 (1939); Tebbitt, Brittonia 55(1): 24 (2003). – Type: China, Yunnan, Szemao, *Henry* 12251A (holo B *n.v.*; iso E *n.v.*, K *n.v.*).
- **BURMA:** Myelleyiana, 1 iv 1953, *Tha Hla, Chit Koko* 3724 (K *n.v.*); Kachin Hills, *S.M. Toppin* 4299 (K *n.v.*).
- **IUCN category: LC.** A very widespread species.

Begonia acuminatissima Merr. [§ Diploclinium], Philipp. J. Sci. 6: 395 ('1911', 1912); Merrill, Enum. Philipp. Fl. Pl. 3: 119 (1923); Smith & Wasshausen, Phytologia 54(7): 466 (1984). – Type: Philippines, Mindanao, Balut Island, 8 x 1906, *E.D. Merrill* 5419 (syn not located); Philippines, Mindanao, Butuan, Agusan River, *E.D. Merrill* 7306 (syn B, BM).

Begonia camiguinensis Elmer, Leafl. Philipp. Bot. 7: 2553 (1915); Merrill, Enum. Philipp. Fl. Pl. 3: 119 (1923); Smith & Wasshausen, Phytologia 54(7): 466 (1984). – Type: Philippines, Mindanao, Camiguin Island, Mambajao, xi 1912, *A.D.E. Elmer* 14222 (syn B, BM, E, FI, K, L, NY, P, U).

- **PHILIPPINES: Mindanao:** Camiguin Island, Katibawasan Falls, *R. Rubite* 71 (PNH *n.v.*); Agusan, San Antonio, Tagnote Falls, *R. Rubite* 321 (PNH *n.v.*); Balut Island, 8 x 1906, *E.D. Merrill* 5419 (syn not located) – Type of *Begonia acuminatissima* Merr.; Butuan, Agusan River, *E.D. Merrill* 7306 (syn B, BM) – Type of *Begonia acuminatissima* Merr.; Camiguin Island, Mambajao, xi 1912, *A.D.E. Elmer* 14222 (syn B, BM, E, FI, K, L, NY, P, U) – Type of *Begonia camiguinensis* Elmer.
- **IUCN category: VU D2.** This species is restricted to c. three localities in the lowland forests of Mindanao (c. 400 m). These forests are highly reduced in area and fragmented, and show evidence of recent logging.

Begonia adenodes Irmsch. [§ Petermannia], Webbia 9: 478 (1953). – Type: Borneo, Sarawak, Sakarrang, x 1867, *O. Beccari* PB3843 (syn FI); Borneo, Sarawak, Sakarrang, x 1867, *O. Beccari* PB3859 (syn FI *n.v.*).
- **BORNEO: Sarawak:** Sakarrang, x 1867, *O. Beccari* PB3843 (syn FI) – Type of *Begonia adenodes* Irmsch.; Sakarrang, x 1867, *O. Beccari* PB3859 (syn FI *n.v.*) – Type of *Begonia adenodes* Irmsch.
- **IUCN category: DD.** Insufficient specimens could be georeferenced with certainty.

Begonia adenopoda Lem. [§ Lauchea], Jard. Fleur. 2: 17 (1852); Candolle, Prodr. 15(1): 353 (1864); Golding & Karegeannes, Phytologia 54(7): 499 (1984). – Type: not located.
    Begonia verticillata Hook. Later homonym, Icon. Pl. t. 811 (1851); Candolle, Prodr. 15(1): 353 (1864); Golding & Karegeannes, Phytologia 54(7): 499 (1984). – Type: Burma, Moulmein, *W. Lobb* 382 (holo K; iso BM, E, FI).
- **BURMA:** Moulmein, 1862, *C.S.P. Parish s.n.* (B, K, P); Moulmein, 1859, *C.S.P. Parish s.n.* (K); Moulmein, *W. Lobb* 382 (holo K; iso BM, E, FI) – Type of *Begonia verticillata* Hook. Later homonym.
- **Notes:** A very distinctive species with a whorl of triangular leaves subtending the inflorescence.
- **IUCN category: DD.** Insufficient specimens could be georeferenced with certainty.

Begonia adenostegia Stapf [§ Platycentrum], Trans. Linn. Soc. London, Bot., II 4: 164 (1894); Ridley, J. Malayan Branch Roy. Asiat. Soc. 46: 261 (1906); Gibbs, J. Linn. Soc., Bot. 42: 84 (1914); Sands, in Beaman *et al.*, Pl. Mt. Kinabalu 147 (2001). – Type: Borneo, Sabah, Mt. Kinabalu, Kinitaki, *G.D. Haviland* 1270 (holo K; iso BO *n.v.*).
- **BORNEO: Sabah:** Mt. Kinabalu, Kinitaki, *G.D. Haviland* 1270 (holo K; iso BO *n.v.*) – Type of *Begonia adenostegia* Stapf; Mt. Kinabalu, Penibukan, 2 ii 1933, *J. & M.S. Clemens* 30559 (A, BM[2], BO[2]).
- **Notes:** It would seem unlikely that this species belongs in § *Platycentrum*. The 2-locular fruit may well be a secondary acquisition, and its true affinity is probably with § *Petermannia*, but further research is required.
- **IUCN category: VU D2.** Known only from the lower montane forests of Mt. Kinabalu, where it appears to be limited to ultramafic substrate (Sands, 2001).

Begonia adscendens C.B. Clarke [§ Diploclinium], J. Linn. Soc., Bot. 25: 26 (1890). – Type: India, Naga Hills, Jakpho, ix 1937, *C.B. Clarke* 41240 (holo K).
- **BURMA:** Mungku Hkyet, 19 vii 1937, *F.K. Ward* 12977 (BM) [as *Begonia aff. adscendens*]; West Central Burma, Mt. Victoria, 20 x 1953, *F.K. Ward* 21489 (BM); West Central Burma, Mt. Victoria, 5 xi 1956, *F.K. Ward* 22822 (BM).
    LAOS: Champassak Forest, 30 xi 1998, *J. Munzinger* 250 (L, P).
- **Notes:** If correctly identified, the record from southern Laos represents a large disjunction for this species, which is mainly found along the Burma/India border. Further work is needed to clarify the limits between this species and the similar *B. labordei* H. Lév.

- **IUCN category: LC.** Has a distribution spreading c. 700 km along the Burma/India border, and a recent collection possibly extends the distribution to southern Laos.

Begonia aequata A. Gray [§ Petermannia], U.S. Expl. Exped. Phan. 15: 658 (1854); Candolle, Prodr. 15(1): 321 (1864); Fernández-Villar, Noviss. App. 99 (1880); Merrill, Philipp. J. Sci. 3: 84 (1908); Merrill, Philipp. J. Sci. 6: 406 ('1911', 1912); Merrill, Enum. Philipp. Fl. Pl. 3: 119 (1923). – Type: Philippines, Luzon, Mts. near Banos, *unknown s.n.* (photo) (iso K).
- **PHILIPPINES: Luzon:** Mts. near Banos, *unknown s.n.* (photo) (iso K) – Type of *Begonia aequata* A. Gray; Laguna, Makiling, iv 1906, *A. Loher* 6087 (B, K); Rizal, Angilog, iii 1906, *A. Loher* 6093 (K); Laguna, Makiling, 1906, *A. Loher* 6096 (K); Laguna, Makiling, ii 1907, *E.D. Merrill* 6302 (B, P); Laguna, Los Banos, iv 1906, *A.D.E. Elmer* 8324 (E, K, L); Laguna, Makiling, 5 vi 1958, *J. Sinclair* 9510 (E); Laguna, Makiling, 6 xii 1912 – 9 xii 1912, *C.B. Robinson* 17006 (K); Laguna, Makiling, v 1913, *N.E. Brown* 17514 (B, BM[2], L, P, U); Camarines, Mt. Isaro(g), 22 iii 1997, *P. Wilkie et al.* 29140 (E) [as *Begonia aff. aequata*]; Laguna, Mt. Banajao, *M. Ocampo* 29715 (P); Albay Province, Mt. Malinao, 28 i 1956, *G. Edano* 34418 (L).
- **Notes:** This species belongs to a complex along with *B. binuangensis* Merr., *B. edanoi* Merr., *B. gracilipes* Merr., *B. lagunensis* Elmer, *B. sarmentosa* L.B. Sm. & Wassh. and *B. wenzelii* Merr., at once easily recognisable as a whole due to their climbing habit and sub-symmetric leaves, but very difficult to tell apart.
- **IUCN category: DD.** Taxonomic limits between this species and the rest of this species complex need defining.

Begonia aequilateralis Irmsch. [§ Platycentrum], Mitt. Inst. Allg. Bot. Hamburg 8: 134 (1929); Kiew, Begonias Penins. Malaysia 261 (2005). – Type: Peninsular Malaysia, Selangor, Sungai Buloh, x 1899, *J.S. Goodenough* 10564 (holo K).
- **PENINSULAR MALAYSIA:** Selangor, Sungai Buloh, *R. Kiew* RK1594 (KEP *n.v.*); Selangor, Sungai Buloh, *R. Kiew* RK1609 (SING *n.v.*); Selangor, Sungai Buloh, *R. Kiew* RK4715 (SING *n.v.*); Selangor, Sungai Buloh, x 1899, *J.S. Goodenough* 10564 (holo K) – Type of *Begonia aequilateralis* Irmsch.; Selangor, Kepong, Sungai Kroh, *J. Wyatt-Smith* 60831 (KEP *n.v.*).
- **IUCN category: CR B2ab(iii), D1.** Known from one population of 20 individuals, in a locality threatened by development (Kiew, 2005).

Begonia affinis Merr. [§ Petermannia], Philipp. J. Sci. 7: 308 (1912); Merrill, Enum. Philipp. Fl. Pl. 3: 119 (1923). – Type: Philippines, Mindanao, Zamboanga, Sax River Mountains, xi 1911 – xii 1911, *E.D. Merrill* 8251 (syn BM[2], K, L, P).
- **PHILIPPINES: Mindanao:** Zamboanga, Sax River Mountains, 27 xi 1911, *E.D. Merrill* 8248 (para B) – Type of *Begonia affinis* Merr.; Zamboanga, Sax River Mountains, xi 1911 – xii 1911, *E.D. Merrill* 8251 (syn BM[2], K, L, P) – Type of *Begonia affinis* Merr.; Zamboanga, Malangas, x 1919 – xi 1919, *M. Ramos, G. Edano* 36900 (K).
- **IUCN category: DD.** The localities given by Merrill in the protologue are not precise, but it seems likely that the type locality is within the Pasonanca Natural Park near the tip of the Zamboanga peninsula.

Begonia agusanensis Merr. [§ Petermannia], Philipp. J. Sci. 6: 377 ('1911', 1912); Merrill, Enum. Philipp. Fl. Pl. 3: 119 (1923). – Type: Philippines, Mindanao, Butuan, Agusan River, Waloe, x 1910, *E.D. Merrill* 7312 (syn K).
- **PHILIPPINES: Mindanao:** Butuan, Talacogon, iii 1911 – vii 1911, *C.M. Weber* 1209 (para E, K, P) – Type of *Begonia agusanensis* Merr.; Butuan, Agusan River, Waloe, x 1910, *E.D. Merrill* 7312

(syn K) – Type of *Begonia agusanensis* Merr.; Butuan, Tungao, 22 vi 1961, *D.R. Mendoza* 42481 (P).

- **IUCN category: EN B2ab(iii)**. Endemic to low-altitude forests (c. 50 m) of the Agusan River system, which is very heavily developed.

Begonia alba Merr. [§ Diploclinium], Philipp. J. Sci. 10: 45 (1915); Merrill, Enum. Philipp. Fl. Pl. 3: 120 (1923). – Type: Philippines, Luzon, Ifugao, Mt. Polis, ii 1913, *R.C. McGregor* 19963 (syn K).
- **PHILIPPINES: Luzon:** Ifugao, Mt. Polis, ii 1913, *R.C. McGregor* 19963 (syn K) – Type of *Begonia alba* Merr.
- **IUCN category: CR B2ab(iii)**. Known only from the type locality on Mt. Polis, which has little remaining forest cover and receives no legal protection (Collar *et al.*, 1999).

Begonia albobracteata Ridl. [§ Petermannia], Trans. Linn. Soc. London, Bot., II 9: 60 (1916). – Type: New Guinea, Papua, Camp I – Camp III, *C.B. Kloss s.n.* (holo BM).
- **NEW GUINEA: Papua:** Camp I – Camp III, *C.B. Kloss s.n.* (holo BM) – Type of *Begonia albobracteata* Ridl.
- **IUCN category: DD**. Insufficient specimens could be georeferenced with certainty.

Begonia alicida C.B. Clarke in Hook. f. **[§ Alicida]**, Fl. Brit. Ind. 2: 637 (1879); Craib, Fl. Siam. 1: 770 (1931). – Type: Burma, Moulmein, 1859, *C.S.P. Parish s.n.* (holo K).
- **BURMA:** Moulmein, 1859, *C.S.P. Parish s.n.* (holo K) – Type of *Begonia alicida* C.B. Clarke.
  THAILAND: Wangka, Neeckey, *A.J.G.H. Kostermans* (P); Ayuthia, Saraburi, vii 1925, *Nai Nao* 127 (B, BM, L); Kwae Noi River Basin, 29 iv 1946, *G. Hoed* 234 (A, L); Kwae Noi River Basin, Brangkasi, 19 vi 1946 – 22 vi 1946, *A.J.G.H. Kostermans* 682 (A, L, P); Chiangmai, Doi Sahng Liang, vii 1997, *J.F. Maxwell* 97-807 (A); Kwae Noi River Basin, 2 vi 1946, *G. Hoed* 908 (A, L); Kin Sayok, 9 vii 1946, *A.J.G.H. Kostermans* 1039 (A, P); Kin Sayok, 9 vii 1946, *A.J.G.H. Kostermans* 1041 (A, P); Kwae Noi River Basin, Rintin, 31 vii 1946, *A.J.G.H. Kostermans* 1409 (A); Ratburi, 17 vii 1924, *A. Marcan* 1456 (BM[2]); Rachaburi, Kanburi, Sai Yok, 31 vii 1828, *N. Put* 1787 (ABD, BM); Ayuthia, Saraburi, 20 viii 1828, *N. Put* 2422 (ABD, BM); 26 vi 1922, *A.F.G. Kerr* 6163 (ABD, BM); Ayuthia, Saraburi, 6 vi 1923, *A.F.G. Kerr* 7058 (BM); ix 1924, *A.F.G. Kerr* 9093 (ABD, B, BM); Bangkok, 19 vii 1930, *A.F.G. Kerr* 19735 (BM).
- **IUCN category: DD**. Insufficient specimens could be georeferenced with certainty.

Begonia alpina L.B. Sm. & Wassh. [§ Platycentrum], Phytologia 54(7): 469 (1984); Kiew, Begonias Penins. Malaysia 149 (2005). – *Begonia monticola* Ridl., J. Fed. Malay States Mus. 5: 34 (1914); Ridley, Fl. Malay Penins. 1: 862 (1922); Irmscher, Mitt. Inst. Allg. Bot. Hamburg 8: 128 (1929); Smith & Wasshausen, Phytologia 54(7): 469 (1984). – Type: Peninsular Malaysia, Perak, Gunung Mengkuang Lebah, 3 ii 1913, *H.C. Robinson s.n.* (lecto SING *n.v.*; isolecto BM, K).
- **PENINSULAR MALAYSIA:** Perak, Gunung Mengkuang Lebah, 3 ii 1913, *H.C. Robinson s.n.* (lecto SING *n.v.*; isolecto BM, K) – Type of *Begonia monticola* Ridl.
- **IUCN category: DD**. Insufficient specimens could be georeferenced with certainty.

Begonia altissima Ridl. [§ Platycentrum], J. Fed. Malay States Mus. 8(4): 39 (1917). – Type: Sumatra, Korinchi, Purdock, 23 iv 1914, *H.C. Robinson, C.B. Kloss* (syn BM[2], K).
- **SUMATRA:** Korinchi, Purdock, 23 iv 1914, *H.C. Robinson, C.B. Kloss* (syn BM[2], K) – Type of *Begonia altissima* Ridl.
- **Notes:** Originally placed in § *Petermannia* in error; this species patently belongs in § *Platycentrum* and is close to *B. teysmanniana* and *B. laevis*.

■ **IUCN category: LC**. Endemic to the higher altitude regions of Mt. Kerinci, and hence in the Kerinci-Seblat National Park.

Begonia alvarezii Merr. [§ Diploclinium], Philipp. J. Sci. 6: 405 ('1911', 1912); Merrill, Enum. Philipp. Fl. Pl. 3: 120 (1923). – Type: Philippines, Luzon, Nueva Ecija Province, Santor River, ii 1911, *R.J. Alvarez* FB22446 (syn not located).
■ **PHILIPPINES: Luzon:** Apayao, v 1917, *E. Fenix s.n.* (K); Tayabas, Mt. Pular, i 1917, *M. Ramos* 19452 (K); Bulacan Province, Angat, ix 1913, *M. Ramos* 21756 (K); Nueva Ecija Province, Santor River, ii 1911, *R.J. Alvarez* FB22446 (syn not located) – Type of *Begonia alvarezii* Merr.
■ **IUCN category: VU D2**. Known only from four collections, each from a different locality in Luzon. Although none of these can be georeferenced with certainty, the number of locations and the fact that forest cover is so low in Luzon puts this species firmly in the VU category.

Begonia alveolata T.T. Yu [§ Diploclinium], Bull. Fan. Mem. Inst. Biol. n.s. 1: 121 (1948). – Type: China, Yunnan, Ping Bien, *C.W. Wang* 82780 (Yun. Bot. Instit. *n.v.*).
  Begonia pingbiensis C.Y. Wu, Acta Phytotax. Sin. 33: 258 (1995); Shui & Peng, Bot. Bull. Acad. Sin. 43: 313 (2002). – Type: China, Yunnan, Pingbian, 30 viii 1934, *H-T. Tsai* 61724 (holo KUN *n.v.*).
■ **VIETNAM:** Laoke, Chapa, 10 xii 1964, *Sino-Vietnam expedition* 295 (holo KUN *n.v.*) – Type of *Begonia pingbiensis* var. *angustior* C.Y. Wu.
■ **IUCN category: DD**. Insufficient specimens could be georeferenced with certainty.

Begonia amphioxus Sands [§ Petermannia], Kew Mag. 7: 77 (1990); Kiew, Gard. Bull. Singapore 53: 247 (2001); Tebbitt, Begonias 79 (2005). – Type: Borneo, Sabah, Pensiangan District, Batu Punggul, Sungei Sansiang, 29 iv 1984, *M.J.S. Sands* 4045 (holo K *n.v.*).
■ **BORNEO: Sabah:** Pensiangan District, Batu Punggul, Sungei Sansiang, 29 iv 1984, *M.J.S. Sands* 4045 (holo K *n.v.*) – Type of *Begonia amphioxus* Sands; Pensiangan District, Batu Punggul, 15 v 1997, *R. Kiew, S. Anthonysamy* 4379 (E).
■ **IUCN category: VU B2ab(iii)**. Known only from two nearby limestone localities in Sabah.

Begonia anceps Irmsch. [§ Diploclinium], Notes Roy. Bot. Gard. Edinburgh 21: 35 (1951). – Type: China, Yunnan, Ma-li-po, *C.W. Wang* 86714 (holo KUN *n.v.*); China, Yunnan, Ma-li-po, *C.W. Wang* 86714 (mero B).
■ **VIETNAM:** Laoke, Chapa, iv 1925, *P.A. Petelot* 7097 (B).
■ **Notes:** Some authors regard this as a synonym of *B. morifolia* T.T. Yu (Shui *et al.*, 2002).
■ **IUCN category: DD**. Taxonomically uncertain. The limits between this species and *B. alveolata* and *B. morifolia* are not clear (R. Kiew, pers. comm.).

Begonia angilogensis Merr. [not placed to section], Philipp. J. Sci. 26: 477 (1925). – Type: Philippines, Luzon, Rizal Province, Mt. Angilog, iv 1922, *M. Ramos* 40777 (syn B, K, L, P).
■ **PHILIPPINES: Luzon:** Rizal Province, Mt. Angilog, iv 1922, *M. Ramos* 40777 (syn B, K, L, P) – Type of *Begonia angilogensis* Merr.
■ **IUCN category: DD**. Insufficient specimens could be georeferenced with certainty.

Begonia angustilimba Merr. [§ Petermannia], J. Straits Branch Roy. Asiat. Soc. 86: 344 (1922). – Type: Borneo, Sabah, Sandakan District, 1920, *M. Ramos* 1388 (syn BM, K, L, P).
■ **BORNEO: Sabah:** Sandakan District, 1920, *M. Ramos* 1388 (syn BM, K, L, P) – Type of *Begonia angustilimba* Merr.
■ **IUCN category: DD**. Insufficient specimens could be georeferenced with certainty.

Begonia anisoptera Merr. [§ Diploclinium], Philipp. J. Sci. 6: 398 ('1911', 1912); Merrill, Enum. Philipp. Fl. Pl. 3: 120 (1923). – Type: Philippines, Mindanao, Zamboanga, 10 x 1906, *E.D. Merrill* 5482 (syn B).

- **PHILIPPINES: Mindanao:** Zamboanga, 10 x 1906, *E.D. Merrill* 5482 (syn B) – Type of *Begonia anisoptera* Merr.; Zamboanga, xi 1911 – xii 1911, *E.D. Merrill* 8304 (B, BM[2], K, P); Zamboanga, Port Banga, iv 1908, *W.I. Hutchinson* FB12342 (para not located) – Type of *Begonia anisoptera* Merr.
- **IUCN category: EN B2ab(iii)**. Known only from two lowland localities in the fragmented and denuded Mindanao–Eastern Visayas rain forest eco-region.

Begonia annulata K. Koch [§ Platycentrum], Berliner Allg. Gartenzeitung 10: 76 (1857); Irmscher, Bot. Jahrb. Syst. 78: 191 (1959). – Type: not located.
   Begonia griffithii Hook., Bot. Mag. 83: 4984 (1857); Regel, Gartenflora 8: 15 (1859); Candolle, Prodr. 15(1): 350 (1864); Clarke, Fl. Brit. Ind. 2: 647 (1879); Irmscher, Bot. Jahrb. Syst. 78: 191 (1959). – Type: not located.
   Begonia barbata Wall. ex A. DC., Prodr. 15(1): 348 (1864); Clarke, Fl. Brit. Ind. 2: 646 (1879); Irmscher, Bot. Jahrb. Syst. 78: 191 (1959). – Type: India, Sillet Mts., *N. Wallich* 3679A (holo not located).

- **BURMA:** *M.P. Chin* 102 (B); Southern Shan States, King Jung, ix 1909, *R.W. MacGregor* 783 (E); Upper Cheidwin Distr., Nampakon Drainage, 29 iv 1927, *M.P. Chin* 5847 (B); Toungoo Distr., Thandoung, xii 1937, *F.G. Dickason* 6759 (A).
   VIETNAM: Tonkin Prov., Taai Wong Mo Shan, xi 1936, *H-T. Tsai* 27278 (A); Tonkin Prov., Taai Wong Mo Shan, viii 1939, *Tsang Wai Tak* 29457 (A).
- **IUCN category: DD**. Insufficient specimens could be georeferenced with certainty.

Begonia anthonyi Kiew [§ Petermannia], Gard. Bull. Singapore 53: 248 (2001). – Type: Borneo, Sabah, Pensiangan District, Batu Punggul, *R. Kiew, S. Anthonysamy* RK4352 (holo SAN *n.v.*; iso K *n.v.*, SAR *n.v.*, SING *n.v.*).

- **BORNEO: Sabah:** Pensiangan District, Batu Punggul, *R. Kiew, S. Anthonysamy* RK4352 (holo SAN *n.v.*; iso K *n.v.*, SAR *n.v.*, SING *n.v.*) – Type of *Begonia anthonyi* Kiew.
- **IUCN category: VU B2ab(iii)**. Known only from two nearby limestone localities in Sabah.

Begonia apayaoensis Merr. [§ Petermannia], Philipp. J. Sci. 13: 39 (1918); Merrill, Enum. Philipp. Fl. Pl. 3: 120 (1923). – Type: Philippines, Luzon, Apayao, Mt. Sulu, 22 v 1917, *E. Fenix* 28403 (syn K).

- **PHILIPPINES: Luzon:** Apayao, Mt. Sulu, 22 v 1917, *E. Fenix* 28403 (syn K) – Type of *Begonia apayaoensis* Merr.
- **IUCN category: DD**. Insufficient specimens could be georeferenced with certainty.

Begonia aptera Blume [§ Sphenanthera], Enum. Pl. Javae 1: 97 (1827); Candolle, Prodr. 15(1): 397 (1864); Tebbitt, Brittonia 55(1): 25 (2003). – *Diploclinium apterum* (Blume) Miq., Fl. Ned. Ind. 1(1): 691 (1856); Candolle, Prodr. 15(1): 397 (1864). – Type: Sulawesi, Tondano, *Anon. s.n.* Herb. Lugd. Bat. 898194-39 (lecto L, here designated).
   Begonia cristata Warb. ex L.B. Sm. & Wassh., **syn. nov.**, Phytologia 52: 442 (1983); Tebbitt, Brittonia 55(1): 24 (2003). – Type: Sulawesi, Minahassa, Tomohon, iv 1894, *K.F. & P.B. Sarasin* 288 (lecto K).
   Begonia cristata Warb. ex Koord. *nom. nud.*, Natuurw. Tijdschr. Ned.-Indië 63: 90 (1904); Koorders-Schumacher, Suppl. Fl. Celebes 3: 45 (1922).

- **SULAWESI:** Tondano, *Anon. s.n.* Herb. Lugd. Bat. 898194-39 (lecto L) – Type of *Begonia aptera* Blume; *E.A. Forsten s.n.* (L); vii 1840, *E.A. Forsten s.n.* (B); *unknown s.n.* (B); Gorantalo, Olama River,

9 iv 2002, *M. Mendum, H.J. Atkins, M. Newman, Hendrian, A. Sofyan* 49 (A, E, L); E of Tongoa, 24 ii 1981, *Johansson, H. Nybom, S. Riebe* 58 (L); Goeroepahi, 27 iii 1917, *Kauderns* 61 (L); Tongoa (E of), 25 ii 1981, *Johansson* 76 (K, L); Mt. Sojol, 25 ii 2000, *G.C.G. Argent, M. Mendum, Hendrian* 151 (E); Gorantalo, Gunung Boliohutu, 23 iv 2002, *M. Mendum, H.J. Atkins, M. Newman, Hendrian, A. Sofyan* 186 (E); Tomohon, Gunung Masarang, 22 vi 1956, *L.L. Forman* 207 (K); G. Rantemario, 7 xi 1993, *S. Kofman* 210 (L); Emrekang District, Rantemario, 5 iii 2000, *G.C.G. Argent, M. Mendum, Hendrian* 238 (E); Mt. Klabut, 27 vi 1956, *L.L. Forman* 248 (K, L, P); Minahassa, Tomohon, iv 1894, *K.F. & P.B. Sarasin* 288 (lecto K) – Type of *Begonia cristata* Warb. ex L.B. Sm. & Wassh.; Minahassa, Gunung Lokon, 4 vii 1956, *L.L. Forman* 371 (K, L); Bolaang-Mongondow Distr., Dumoga Bone National Park, 6 i 1996, *M. Kato et al.* 382 (A); 16 v 1894, *K.F. & P.B. Sarasin* 488 (B[2]); iv 1894, *K.F. & P.B. Sarasin* 488 (B[2]); Bukit Dako, Lakatan Distr., Toli-Toli, 25 ii 1985, *Ramlanto, Z. Fanani* 527 (L); B. Watoewila, 25 iii 1929, *G.K. Kjellberg* 1032 (BO); Maboesa-Sae, 21 iii 1937, *P.J. Eyma* 1188 (BO); Pasoei, 9 vi 1929, *G.K. Kjellberg* 1627 (B); *unknown* 1721 (L); Todjambor, 23 vi 1929, *G.K. Kjellberg* 1750 (B); Buntu Area, Kpg Lokkok, 15 ix 2003, *J.J. Vermeulen* 2300 (L); Tanah Toraja, Rantepoa-Palopo divide, 2 ii 2004, *J.J. Vermeulen* 2409 (L); Minahassa, Mt. Soputan, 11 x 1973, *E.F. de Vogel* 2503 (L); G. Potong, 28 ii 2001, *P.J.A. Kessler* 2951 (L); Palolo, Kamarora, 15 iii 2001, *P.J.A. Kessler* 3025 (L); Nr. Pangi, 7 iii 1990, *J.S. Burley et al.* 3717 (K, L); Loewoek distr., G. Loloa – G. Beabis, 27 ix 1938, *P.J. Eyma* 3839 (L); Menado, G. Ngilalaki, 11 vii 1939, *S. Bloembergen* 4125 (L); Sopu Valley, 2 v 1979, *E.F. de Vogel* 5174 (L); Sopu Valley, 24 iii 1979, *E. Hennipman* 5588 (A, K, L) [as *Begonia cf. cristata*]; Sopu Valley, 26 v 1979, *E. Hennipman* 5633 (K, L) [as *Begonia cf. cristata*]; Bolaang-Mongondow Distr., Dumoga Bone National Park, Gunung Mogogonipa, 6 iv 1985, *E.F. de Vogel, J.J. Vermeulen* 7020 (L); Palu – Parigi, 17 iv 1975, *W. Meijer* 9329 (L); Minahassa, Bojong, 1888, *O. Warburg* 15187 (B[2]); Minahassa, Tomohon, 6 vi 1954, *A.H.G. Alston* 15679 (A, BM, L); Minahassa, *S.H. Koorders* 16244B (L); Minahassa, *S.H. Koorders* 16245B (B, L); Minahassa, *S.H. Koorders* 16246B (B).

- **Notes:** This name has often been misapplied to *B. acetosella*, possibly due to Blume stating erroneously in his description that it has 4-locular fruit (*'tetragonis'*). The rest of Blume's description agrees with the delimitation of *B. cristata*, and indeed both *B. aptera* and *B. cristata* are known only from Sulawesi. The latter is included here in synonymy (M.C. Tebbitt, pers. comm.). *Begonia aptera* is a coarser plant than the related *B. longifolia*, distinct in having larger elliptic-oblong leaves and denser inflorescences which surround the main stem.
- **IUCN category: LC.** Widespread in Sulawesi and very frequently collected, including from secondary forests.

Begonia archboldiana Merr. & L.M. Perry [not placed to section], J. Arnold Arbor. 24: 42 (1943). – Type: New Guinea, Papua New Guinea, Bella Vista, xi 1933, *L.J. Brass* 5470 (holo NY *n.v.*; iso A *n.v.*, B).

- **NEW GUINEA: Papua New Guinea:** Bella Vista, xi 1933, *L.J. Brass* 5470 (holo NY *n.v.*; iso A *n.v.*, B) – Type of *Begonia archboldiana* Merr. & L.M. Perry; Papua New Guinea, *C.E. Carr* 14974 (B, L) [as *Begonia cf. archboldiana*].
- **IUCN category: DD.** Insufficient specimens could be georeferenced with certainty.

Begonia areolata Miq. [§ Platycentrum], Pl. Jungh. 4: 417 ('1855', 1857); Candolle, Prodr. 15(1): 397 (1864); Koorders, Exkurs.-Fl. Java 2: 650 (1912). – *Diploclinium areolatum* (Miq.) Miq., Fl. Ned. Ind. 1(1): 689 (1856). – Type: Java, Patoea, *F.W. Junghuhn s.n.* (not located).
  Begonia papillosa Reinw. ex Koord. *nom. nud.*, Exkurs.-Fl. Java 2: 650 (1912). – Type: not located.

Begonia areolata var. areolata
- **SUMATRA:** Sumatra, 5 vi 1904, *R.M. Pringgo Atmodjo* 360 (L); Sumatra, *J.E. Teijsmann* 1102 (holo L; mero B) – Type of *Begonia areolata* var. *minor* Miq.; G. Sibjak, 1923, *T.J. Stomps s.n.* (L); Korinchi, 1877 – 1888, *unknown s.n.* (L); Gajo Coeas, 16 ii 1904, *R.M. Pringgo Atmodjo* 20 (B, L); Mt. Singalan, 1878, *O. Beccari* PS126 (para FI[3]) – Type of *Begonia beccariana* Ridl.; Gunung Kambot – Gunung Gadot, 23 i 1981, *M. Hotta et al.* 246 (A); Mt. Singalan, 1878, *O. Beccari* PS287 (B); Deleng Singkoet, 8 vi 1928, *C. Hanel, Rahmat Si Boeea* 515 (A); Jambi Prov., Gunung Tujuh, 26 vii 2006, *D. Girmansyah, A. Poulsen, I. Hatta, R. Neivita* 786 (E); Gunung Gadut, 28 i 1983, *M. Hotta et al.* 874 (A); Solok, Gunung Talang, 1 viii 1953, *J.v. Borssum Waalkes* 2808 (L); Barisan Range, Air Sirah, 7 iii 1954, *E.F. de Vogel* 2854 (L); Pajakumbuh, Mt. Sago, 14 iv 1955, *W. Meijer* 3181 (L); Solok, Gunung Talang, 15 xi 1988, *H. Nagamasu* 3487 (L); Palembang, Gunung Raja, 2 xi 1929, *C.G.G.J.v. Steenis* 3606 (L); Pajakumbuh, Mt. Sago, 1919, *H.A.B. Bunnemeijer* 3992 (B); Pajakumbuh, Mt. Sago, 6 x 1955, *W. Meijer* 4031 (L); Pajakumbuh, Mt. Sago, *H.A.B. Bunnemeijer* 4375 (B); G. Talang, 1919, *H.A.B. Bunnemeijer* 5091 (BO); *H.A.B. Bunnemeijer* 5267 (B); Berastagi Woods, 28 viii 1918, *J.A. Lorzing* 5969 (L); Atjeh, Boer ui Lintang, 1 ix 1934, *C.G.G.J.v. Steenis* 6323 (A); Pajakumbuh, Mt. Sago, 28 vii 1957, *W. Meijer* 7256 (L) [as *Begonia cf. areolata*]; Barisan Range, Air Sirah, 8 v 1985, *E.F. de Vogel, J.J. Vermeulen* 7528 (L); Pajakumbuh, Mt. Sago, 1953, *W. Meijer* 7636 (L); Asahan, Dolok Si Manoek-manoek, 30 vii 1936, *Rahmat Si Boeea* 9788 (A); Asahan, 12 x 1936 – 2 xii 1936, *Rhamat Si Boeea* 10378 (A); 7 xi 1936, *Rhamat Si Boeea* 10799 (A); Gunung Bandahara, 21 vii 1962, *de Wilde, de Wilde-Duyfies* 13136 (L); Gunung Bandahara, Track from Kampung Seldok, 23 vi 1972, *W.J.J.O. de Wilde* 13213 (L[2]); Dolok-baros, 14 viii 1928, *J.A. Lorzing* 13482 (L); Atjeh, Gunung Leuser Nature Reserve, 18 vii 1972, *de Wilde, de Wilde-Duyfies* 13715 (L); Atjeh, Gunung Leuser Nature Reserve, 16 viii 1972, *de Wilde, de Wilde-Duyfies* 14310 (L[2]); Lae Pondon, 29 iii 1954, *A.H.G. Alston* 14900 (BM, L); Atjeh, Gunung Leuser Nature Reserve, 20 ii 1975, *de Wilde, de Wilde-Duyfies* 15097 (L).
- **JAVA:** Java, *P.W. Korthals* (B); Java, *T. Horsfield s.n.* (BM); Java, *T. Horsfield* 144 (BM); Java, *T. Horsfield* 193 (BM); Java, 1859, *A.F. Jagor* 434 (B[2]); Preanger, Tjadas Malang, 22 i 1917, *R.C. Bakhuizen van den Brink s.n.* (K); Banjoemas, Gunung Slamat, 13 iii 2004, *W.S. Hoover, J.M. Hunter, H. Wiriadinata, D. Girmansyah* 35 (A); Banjoemas, Gunung Slamat, 13 iv 1911, *J.A. Lorzing* 244 (BO); Banjoemas, Gunung Slamat – Batureden, 18 viii 1973, *Murata et al.* 938 (L); Banjoemas, Gunung Slamat – Batureden, 18 viii 1973, *Murata et al.* 940 (L); Mt. Telagabodas, 21 iii 2001, *W.S. Hoover et al.* 941 (A); Rangga, Gunung Semboeng, 18 iii 1914, *J.A. Lorzing* 1167 (BO); Ijadas Malang, 22 i 1917, *R.C. Bakhuizen van den Brink* 1401 (BO, K); West Java, Mt. Cikurai, 14 iii 2002, *H. Wiriadinata, D. Girmansyah* 10515 (A); Banjoemas, Gunung Slamat, 19 iii 2004, *H. Wiriadinata et al.* 11348 (A); Banjoemas, Gunung Slamat, 19 iii 2004, *H. Wiriadinata et al.* 11356 (A); G. Tjarame (Tjeremai), 22 xii 1940, *C.G.G.J.v. Steenis* 12835 (K, L); Pekalongan, Jororedjo, 17 ix 1914, *C.A.B. Backer* 16241 (BO).

Begonia areolata var. minor Miq., Fl. Ned. Ind., Suppl. 1: 1091 (1861). – Type: Sumatra, *J.E. Teijsmann* 1102 (holo L); Sumatra, *J.E. Teijsmann* 1102 (mero B).
- **SUMATRA:** *J.E. Teijsmann* 1102 (holo L; mero B) – Type of *Begonia areolata* var. *minor* Miq.
- **Notes:** See notes under *B. beccariana* Ridl.
- **IUCN category: LC.** Widespread in Java and Sumatra.

Begonia arfakensis (Gibbs) L.L. Forrest & Hollingsw. [§ Symbegonia], Plant Syst. Evol. 241: 208 (2003). – *Symbegonia arfakensis* Gibbs, Fl. Arfak Mts. 149 (1917); Forrest & Hollingsworth, Plant Syst. Evol. 241: 208 (2003). – Type: New Guinea, Papua, Vogelkopf, Arfak Mts., Angi Lakes, xii 1913, *L.S. Gibbs* 5953 '6953' (holo BM).

- **NEW GUINEA: Papua:** Vogelkopf, Arfak Mts., Angi Lakes, 9 x 1948 – 22 x 1948, *A.J.G.H. Kostermans* 2061 (BO, L); Vogelkopf, Arfak Mts., Angi Lakes, 9 x 1948 – 22 x 1948, *A.J.G.H. Kostermans* 2273 (BO); Wissel Lakes, 11 iv 1939, *P.J. Eyma* 4848 (BO) [as *Begonia* aff. *arfakensis*]; Vogelkopf, Arfak Mts., Angi Lakes, xii 1913, *L.S. Gibbs* 5953 '6953' (holo BM) – Type of *Symbegonia arfakensis* Gibbs; Vogelkopf, Ije River Valley, 8 xi 1961, *P.v. Royen, H. Sleumer* 7789 (BO, K, L); Vogelkopf, Arfak Mts., Minjambau, 19 v 1962, *C. Koster* BW13859 (L[2]).
- **IUCN category: LC**. Although it is not possible to georeference many of the collections with certainty, the forests are reasonably undisturbed in the vicinity of the type collection.

Begonia argenteomarginata Tebbitt [§ Symbegonia], Edinburgh J. Bot. 61: 98 ('2004', 2005). – Type: New Guinea, Papua New Guinea, Chimbu Province, Chuave District, Mt. Elimbari, xii 1981, *T.M. Reeve* 595 (holo E; iso K, L *n.v.*).
- **NEW GUINEA: Papua New Guinea:** Chimbu Province, Chuave District, Mt. Elimbari, xii 1981, *T.M. Reeve* 595 (holo E; iso K, L *n.v.*) – Type of *Begonia argenteomarginata* Tebbitt.
- **IUCN category: DD**. Insufficient specimens could be georeferenced with certainty.

Begonia articulata Irmsch. [§ Petermannia], Webbia 9: 497 (1953). – Type: Borneo, Sarawak, Sakarrang, x 1867, *O. Beccari* PB3866 (holo FI).
- **BORNEO: Sarawak:** Sakarrang, x 1867, *O. Beccari* PB3866 (holo FI) – Type of *Begonia articulata* Irmsch.
- **IUCN category: DD**. Insufficient specimens could be georeferenced with certainty.

Begonia artior Irmsch. [§ Petermannia], Webbia 9: 484 (1953). – Type: Borneo, Sarawak, Pinindgiao, xi 1865, *O. Beccari* PB1012 (holo FI).
- **BORNEO: Sarawak:** Pinindgiao, xi 1865, *O. Beccari* PB1012 (holo FI) – Type of *Begonia artior* Irmsch.
- **IUCN category: DD**. Insufficient specimens could be georeferenced with certainty.

Begonia atricha (Miq.) A. DC. [§ Petermannia], Prodr. 15(1): 321 (1864). – *Diploclinium atrichum* Miq., Fl. Ned. Ind. 1(1): 1091 (1856); Miquel, Fl. Ned. Ind., Suppl. 1: 332 (1861); Candolle, Prodr. 15(1): 321 (1864). – Type: Sumatra, Palembajan, *J.E. Teijsmann* 1100 (holo L; mero B).
- **SUMATRA:** Palembajan, *J.E. Teijsmann* 1100 (holo L; mero B) – Type of *Diploclinium atrichum* Miq.; Pajakumbuh, Mt. Sago, 7 iv 1983, *S. Danimihardja* 2331 (L) [as *Begonia atricha* ?]; Bukit Khsang, Sibolangit, 5 viii 1917, *J.A. Lorzing* 5237 (L); Pajakumbuh, Mt. Sago, 18 v 1957, *W. Meijer* 5816 (L); Pajakumbuh, Mt. Sago, 30 ix 1956, *W. Meijer* 8336 (L); Sipora Island, 9 x 1924, *C.B. Kloss* 14653 (BO, K).
  JAVA: Oengaran, 27 ix 1910, *Doctors v. Leeuwen-Reynvaan* 103 (BO).
- **Notes:** Can be recognised by its distinctive large fruit, which tapers gradually towards a very thin pedicel.
- **IUCN category: LC**. Reasonably widespread on Sumatra and also found on Java.

Begonia augustae Irmsch. [§ Petermannia], Bot. Jahrb. Syst. 50: 350 (1913); Merrill & Perry, J. Arnold Arbor. 24: 58 (1943). – Type: New Guinea, Papua New Guinea, Sepik River, 27 x 1910, *L. Schulze Jena* 226 (holo B).
- **NEW GUINEA: Papua:** Batanta Island, Marchesa Bay, 31 iii 1954, *P.v. Royen* 3215 (FI, L); Gunung Top valley, 8 vi 1993, *McDonald, Ismail* 3833 (L); Vogelkopf, Tohkiri Range, 26 x 1961, *P.v. Royen, H. Sleumer* 7104 (L). **Papua New Guinea:** Mt. Hagen, 11 iv 1969, *Slopp s.n.* (L) [as *Begonia augustae* ?]; Sepik River, 27 x 1910, *L. Schulze Jena* 226 (holo B) – Type of *Begonia augustae* Irmsch.; Fly

River, viii 1936, *L.J. Brass* 7392 (BO); Fly River, viii 1936, *L.J. Brass* 7417 (BM, L); Waro airstrip, 14 x 1973, *M. Jacobs* 9255 (L); Bismarck Mts., 21 x 1995, *W. Takeuchi et al.* 10992 (L) [as *Begonia cf. augustae*]; Morobe Distr., Oomsis, 15 iii 1960, *Henry* 11953 (L); Chimbu Province, Crater Mt. Wildlife Management area, 27 vii 1998, *W. Takeuchi* 12628 (L) [as *Begonia augustae* ?]; Morobe Distr., Kipu, 7 i 1966, *H. Streimann* 26118 (L); Misima Island, Mt. Sisa, 28 vii 1956, *L.J. Brass* 27557 (L); West Sepik Province, Bewani, 17 ix 1977, *J. Wiakabu, S. Feni* 70487 (E).
- **IUCN category: LC.** Widespread in Papua and Papua New Guinea.

Begonia awongii Sands [§ Petermannia], in Coode *et al.*, Checkl. Fl. Pl. Gymnosperms Brunei Darussalam App. 2: 432 (1997). – Type: Borneo, Brunei, Temburong, 24 iii 1991, 4°30'N, 115°8'E, *M.J.S. Sands* 5568 (holo K; iso BRUN *n.v.*).
- **BORNEO: Brunei:** Batu Apoi Forest Reserve, 20 xi 1991, *C. Hansen* 1591 (L); Temburong, 24 iii 1991, *M.J.S. Sands* 5568 (holo K; iso BRUN *n.v.*) – Type of *Begonia awongii* Sands; Kuala Belalong Field Centre, 22 ii 1991, *G.C.G. Argent, D. Mitchell* 9188 (E, L).
- **IUCN category: LC.** Occurs in the protected Batu Apoi reserve. Forest cover is high in Brunei, and clearance rates are currently low.

Begonia axillaris Ridl. [§ Petermannia], J. Straits Branch Roy. Asiat. Soc. 46: 249 (1906). – Type: Sumatra, Lingga Arch, 17 vi 1893, *R.W. Hullett* 5707 (SING).
- **SUMATRA:** Lingga Arch, 17 vi 1893, *R.W. Hullett* 5707 (SING) – Type of *Begonia axillaris* Ridl.
- **IUCN category: DD.** Insufficient specimens could be georeferenced with certainty.

Begonia axillipara Ridl. [§ Petermannia], Trans. Linn. Soc. London, Bot., II 9: 60 (1916); Tebbitt & Dickson, Brittonia 52(1): 114 (2000). – Type: New Guinea, Papua, Canoe Camp, *C.B. Kloss s.n.* (holo BM).
- **NEW GUINEA: Papua:** Canoe Camp, *C.B. Kloss s.n.* (holo BM) – Type of *Begonia axillipara* Ridl.
- **IUCN category: DD.** Insufficient specimens could be georeferenced with certainty.

Begonia bahakensis Sands [§ Petermannia], in Coode *et al.*, Checkl. Fl. Pl. Gymnosperms Brunei Darussalam App. 2: 432 (1997). – Type: Borneo, Brunei, Ulu Tutong, 4°22'52"N, 114°50'14"E, *Kirkup* 503 (holo K; iso BRUN *n.v.*).
- **BORNEO: Brunei:** Ulu Tutong, *Kirkup* 503 (holo K; iso BRUN *n.v.*) – Type of *Begonia bahakensis* Sands.
- **IUCN category: LC.** Although currently known only from the type, forest cover in Brunei is such that none of the criteria of threat are met.

Begonia balansana Gagnep. [not placed to section], Bull. Mus. Hist. Nat. (Paris) 25: 194 (1919); Gagnepain, in Lecomte (ed.), Fl. Indo-Chine 2: 1114 (1921); Shui & Huang, Acta Bot. Yunnan. 21(1): 12 (1999); Tebbitt, Edinburgh J. Bot. 60: 3 (2003). – Type: Vietnam, Tonkin, Mt. Bavi, i 1885, *B. Balansa* 3758 (syn P); Vietnam, Tonkin, Lankok Valley, x 1887, *B. Balansa* 3764 (syn P).
  Begonia handelii auct. non Irmsch.: Shui & Huang, Acta Bot. Yunnan. 21(1): 11 (1999).
- **VIETNAM:** Tonkin, Mt. Bavi, i 1885, *B. Balansa* 3758 (syn P) – Type of *Begonia balansana* Gagnep.; Tonkin, Lankok Valley, x 1887, *B. Balansa* 3764 (syn P) – Type of *Begonia balansana* Gagnep.
- **IUCN category: CR B2ab(v).** The showy nature of this narrowly endemic species causes it to be uprooted in large numbers by tourists (Tebbitt, 2003a).

Begonia baramensis Merr. [§ Petermannia], Sarawak Mus. J. 3: 529 (1928). – Type: Borneo, Sarawak, Baram River (Upper), 1920, *J.C. Moulton s.n.* (syn B, K).

- **BORNEO: Sarawak:** Baram River (Upper), 1920, *J.C. Moulton s.n.* (syn B, K) – Type of *Begonia baramensis* Merr.; Gunung Mulu National Park, 1977 – 1978, *C. Hansen* 38 (L); Baram, Sungai Segelam, 31 iii 1978, *S.C. Chin* 2859 (L); Maputi, 21 vi 1955, *W.M.A. Brooke* 10096 (BM, L); Maputi, 5 vii 1954, *W.M.A. Brooke* 10183a (L); Bario, Bt. Lawi, 23 i 1988, *D. Awa, B. Lee* 20928 (L); Ulu Lawas, Sg. Telau, 20 x 1971, *P. Chai, I. Paie* 27940 (L); Ulu Lawas, Kenaya F.R., 26 x 1971, *P. Chai, I. Paie* 31503 (L); Bukit Tebunan, vii 1989, *B. Lee* 52444 (L). **Sabah:** Mt. Kinabalu, Marai Parai, 18 ix 1993, *K.M. Wong* 2433 (L) [as *Begonia cf. baramensis*]. **Brunei:** Ulu Temburong, 8 xi 1957, *P.S. Ashton* 387 (L); Kuala Belalong Field Centre, 22 vi 1989, *K.M. Wong* 1207 (L); Temburong, Bukit Belalong, 20 vii 1989, *K.M. Wong* 1381 (L); Batu Apoi Forest Reserve, 20 xi 1991, *C. Hansen* 1589 (L); Batu Apoi Forest Reserve, 23 xi 1991, *C. Hansen* 1609 (L); Bt. Banger Banger, 19 i 1964, *M. Hotta* 13141 (L); Nyamokning, 1 vi 1996, *A. Kalat et al.* 17555 (K, L). **Kalimantan:** Mt. Kemoll, 15 x 1925, *F.H. Endert* 4109 (L); Krayan, Long Umung – Pa Raya, 16 ix 1990, *M. Kato et al.* 23431 (L).
- **IUCN category: LC.** Occurs in Brunei, Sarawak and Kalimantan, and within protected areas.

Begonia barbellata Ridl. [§ Petermannia], J. Fed. Malay States Mus. 10: 135 (1920); Ridley, Fl. Malay Penins. 1: 856 (1922); Irmscher, Mitt. Inst. Allg. Bot. Hamburg 8: 116 (1929); Kiew, Begonias Penins. Malaysia 129 (2005). – Type: Peninsular Malaysia, Kelantan, Chaning Woods, ii 1917, *H.N. Ridley s.n.* (holo K *n.v.*).
- **PENINSULAR MALAYSIA:** Kelantan, Chaning Woods, ii 1917, *H.N. Ridley s.n.* (holo K *n.v.*) – Type of *Begonia barbellata* Ridl.; Kelantan, Kuala Aring, *B.H. Kiew* 23 (L); Kelantan, Aring Forest Reserve, 1992, *S. Anthonysamy* 1096 (L); Johore, Bukit Pengantin, 30 i 1971, *M. Shah, A. Shukor* 2258 (L); Trengganu, Batu Biwa, 22 x 1986, *R. Kiew* 2300 (L); Johore, Kota Tinggi, 13 i 1960, *J. Sinclair* 10155 (E, L); Pahang, Bukit Rengit, 10 xi 1999, *S. Damahuri et al.* 15311 (L); Kelantan, Bukit Batu Papan, 6 vii 1935, *M.R. Henderson* 29566 (L); Trengganu, Ulu Brang, vii 1937, *L. Moysey, Kiah* 33816 (L). **SUMATRA:** Riau Province, Bukit Tiga Puluh, 30 vii 2006, *D. Girmansyah, A. Poulsen, I. Hatta, F. Antoni* 794 (E); Riau Province, Bukit Tiga Puluh, 31 vii 2006, *D. Girmansyah, A. Poulsen, I. Hatta, F. Antoni* 797 (E).
- **Notes:** Previously recorded as endemic to Peninsular Malaysia (Kiew, 2005), but recent collections have extended the distribution to Riau Province, Sumatra.
- **IUCN category: LC.** Is found in Peninsular Malaysia, including in a protected area (Ulu Trengganu), and Sumatra (Proposed Seberida Nature Reserve).

Begonia bartlettiana Merr. & L.M. Perry [§ Diploclinium], J. Arnold Arbor. 29: 160 (1948). – Type: New Guinea, Papua New Guinea, Morobe Distr., Kajabit Mission, x 1939, *J. & M.S. Clemens* 10762 (holo A *n.v.*).
- **NEW GUINEA: Papua New Guinea:** Morobe Distr., Kajabit Mission, x 1939, *J. & M.S. Clemens* 10762 (holo A *n.v.*) – Type of *Begonia bartlettiana* Merr. & L.M. Perry.
- **IUCN category: DD.** Insufficient specimens could be georeferenced with certainty.

Begonia bataiensis Kiew [§ Leprosae], Gard. Bull. Singapore 57: 20 (2005). – Type: Vietnam, Kien Giang Prov., Ba Tai Hill, x 2004, *J.J. Vermeulen* 2586 (holo HCMC *n.v.*; iso HN *n.v.*, SING *n.v.*).
- **VIETNAM:** Kien Giang Prov., Ba Voi (Mo So) Hill, *Truong Quang Tam* MS56 (para BISH *n.v.*) – Type of *Begonia bataiensis* Kiew; Kien Giang Prov., Ba Tai Hill, x 2004, *J.J. Vermeulen* 2586 (holo HCMC *n.v.*; iso HN *n.v.*, SING *n.v.*) – Type of *Begonia bataiensis* Kiew; Kien Giang Prov., Ba Tai Hill, *J.J. Vermeulen* 2587 (para SING *n.v.*) – Type of *Begonia bataiensis* Kiew.
- **IUCN category: EN B2ab(iii).** Known only from two limestone hills, one of which (Ba Voi) is scheduled for quarrying.

Begonia baturongensis Kiew [§ Petermannia], Gard. Bull. Singapore 53: 251 (2001). – Type: Borneo, Sabah, Lahad Datu District, Bukit Baturong, *R. Kiew* 5026 (holo SAN *n.v.*; iso K *n.v.*, SAR *n.v.*, SING *n.v.*).

- **BORNEO: Sabah:** Lahad Datu District, Bukit Baturong, *R. Kiew* 5026 (holo SAN; iso K *n.v.*, SAR *n.v.*, SING *n.v.*) – Type of *Begonia baturongensis* Kiew.
- **IUCN category: VU D2**. Known only from a single locality which is not under formal protection.

Begonia baviensis Gagnep. [§ Platycentrum], Bull. Mus. Hist. Nat. (Paris) 25: 195 (1919); Gagnep. in Lecomte (ed.), Fl. Indo-Chine 2: 1109 (1921); Hô, Ill. Fl. Vietnam 1(2): 730 (1991). – Type: Vietnam, Tonkin, Phuong-Lam, Phuong-Lam – Doi Pu Ka, 12 xi 1887, *B. Balansa* 3761 (syn P); Vietnam, Tonkin, Lankok Valley, Lankok Valley – Hainan, *B. Balansa* 3762 (syn not located); Vietnam, Tonkin, Lankok Valley, Lankok Valley – Hainan, x 1887, *B. Balansa* 3767 (syn P); Vietnam, Muong-xen, *P.H. Lecomte, A.E. Finet* 422 (syn P); Vietnam, Muong-xen, *P.H. Lecomte, A.E. Finet* 430 (syn P); Vietnam, Muong-xen, *P.H. Lecomte, A.E. Finet* 473 (syn P).

- **VIETNAM:** Annam, Honba Massif, 15 ix 1918, *A.J.B. Chevalier s.n.* (P); Hoa Bin Province, Quyet Chien mun, 30 xi 2003, *S.G. Wu et al. s.n.* (A); Muong-xen, *P.H. Lecomte, A.E. Finet* 422 (syn P) – Type of *Begonia baviensis* Gagnep.; Muong-xen, *P.H. Lecomte, A.E. Finet* 430 (syn P) – Type of *Begonia baviensis* Gagnep.; Muong-xen, *P.H. Lecomte, A.E. Finet* 473 (syn P) – Type of *Begonia baviensis* Gagnep.; Tonkin, Mt. Bavi, 2 x 1940, *P.A. Petelot* 2587 (A); Cuc Phuong National Park, 17 ix 1999, *N.T. Hiep* 3061 (L); Tonkin, Phuong-Lam – Cho-bo, 12 xi 1887, *B. Balansa* 3761 (syn P) – Type of *Begonia baviensis* Gagnep.; Tonkin, Lankok Valley – Mt. Bavi, *B. Balansa* 3762 (syn not located) – Type of *Begonia baviensis* Gagnep.; Tonkin, valley of Bantan, xii 1807, *B. Balansa* 3766 (K, P); Tonkin, Lankok Valley – Mt. Bavi, x 1887, *B. Balansa* 3767 (syn P) – Type of *Begonia baviensis* Gagnep.; Cuc Phuong National Park, 15 xi 2000, *unknown* NTH 4245 (CPNP *n.v.*); Annam, Honba Massif, 18 ix 1918 – 20 ix 1918, *A.J.B. Chevalier* 13641 (P); Muong-xen – Lao-kay, 5 xii 1913, *A.J.B. Chevalier* 29367 (P); Muong-xen – Lao-kay, 5 xii 1913, *A.J.B. Chevalier* 29368 (P); Annam, Honba Massif, 18 ix 1918 – 20 ix 1918, *A.J.B. Chevalier* 38641 (P).
- **IUCN category: NT**. Although this species has a narrow distribution, it is well represented in collections (including from the Cuc Phuong and Ba Vi National Parks), which indicates it is reasonably abundant.

Begonia beccariana Ridl. [§ Platycentrum], J. Malayan Branch Roy. Asiat. Soc. 1(87): 62 (1923). – Type: Sumatra, Berastagi Woods, ii 1921, *H.N. Ridley s.n.* (syn K[2]).

- **SUMATRA:** Berastagi Woods, ii 1921, *H.N. Ridley s.n.* (syn K[2]) – Type of *Begonia beccariana* Ridl.; Mt. Singalan, 21 ii 2004, *D. Girmansyah et al.* 9 (E); Atjeh, Bur ui Papandji, 22 vi 1930, *Frey-Wyssling* 47 (BO); Mt. Singalan, 1878, *O. Beccari* PS126 (para B, K) – Type of *Begonia beccariana* Ridl.; Liwa Benkoelen, 26 iv 1928, *C.N.A. de Voogd* 129 (BO, L); Lampongs, Tangamoes, 2 vii 1928, *C.N.A. de Voogd* 175 (BO, L); Mt. Singalan, 1878, *O. Beccari* PS287 (FI[2], K); Bukit Palelawan Nature Reserve, 10 ii 1983, *J.J. Afriastini* 716 (L); Sibayak Volcano, 15 ii 1932, *Bangham* 1017 (A); Deli Gilibajak, iv 1927, *J.G.B. Beumée* 1571 (BO); Atjeh, 25 ii 1980, *S. Prawiroatmodjo* 2435 (L); Palembang, Ranaumeer, 1929, *C.G.G.J.v. Steenis* 3712 (BO); Mt. Tandikat, 24 vii 1955, *W. Meijer* 3916 (L); Atjeh, Boer ui Lintang, 1 ix 1934, *C.G.G.J.v. Steenis* 6323 (BO); Karoland, Deleng Baroes, 21 vi 1927, *H.H. Bartlett* 8515 (L); Atjeh, Gajolanden, 4 iii 1937, *C.G.G.J.v. Steenis* 9467 (L); Asahan, 12 x 1936 – 2 xii 1936, *Rhamat Si Boeea* 10378 (L); 7 xi 1936, *Rhamat Si Boeea* 10799 (L); Gunung Bandahara, 19 vi 1972, *de Wilde, de Wilde-Duyfies* 13046 (L); Gunung Bandahara, Track from Kampung Seldok, 21 vi 1972, *W.J.J.O. de Wilde* 13136 (BO); Gunung Bandahara, Track from Kampung Seldok, 23 vi 1972, *W.J.J.O. de Wilde* 13213 (BO, L); Atjeh, 10 ii 1975, *de Wilde, de Wilde-*

Duyfies 14702 (L); Lae Pondon, 29 iii 1954, *A.H.G. Alston* 14900 (BO); Atjeh, Gunung Leuser Nature Reserve, 22 vi 1979, *de Wilde, de Wilde-Duyfies* 18328 (L); Atjeh, Gunung Leuser Nature Reserve, 22 vi 1979, *de Wilde, de Wilde-Duyfies* 18365 (L).

- **Notes:** Confirmed here as belonging to § *Platycentrum* (previously doubtfully assigned to this section in Doorenbos *et al.*, 1998). Probably synonymous with *B. areolata* Miq. to form a variable species distributed across Java and Sumatra, and easily recognised through having paired leaves subtending the inflorescence. See also notes under *B. bifolia* Ridl.
- **IUCN category: DD**. Not evaluated due to its unclear distinction from the earlier described *B. areolata* Miq.

Begonia beccarii Warb. [not placed to section], Repert. Spec. Nov. Regni Veg. 18: 329 (1922). – Type: Borneo, Sarawak, *O. Beccari* PB1013 (holo B; iso FI[2], P).

- **BORNEO: Sarawak:** *O. Beccari* PB1013 (holo B; iso FI[2], P) – Type of *Begonia beccarii* Warb.
- **Notes:** Some duplicates of the type collection have been determined as *Begonia promethea* Ridl. by Irmscher, and it would seem that this species could be synonymised after further investigation.
- **IUCN category: DD**. Insufficient specimens could be georeferenced with certainty.

Begonia berhamanii Kiew [§ Petermannia], Gard. Bull. Singapore 53: 253 (2001). – Type: Borneo, Sabah, Pensiangan District, Batu Punggul, *R. Kiew, S. Anthonysamy* RK5046 (holo SAN *n.v.*; iso SING *n.v.*).

- **BORNEO: Sabah:** Pensiangan District, Batu Punggul, *R. Kiew, S. Anthonysamy* RK5046 (holo SAN *n.v.*; iso SING *n.v.*) – Type of *Begonia berhamanii* Kiew.
- **IUCN category: VU D2**. Known only from a single locality which is not under formal protection.

Begonia beryllae Ridl. [§ Petermannia], Sarawak Mus. J. 2(6): 177 (1915); Sands, Pl. Mt. Kinabalu 147 (2001). – Type: Borneo, Sabah, Mt. Kinabalu, *J.C. Moulton s.n.* (cult) (holo K).

- **BORNEO: Sabah:** Mt. Kinabalu, *J.C. Moulton s.n.* (cult) (holo K) – Type of *Begonia beryllae* Ridl.; Mt. Kinabalu, 4 ix 1913, *unknown* 79 (E); Mt. Kinabalu, 25 iii 1982, *J. Sinclair* 193 (E); Mt. Kinabalu, Tenompok F.R., 11 v 1967, *W.R. Price* 231 (K); Tenom, Crocker Range, 10 iii 1991, *A. Lamb* 407/91 (E[2]); Mt. Kinabalu, 22 i 1976, *P.F. Stevens* 620 (L); G. Alab, 26 ii 1969, *H.P. Nooteboom* 1045 (L); Penampang, 11 xii 1968, *Kokawa, Hotta* 1737 (L); Penampang, Alab Mts., 14 xii 1968, *Kokawa, Hotta* 2172 (L); Mt. Kinabalu, Liwagu River, 3 i 1969, *Kokawa, Hotta* 2860 (L); Mt. Kinabalu, *W.L. Chew, E.J.H. Corner* 4063 (K) [as *Begonia cf. beryllae*]; Mt. Kinabalu, Sungei Mamut, 3 ii 1969, *S.H. Koorders* 4880 (L); Mt. Kinabalu, Sungai Berumbang, 14 ii 1969, *Kokawa, Hotta* 5716 (L); Mt. Kinabalu, Mamut Ridge – Ulu Berambung, 19 ii 1954, *Kokawa, Hotta* 5874 (L); Tenom, Crocker Range, 9 x 1983, *J.H. Beaman* 7158 (L); Mt. Kinabalu, 13 x 1983, *J.H. Beaman* 7211 (L); Pinosuk Plateau, 3 x 1986, *J.M. Huisman, J.J. Vermeulen, E.F. de Vogel, P.C.v. Welzen* 8006 (L); Pinosuk Plateau, 22 iii 1984, *J.H. Beaman* 8995 (L); Pinosuk Plateau, 14 iv 1984, *J.H. Beaman* 9371 (L); Mt. Kinabalu, Tenompok F.R., 21 iv 1984, *J.H. Beaman* 9439 (L); Mt. Kinabalu, Sungai Berumbang, 25 viii 1976, *M. Hotta* 20326 (L); Mt. Kinabalu, Kambarangoh Relay Stn, 15 vii 1963, *H.P. Fuchs* 21043 (K, L); Mt. Kinabalu, Tenompok F.R., 21 i 1932, *J. & M.S. Clemens* 28392 (BM, K, L); Mt. Kinabalu, Sosopodon, 21 ii 1962, *W. Meijer* 29243 (K, L); Ransu, Bukit Kulimpisan, vii 1961, *W. Meijer* 34630 (K); Mt. Kinabalu, Penibukan, Pinokkok Falls, *J. & M.S. Clemens* 50005 (BM); Tambunan, Alab range, Gunung Alab, 20 vii 1984, *Amin et al.* 60304 (K, L); Ranau District, 12 iii 1986, *Amin et al.* 114225 (L); Tenom, Crocker Range, 25 viii 1986, *F. Krispinum* 120576 (E); Kota Belud, 15 iv 1989, *K. Julius* 124869 (L); Tambunan, Gunung Trusmadi, 26 viii 1988, *Fidilis et al.* 125597 (K).

- **IUCN category: LC.** The bulk of the known range of this species is encompassed by the Crocker Range National Park and the Kinabalu Park.

Begonia bifolia Ridl. [§ Platycentrum], J. Fed. Malay States Mus. 8(4): 40 (1917). – Type: Sumatra, Korinchi, Barong Baru, 7 vi 1914, *H.C. Robinson, C.B. Kloss* (syn BM, K).
- **SUMATRA:** Korinchi, Barong Baru, 7 vi 1914, *H.C. Robinson, C.B. Kloss* (syn BM, K) – Type of *Begonia bifolia* Ridl.
- **Notes:** Ridley speculates in his notes to this species that it may be the same as *B. areolata* Miq., although he stated he did not have access to any specimens of the latter to confirm this. Examination of the type shows it to resemble *B. areolata*, with which it shares the paired leaves below the inflorescence. See also notes under *B. beccariana* Ridl.
- **IUCN category: DD.** Not evaluated due to its unclear distinction from the earlier described *B. areolata* Miq.

Begonia biliranensis Merr. [§ Diploclinium], Philipp. J. Sci. 10: 46 (1915); Merrill, Enum. Philipp. Fl. Pl. 3: 120 (1923). – Type: Philippines, Leyte, Biliran, vi 1914, *R.C. McGregor* 18822 (syn BM, K, P).
- **PHILIPPINES: Panay:** 27 xii 1912 – 31 xii 1912, *C.B. Robinson* 18041 (BM[2], P). **Leyte:** Biliran, vi 1914, *R.C. McGregor* 18544 (para P) – Type of *Begonia biliranensis* Merr.; Biliran, vi 1914, *R.C. McGregor* 18760 (para not located) – Type of *Begonia biliranensis* Merr.; Biliran, vi 1914, *R.C. McGregor* 18822 (syn BM, K, P) – Type of *Begonia biliranensis* Merr. **Samar:** Catubig River, ii 1916 – iii 1916, *M. Ramos* 24478 (K, P) [as *Begonia cf. biliranensis*].
- **IUCN category: EN B2ab(iii).** Endemic to Biliran (although possibly extended to the Catubig River in northern Samar; however, this is a tentative determination), where there are no formally protected areas.

Begonia binuangensis Merr. [§ Petermannia], Philipp. J. Sci. 13: 40 (1918); Merrill, Enum. Philipp. Fl. Pl. 3: 120 (1923). – Type: Philippines, Luzon, Tayabas, Mt. Binuang, v 1917, 14°46′29″N, 121°33′30″E, *M. Ramos, G. Edano* 28813 (syn K).
- **PHILIPPINES:** *A. Loher* 6088 (B). **Luzon:** Tayabas, Mt. Alzapan, v 1925 – vi 1925, *M. Ramos, G. Edano* 45??? (P); Sorsogon, Mt. Bulusan, x 1915, *A.D.E. Elmer* 14366 (B, BM[2], K, L, P, U) – Type of *Begonia hemicardia* Elmer ex Merr. *nom. nud.*; Sorsogon, Mt. Bulusan, xi 1916, *A.D.E. Elmer* 15262 (B, BM[2], K, L, P, U); Sorsogon, Mt. Bulusan, v 1916, *A.D.E. Elmer* 16083 (B, BM[2], K, L, P, U) – Type of *Begonia bulusanensis* Elmer ex Merr. *nom. nud.*; Sorsogon, vii 1915 – viii 1915, *M. Ramos* 23374 (K); Tayabas, Mt. Binuang, v 1917, *M. Ramos, G. Edano* 28813 (syn K) – Type of *Begonia binuangensis* Merr.; Catanduanes, xi 1917 – xii 1917, *M. Ramos* 30436 (K); Sorsogon, Mt. Bulusan, v 1957, *G. Edano, H. Gutierrez* 37847 (BM); Rizal Province, Mt. Irig, ii 1923, *M. Ramos* 41882 (B, K, L, P). **Mindoro:** Mt. Halcon, iii 1922, *M. Ramos, G. Edano* 40803 (K). **Panay:** Capiz, Mt. Macosolon, iv 1910 – v 1910, *M. Ramos, G. Edano* 30742 (L). **Negros:** Negros Occidental, Mt. Canlaon, iv 1954, *G. Edano* 21925 (K, L).
- **Notes:** There are two unpublished names from Elmer considered by Merrill to be synonymous with this species (*B. hemicardia* and *B. bulusanensis*). See notes under *B. aequata* A. Gray.
- **IUCN category: DD.** Taxonomic limits between this species and the rest of the complex need defining.

Begonia bipinnatifida J.J. Sm. [§ Petermannia], Bull. Dep. Agric. Indes Neerl. 2: 47 (1906). – Type: New Guinea, Papua, *unknown s.n.* (holo BO; iso BO).
- **NEW GUINEA: Papua:** *unknown s.n.* (holo BO; iso BO) – Type of *Begonia bipinnatifida* J.J. Sm.; Biak, 1946, *D.T.E.v.d. Ploeg s.n.* (L); Saroerai, 22 vii 1939, *L.J.V. Dijk* 10 (L); Jappen-Biak, 12 vii 1939,

L.J.V. Dijk 62 (L); 19 viii 1966, *A.J.G.H. Kostermans* 428 (L); Napan Distr., Wati, 17 iv 1943, *S. Ijiri, T. Niimura* 621 (L); Jappen-Biak, 16 ix 1939, *Aet, Idjan* 801 (BO) [as *Begonia cf. bipinnatifida*]; Biak, 14 vi 1959, *J.J.F.E. de Wilde* 1165 (L); Cycloop Mts., Anafre River, 15 iii 1957, *C. Koster* 4283 (L); Kebar Valley, 15 xi 1954, *P.v. Royen* 5065 (L); Cycloop Mts., Faika River, 8 vi 1961, *P.v. Royen, H. Sleumer* 5750 (L); Cycloop Mts., 27 vi 1961, *P.v. Royen, H. Sleumer* 6056 (L); Cycloop Mts., 8 viii 1961, *P.v. Royen, H. Sleumer* 6534 (L); Rouffaer River, viii 1926, *Doctors v. Leeuwen-Reynvaan* 9676 (L); Rouffaer River, ix 1926, *Doctors v. Leeuwen-Reynvaan* 10364 (L); Jappen-Biak, 29 ix 1939, *L.J.V. Dijk* 62928 (L). **Papua New Guinea:** Telefomin Subdist., Star Mts., 12 v 1959, *C. Kalkman, M.O. Tissing* 4032 (L); Pieni River, 23 vi 1961, *P.J. Darbyshire, R.D. Hoogland* 8006 (L); West Sepik Province, Bewani, 31 viii 1982, *J. Wiakabu* 73800 (L).

- **Notes:** Whether this species is distinct from the earlier described *B. warburgii* K. Schum. & Lauterb. needs further investigation.
- **IUCN category: DD**. Delimitation of this species relative to *B. warburgii* is not clear.

Begonia boisiana Gagnep. [not placed to section], Bull. Mus. Hist. Nat. (Paris) 25: 195 (1919); Gagnepain, Bull. Mus. Hist. Nat. (Paris) 25: 281 (1919); Gagnepain, in Lecomte (ed.), Fl. Indo-Chine 2: 1102 (1921); Hô, Ill. Fl. Vietnam 1(2): 730 (1991); Tebbitt, Begonias 87 (2005). – Type: Vietnam, Kien-Khe, *Bon* 2276 (syn P); Vietnam, Tonkin, Surprise Island, *J.O. Debeaux* 293 (syn not located); Vietnam, Tonkin, Surprise Island, 1911, *P.H. Lecomte, A.E. Finet* 735 (syn P); Vietnam, Tonkin, Surprise Island, *P.H. Lecomte, A.E. Finet* 742 (syn not located); Vietnam, Tonkin, Surprise Island, *P.H. Lecomte, A.E. Finet* 758 (syn P); Vietnam, Tonkin, Along Bay, *D.G.J.M. Bois s.n.* (syn not located).

- **VIETNAM:** Tonkin, Along Bay, *D.G.J.M. Bois s.n.* (syn not located) – Type of *Begonia boisiana* Gagnep.; Tonkin, xii 1908, *D'Alleizette s.n.* (L); Tonkin, v 1909, *D'Alleizette s.n.* (L); Cuc Phuong National Park, 1 vii 1999, *N.M. Cuong* 222 (L); Tonkin, Surprise Island, *J.O. Debeaux* 293 (syn not located) – Type of *Begonia boisiana* Gagnep.; Tonkin, Surprise Island, 1911, *P.H. Lecomte, A.E. Finet* 735 (syn P) – Type of *Begonia boisiana* Gagnep.; Tonkin, Surprise Island, *P.H. Lecomte, A.E. Finet* 742 (syn not located) – Type of *Begonia boisiana* Gagnep.; Annam, Thua-thien Province, Mt. Bach Ma, 12 viii 1942, *J.E. Vidal* 756A (P); Tonkin, Surprise Island, *P.H. Lecomte, A.E. Finet* 758 (syn P) – Type of *Begonia boisiana* Gagnep.; Cuc Phuong National Park, 13 iv 1970, *unknown* CPNP 996 (CPNP *n.v.*); Annam, Mt. Bani (Bana), 6 vi 1920, *E. Poilane* 1530 (P); Kien-Khe, *Bon* 2276 (syn P) – Type of *Begonia boisiana* Gagnep.; Cuc Phuong National Park, 16 ix 1999, *N.T. Hiep* 3016 (L); Cuc Phuong National Park, 17 ix 1999, *N.T. Hiep* 3069 (L); Cuc Phuong National Park, 1 ii 2000, *N.T. Hiep* 3089 (L); Annam, Mt. Bani (Bana), vii 1927, *J. & M.S. Clemens* 3898 (P); Annam, Mt. Bani (Bana), 8 vii 1923, *E. Poilane* 6961 (P); Annam, 13 ix 1938, *E. Poilane* 27768 (P); Annam, 15 iv 1939, *E. Poilane* 29732 (P).
- **IUCN category: LC**. Fairly widespread in Vietnam, from sea level to 1400 m. The Cuc Phuong and Surprise Island localities are protected.

Begonia bolsteri Merr. [§ Petermannia], Philipp. J. Sci. 6: 387 ('1911', 1912); Merrill, Enum. Philipp. Fl. Pl. 3: 120 (1923). – Type: Philippines, Mindanao, Surigao, iv 1906, *F.H. Bolster* 310 (syn B, K).

- **PHILIPPINES: Luzon:** Sorsogon, Mt. Bulusan, iv 1916, *A.D.E. Elmer* 15886 (BM, BO, K, L). **Leyte:** Cabalian, v 1922, *G. Lopez* 40807 (P). **Mindanao:** Surigao, iv 1906, *F.H. Bolster* 310 (syn B, K) – Type of *Begonia bolsteri* Merr.; Surigao, iv 1925, *C.A. Wenzel* 3230 (K); Surigao, iv 1918, *M. Ramos* 34458 (B, BM, K, P).
- **IUCN category: EN B2ab(iii)**. Although originally said to be common in suitable shady habitat (75 m) by the collector (1906), there are no protected areas in Surigao covering forests of this altitude, and the Luzon rain forest eco-region is highly fragmented and modified. It is known

from only three locations. However, the species is capable of surviving in even small pockets of forest (Sands, 2001).

Begonia bonii Gagnep. [§ Reichenheimia], Bull. Mus. Hist. Nat. (Paris) 25: 196 (1919); Gagnepain, in Lecomte (ed.), Fl. Indo-Chine 2: 1115 (1921). – Type: Vietnam, Kien-Khe, Col de Dong-bau, *Bon* 2872 (holo P); Vietnam, Kien-Khe, Col de Dong-bau, *Bon* 2872 (mero B).
- **VIETNAM:** Cuc Phuong National Park, 18 x 1963, *Pocs, Kornas* 924 (P); Kien-Khe, Col de Dong-bau, *Bon* 2872 (holo P; mero B) – Type of *Begonia bonii* Gagnep.
- **IUCN category: DD**. The type locality is defined only as Dong-Bau Mts., which cannot be georeferenced with certainty. Apart from the type, there is only one other collection referred to this name, which is without flowers or fruits and hence cannot be confirmed with certainty.

Begonia bonthainensis Hemsl. [§ Petermannia], Bull. Misc. Inform. Kew 1896: 37 (1896). – Type: Sulawesi, Bonthain Peak, x 1894, *A.H. Everett* 34 (holo K).
- **SULAWESI:** Bonthain Peak, x 1894, *A.H. Everett* 34 (holo K) – Type of *Begonia bonthainensis* Hemsl.; Bonthain Peak, 12 vii 1976, *W. Meijer* 11036 (L); Bonthain Peak, 18 v 1921, *H.A.B. Bunnemeijer* 11607 (BO); Lombasang, 26 v 1921, *H.A.B. Bunnemeijer* 11729 (BO); Bonthain Peak, 10 vi 1921, *H.A.B. Bunnemeijer* 12030 (BO).
- **IUCN category: VU D2**. The Lompobatang massif is densely populated, and the forests below 1700 m are heavily disturbed. Although the distribution of this species is in the 2000–3000 m range, and hence within a protected area, given pressures on forests in this region this species must be considered as vulnerable.

Begonia borneensis A. DC. [§ Petermannia], Ann. Sci. Nat. Bot. 11: 128 (1859); Merrill, Sarawak Mus. J. 3: 530 (1928); Ridley, J. Straits Branch Roy. Asiat. Soc. 46: 250 (1906); Merrill, Philipp. J. Sci. 6: 406 ('1911', 1912); Gibbs, J. Linn. Soc., Bot. 42: 84 (1914); Merrill, Philipp. J. Sci. 29: 403 (1926); Merrill, J. Straits Branch Roy. Asiat. Soc. 414 (1929). – Type: Borneo, *E.S. Barber* 329 (syn K); Borneo, Sabah, Labuan, *J. Motley s.n.* (syn K).
- **BORNEO:** *E.S. Barber* 329 (K) – Type of *Begonia borneensis* A. DC.; *G.D. Haviland* 1189 (not located). **Sarawak:** 1924, *E.G. Mjoberg* 186 (K); 1865 – 1868, *O. Beccari* PB1195 (K); ii 1914 – vi 1914, *Native collector* 2410 (K). **Sabah:** Labuan, *J. Motley s.n.* (K) – Type of *Begonia borneensis* A. DC.; Banguey Island, *P. Castro, F. Melegrito* 1362 (B[2], P); Mt. Kinabalu, *G.D. Haviland* 1707 (not located); Sandakan District, Bettotan River, 26 vii 1927, *C.B. Kloss* 18986 (K).
- **IUCN category: DD**. Taxonomic uncertainty; the name appears to be applied to herbarium specimens fairly indiscriminately.

Begonia brachybotrys Merr. & L.M. Perry [§ Petermannia], J. Arnold Arbor. 24: 56 (1943); Tebbitt & Dickson, Brittonia 52(1): 114 (2000). – Type: New Guinea, Papua, Idenburg River, iv 1939, *L.J. Brass* 14112 (holo A *n.v.*; iso L).
   Begonia brachyptera Merr. & L.M. Perry, J. Arnold Arbor. 29: 160 (1948); Tebbitt & Dickson, Brittonia 52(1): 114 (2000). – Type: New Guinea, Papua New Guinea, Morobe Distr., Wantoat, 1940, *J. & M.S. Clemens* 40896 (holo A *n.v.*).
- **LESSER SUNDA ISLANDS: Aru Islands:** Pulau Wokam, Dosinamaloe, 29 v 1938, *P. Buwalda* 5069 (L).
   MOLUCCAS: Sula Islands: Mangoli, 26 ix 1939, *S. Bloembergen* 4689 (L). **Buru:** Waekosi, 4 xi 1984, *M.M.J.v. Balgooy* 4596 (L); Wae Duna River, 25 xi 1984, *M.M.J.v. Balgooy* 4911 (L); Wae Duna River, 26 xi 1984, *M.M.J.v. Balgooy* 4917 (L); 1 xii 1984, *M.M.J.v. Balgooy* 5048 (L); Wae Langa, 9 xi 1984, *H.P. Nooteboom* 5087 (L); Wae Langa, 9 xi 1984, *H.P. Nooteboom* 5089 (L); Wae Duna River, 23

xi 1984, *H.P. Nooteboom* 5222 (L); Wae Duna River, 25 xi 1984, *H.P. Nooteboom* 5277 (L). **Seram:** Sikeu Walala – Wae Tapakasitam, 20 xii 1996, *M. Kato et al.* 1114 (L); Lelesiru, 24 xii 1996, *M. Kato* 1167 (L); Mansuela National Park, Maraina, 2 i 1985, *K. Udea, M. Okamoto, U.W. Mahjar* 2834 (L); Roemoga – Tasikmi, 2 ix 1938 – 4 ix 1938, *P. Buwalda* 5945 (L); Buria – Wae River, 3 ii 1985, *M. Kato et al.* 6001 (L); Mansuela National Park, 15 ii 1985, *M. Kato et al.* 6673 (A, L); Mansuela National Park, Hatumete – Hoale Pass, 20 ii 1985, *M. Kato et al.* 6947 (L).

NEW GUINEA: Papua: Jappen-Biak, G. Wawah, 22 vii 1939, *L.J.V. Dijk* 44 (L); Mamberamo, 28 viii 1953, *C. Versteegh* 47 (L); Jappen-Biak, G. Wawah, 22 vii 1939, *L.J.V. Dijk* 215 (L); vi 1928 – viii 1928, *E. Mayr* 546 (B); Mt. Jaya, 18 xi 2000, *S. Atkins et al.* 640 (L); Mamberamo, 3 iv 1914, *A.C.T. Thomson* 643 (L); Koode River, 16 viii 1938, *E. Meijer Drees* 653 (L); Jappen-Biak, 16 ix 1939, *L.J.V. Dijk* 802 (L); Jappen-Biak, Arijom, 29 ix 1939, *L.J.V. Dijk* 923 (L); Cycloop Mts., Faika River – Gawesar River, 10 vi 1961, *P.v. Royen, H. Sleumer* 5767 (L); Cycloop Mts., Balmungun Creek – Klifon River, 8 viii 1961, *P.v. Royen, H. Sleumer* 6536 (L); Mt. Jaya, 7 xii 1998, *M.J.S. Sands* 7210 (L); Vogelkopf, Ije River Valley, 2 xi 1961, *P.v. Royen, H. Sleumer* 7650 (L); Rouffaer River, viii 1926, *Doctors v. Leeuwen-Reynvaan* 9743 (L); Rouffaer River, viii 1926, *Doctors v. Leeuwen-Reynvaan* 9744 (L); Rouffaer River, viii 1926, *Doctors v. Leeuwen-Reynvaan* 9905 (L); Mt. Jaya, 7 v 2000, *T.M.A. Utteridge et al.* 10615 (L); Vogelkopf, Arfak Mts., Minjambau, 22 v 1962, *C. Versteegh* 12685 (L); Idenburg River, iv 1939, *L.J. Brass* 14112 (holo A *n.v.*, BM; iso L) – Type of *Begonia brachybotrys* Merr. & L.M. Perry; Sentani, 9 iii 1973, *J. Raynal* 16685 (P); 9 v 1973, *J. Raynal* 17645 (P); Upper Ramu, ix 1939, *J. & M.S. Clemens* 40750 (para not located) – Type of *Begonia brachyptera* Merr. & L.M. Perry. **Papua New Guinea:** 1936, *C.E. Carr* 14975? (B); Morobe Distr., Sattelberg, 8 x 1935, *J. & M.S. Clemens* 379a (B); Sorong, 19 viii 1948, *D.R. Pleyte* 617 (L); Morobe Distr., Sattelberg, 31 x 1935, *J. & M.S. Clemens* 655 (B[2]); Sepik District, Maprik Subdistr., Prince Alexander Range, Mt. Turu, 24 viii 1959, *R. Pullen* 1582 (L); Morobe Distr., Wafi River – Watut Mt., 4 iii 1985, *B.J. Conn* 1800 (L); Morobe Distr., Mt. Kaindi, 29 iv 1992, *R. Hoft* 2186 (L); Morobe Distr., Sattelberg, 28 iii 1936, *J. & M.S. Clemens* 2244 (B); Morobe Distr., Yungziang, 11 vi 1936, *J. & M.S. Clemens* 3281 (B); Baiyer River, 28 vi 1981, *A.N. Vinas* 3286 (L); Yodda Valley, 19 ix 1953, *R.D. Hoogland* 3941 (L); Morobe Distr., Bulolo, Dengalu Village, *Floyd* 5245 (L); Central Division, Mafulu, xi 1933, *L.J. Brass* 5429 (para B) – Type of *Begonia brachybotrys* Merr. & L.M. Perry; Sangwep Logging Area, 13 iii 1975, *J.F. Veldkamp* 6168 (L); Busu River, v 1937, *J. & M.S. Clemens* 6327 (para A *n.v.*) – Type of *Begonia brachyptera* Merr. & L.M. Perry; Aitape Subdistr., Pes Village, 8 vi 1961, *P.J. Darbyshire, R.D. Hoogland* 7896 (L); Morobe Distr., Garaina, 10 v 1971, *B.C. Stone* 10169 (L); Morobe Distr., Sopa, 19 vii 1962, *T.G. Harley* 10354 (L); Nassau Mts., x 1926, *Doctors v. Leeuwen-Reynvaan* 10545 (L); Chimbu Province, Crater Mt. Wildlife Management area, 12 ix 1996, *W. Takeuchi* 11125 (A, L); Morobe Distr., Wantoat, 9 viii 1968, *A.N. Millar* 12139 (L); Josephstaal, 28 vii 1999, *W. Takeuchi et al.* 13420 (A, L); Josephstaal, 22 xii 1999, *W. Takeuchi et al.* 13422 (L); Morobe Distr., Bumbu Logging Area, ix 1961, *Henry* 14321 (L); Bewepi Creek, 3 ix 1962, *Henry* 14820 (E[2], L); *C.E. Carr* 14995 (BM, L); Morobe Distr., Atzera Range, 10 vii 2001, *W. Takeuchi et al.* 15424 (L); Isuarava, *C.E. Carr* 15598 (B, L); Isuarava, 15 ii 1936, *C.E. Carr* 15598 (B, L); Morobe Distr., Bubia, xii 2001, *W. Takeuchi et al.* 15806 (L); Kokoda, 18 iii 1936, *C.E. Carr* 16124 (B, L); Kaiser-Wilhelmsland, 24 vi 1927, *F.R.R. Schlechter* 16175 (E); Morobe Distr., Bubia, 17 vii 2001, *W. Takeuchi et al.* 16398 (L); Madang Prov., Saidor Subdistr., Wumundi, 11 viii 1984, *C.D. Sayers* 19778 (L); Morobe Distr., Bupu, 4 iii 1964, *A.N. Millar* 23329 (L); Kokoda, 25 vii 1964, *A.N. Millar* 23496 (L); Morobe Distr., Kipu, 13 xi 1968, *A. Gillison, A. Kairo* 25692 (L); Morobe Distr., Bulolo, Crooked Logging Area, 5 ix 1966, *H. Streimann, A. Kairo* 27900 (L); Eastern Highlands Distr., Arau, 9 x 1959, *L.J. Brass* 31981 (L); Morobe Distr., Sankwep logging area, 23 iv 1968, *J.S. Womersley* 37119 (L); Western Highland Province, Hagen Subdistrict, Trauna Creek, 7 vii 1968, *A.N. Millar* 37627 (L); Kar Kar Island, 16 vii 1968, *A.N. Millar* 37750 (L); Morobe Distr., Wantoat, 1940, *J. & M.S. Clemens* 40896 (holo A *n.v.*) – Type of *Begonia brachyptera* Merr. & L.M. Perry; Morobe Distr., Wantoat, 1940, *J. & M.S. Clemens* 41205 (para A *n.v.*) – Type of *Begonia brachyptera* Merr. & L.M. Perry; West Sepik Province, Bewani, Mt. Yungat, 20 ix 1982, *J. Wiakabu* 50578 (L); West Sepik Province, Vanimo, 30 xi 1971, *H. Streimann* 52967 (L); West Sepik Province, Bewani, 31 viii 1982, *K. Kerenga* 55469 (L); West Sepik Province, Wutung Patrol Post, 8 ix 1982, *K. Kerenga* 56405 (L); Goroka Subprov., Mt. Hozeke, 29 xi 1984, *K. Kerenga, C. Baker* 56917 (L); Mt. Bosavi, 27 viii 1986, *O. Gideon* 57420 (L); Morobe Distr., Lae Subprovince, Musom Village, 19 vi 1978, *P. Katik* 70827 (L); East Sepik Province, Maprik subprovince, 24 vii 1980, *J. Wiakabu* 73491 (L); Madang Prov., Walium Patrol Post, 19 iv 1979, *Sohmer, P. Katik* 75116 (L).

BISMARCK ARCHIPELAGO: New Britain: Kandrian Subdistr., Alimbit River, 8 x 1965, *A. Gillison* 22466 (L); Ulamona Mission, 6 vi 1973, *R.S. Isles, A.N. Vinas* 32353 (L); Hoskins Subdistr., Dakamanu, 21 xii 1967, *M. Coode* 32645 (L); Kokopo Subdistr., Valilie River, 14 viii 1969, *H. Streimann* 44374 (L); Pomio Subdistr., Mt. Lululua, 4 v 1973, *P.F. Stevens, Y. Leiean* 58247 (E, L); 20 ii 1989, *P. Katik* 64088 (L). **New Ireland:** Ambitle Island, 10 xi 2003, *W. Takeuchi* 16731 (L); Kavieng Subdistr., Schieinitz Range, Logagon Village, 24 x 1974, *J. Croft, Y. Leiean* 65630 (L); Konos Province, NE of Lelet Farm, 26 x 1984, *O. Gideon* 77137 (L). **Manus:** Mt. Dremsel, 26 x 1974, *D. Foreman, P. Katik* 59159 (E, L); Buyang, 11 iii 1981, *K. Kerenga, J. Croft* 77357 (L); Rambutyo Island, 16 iii 1981, *K. Kerenga, J. Croft* 77376 (L).

■ **Notes:** See notes under *B. rieckei* Warb. This is the only *Begonia* reported from the Aru Islands.
■ **IUCN category: LC**. Widespread.

Begonia bracteata Jack [§ Bracteibegonia], Malayan Misc. 2(7): 13 (1822); Miquel, Pl. Jungh. 417 ('1855', 1857); Candolle, Prodr. 15(1): 316 (1864); Koorders, Exkurs.-Fl. Java 2: 645 (1912); Golding, Phytologia 54(7): 494 (1984). – *Diploclinium bracteatum* (Jack) Miq., Fl. Ned. Ind. 1(1): 688 (1856); Miquel, Fl. Ned. Ind., Suppl. 1: 332 (1861); Candolle, Prodr. 15(1): 316 (1864). – Type: not located.

Begonia bracteata var. bracteata
■ **SUMATRA:** Sumatra, 7 iv 1917, *C.A.B. Backer* 5062 (B); Sumatra, *P.W. Korthals s.n.* (L); Mt. Singalan, *unknown s.n.* (L); Sipora Island, 8 ix 1924, *Iboet* 10 (L); Palembang, 11 viii 1932, *C.N.A. de Voogd* 1469 (L); Sibual-buali, 27 v 1993, *J.J. Afriastini* 2530 (L); Siberut Island, 21 v 1994, *J.J. Afriastini* 2682 (L); Asahan, Loemban Ria, 5 ii 1934 – 12 iv 1934, *Rahmat Si Boeea* 7567 (A); Asahan, Haboke, 20 x 1935 – 29 x 1935, *Rahmat Si Boeea* 8419 (A); Sipora Island, 10 x 1924, *C.B. Kloss* 14685 (K). **JAVA:** Java, *unknown s.n.* (L; 7 specimens); Java, *C.L. von Blume s.n.* (L[3]); Java, *P.W. Korthals s.n.* (L[2]); Java, *P.W. Korthals* 82 (L); Java, *H.C. van Hall s.n.* (L); Java, *J.C. Ploem s.n.* (L); Java, *Dr. A.J.D. Steenstra Toussaint* 122 (L); Java, *L.R. Lanjouw* 172 (L); Java, *L.R. Lanjouw* 197 (L); Tjidadap, iii 1918, *W.F. Winckel s.n.* (L); Priangan, G. Pangrango, 11 iv 2001, *W.S. Hoover, J.M. Hunter, H. Wiriadinata, D. Girmansyah* 994 (A); Gunung Tjibodas, 10 iii 1923, *W.F. Winckel* 1173 (L); *C.G.C. Reinwardt* 1721 (L); Tjidadap, 10 xi 1923, *W.F. Winckel* 1825B (L); Priangan, G. Pangrango, 7 iv 1894, *V.F. Schiffner* 2267 (A, L); Gunong Salak, *C.L. von Blume* 6070 (L); Batavia, G. Tjipoetik, *R.C. Bakhuizen van den Brink* 7177 (L); West Java, Mt. Bodas, 17 iv 2001, *W.S. Hoover, J.M. Hunter, H. Wiriadinata, D. Girmansyah* 10005 (A); Mt. Menapa, Nanggoeng, 18 xii 1940, *C.G.G.J.v. Steenis* 17406 (L).

Begonia bracteata var. gedeana A. DC., Prodr. 15(1): 316 (1864). – Type: not located.
■ **JAVA:** *Fide* de Candolle (1864).
■ **Notes:** Many specimens are annotated with both *B. bracteata* Jack and *B. lepida* Blume, which are widely considered to be synonyms. Until these names are lectotypified, this question cannot be answered with certainty.
■ **IUCN category: LC**. Widespread in Java and Sumatra.

Begonia brandisiana Kurz [§ Reichenheimia], J. Asiat. Soc. Bengal 40(2): 58 (1871); Kurz, Flora 54: 295 (1871); Clarke, Fl. Brit. Ind. 654 (1879); Craib, Fl. Siam. 1: 772 (1931). – Type: Burma, Martaban, Attaran Valley, *D. Brandis* 1327 (syn K[2]).
- **BURMA:** Martaban, Attaran Valley, *D. Brandis* 1327 (syn K[2]) – Type of *Begonia brandisiana* Kurz.
  LAOS: Khammouane, Nakai, 7 ix 2004, *A. Kool, H.J. de Boer, L. Bjork, V. Lamxay* 166 (E).
- **Notes:** The symmetric leaves (when small and not so lobed) are reminiscent of larger specimens of *B. demissa* Craib. The fruit shape and androecium are also similar, suggesting a previously unsuspected affinity between these two species which will have bearings on their sectional placement. The two taxa are, however, very distinct. See also notes under *B. cardiophora* Irmsch.
- **IUCN category: DD.** Insufficient specimens could be georeferenced with certainty.

Begonia brassii Merr. & L.M. Perry [§ Diploclinium], J. Arnold Arbor. 24: 43 (1943). – Type: New Guinea, Papua, Bele River Valley, xi 1938, *L.J. Brass* 11228 (holo A *n.v.*; iso BM, L).
- **NEW GUINEA: Papua:** Bele River Valley, xi 1938, *L.J. Brass* 11228 (holo A *n.v.*; iso BM, L) – Type of *Begonia brassii* Merr. & L.M. Perry.
- **IUCN category: DD.** Insufficient specimens could be georeferenced with certainty.

Begonia brevipedunculata Y.M. Shui [§ Platycentrum], Novon 16: 269 (2006). – Type: Vietnam, Tonkin, Laocai Prov., 21 xii 1964, *Sino-Vietnam expedition* 671 (holo PE *n.v.*; iso IBSC *n.v.*).
- **VIETNAM:** Tonkin, Laocai Prov., 21 xii 1964, *Sino-Vietnam expedition* 671 (holo PE *n.v.*; iso IBSC *n.v.*) – Type of *Begonia brevipedunculata* Y.M. Shui.
- **IUCN category: DD.** Insufficient specimens could be georeferenced with certainty.

Begonia brevipes Merr. [§ Petermannia], Philipp. J. Sci. 6: 378 ('1911', 1912); Merrill, Enum. Philipp. Fl. Pl. 3: 120 (1923). – Type: Philippines, Luzon, Cagayan Province, Pamplona, 18°27'58"N, 121°20'27"E, *M. Ramos* 7431 (syn K).
- **PHILIPPINES: Luzon:** Cagayan Province, Pamplona, *M. Ramos* 7431 (syn K) – Type of *Begonia brevipes* Merr.
- **IUCN category: DD.** Insufficient specimens could be georeferenced with certainty.

Begonia brevirimosa Irmsch. [§ Petermannia], Bot. Jahrb. Syst. 50: 358 (1913); Tebbitt, Begonias 93 (2005). – Type: New Guinea, Papua New Guinea, 11 vii 1907, *F.R.R. Schlechter* 16240 (holo B; iso B, P).

Begonia brevirimosa subsp. brevirimosa
- **NEW GUINEA: Papua New Guinea:** Northeast New Guinea, 11 vii 1907, *F.R.R. Schlechter* 16240 (holo B; iso B, P) – Type of *Begonia brevirimosa* Irmsch.
  BISMARCK ARCHIPELAGO: New Britain: West New Britain. *Fide* Tebbitt (2005).

Begonia brevirimosa subsp. exotica Tebbitt, Edinburgh J. Bot. 61: 104 ('2004', 2005). – Type: New Guinea, Papua New Guinea, Baiyer River, 26 xi 1954, *Floyd, J.S. Womersley* 6845 (holo A *n.v.*). *Begonia exotica* A.B. Graf *nom. nud.*, Exotica 3: 295 (1963); Golding, Begonian 71: 168 (2004).
- **NEW GUINEA: Papua New Guinea:** Western Highland Province, Korombi, ii 1978, *T.M. Reeve* 142 (para E[2]) – Type of *Begonia brevirimosa* subsp. *exotica* Tebbitt; Baiyer River, 26 xi 1954, *Floyd, J.S. Womersley* 6845 (holo A *n.v.*) – Type of *Begonia brevirimosa* subsp. *exotica* Tebbitt.
- **IUCN category: LC.** Occurs in the Baiyer River Sanctuary and the Wahgi River system.

Begonia bruneiana Sands [§ Petermannia], in Coode *et al.*, Checkl. Fl. Pl. Gymnosperms Brunei Darussalam App. 2:433 (1997). – Type: Borneo, Brunei, Lamunin, Ladan Hills Forest Reserve, 30 iii 1991, *M.J.S. Sands* 5735 (holo K; iso BRUN *n.v.*).

Begonia bruneiana subsp. bruneiana
- **BORNEO: Brunei:** Lamunin, Ladan Hills Forest Reserve, 30 iii 1991, *M.J.S. Sands* 5735 (holo K; iso BRUN *n.v.*) – Type of *Begonia bruneiana* Sands.

Begonia bruneiana subsp. angustifolia Sands, in Coode *et al.*, Checkl. Fl. Pl. Gymnosperms Brunei Darussalam App. 2: 433 (1997). – Type: Borneo, Brunei, Ulu Tutong, 7 v 1992, 4°25′N, 114°50′E, *R.J. Johns* 7510 (holo K; iso BRUN *n.v.*).
- **BORNEO: Brunei:** Ulu Tutong, 7 v 1992, *R.J. Johns* 7510 (holo K; iso BRUN *n.v.*) – Type of *Begonia bruneiana* subsp. *angustifolia* Sands.

Begonia bruneiana subsp. labiensis Sands, in Coode *et al.*, Checkl. Fl. Pl. Gymnosperms Brunei Darussalam App. 2: 433 (1997). – Type: Borneo, Brunei, Belait, 20 iii 1993, 4°22′N, 114°38′E, *M. Coode* 7293 (holo K; iso BRUN *n.v.*).
- **BORNEO: Brunei:** Belait, 20 iii 1993, *M. Coode* 7293 (holo K; iso BRUN *n.v.*) – Type of *Begonia bruneiana* subsp. *labiensis* Sands.

Begonia bruneiana subsp. retakensis Sands, in Coode *et al.*, Checkl. Fl. Pl. Gymnosperms Brunei Darussalam App. 2: 433 (1997). – Type: Borneo, Brunei, Temburong, 10 iii 1991, 4°22′N, 115°17′E, *M.J.S. Sands* 5316 (holo K; iso BRUN *n.v.*).
- **BORNEO: Brunei:** Temburong, 10 iii 1991, *M.J.S. Sands* 5316 (holo K; iso BRUN *n.v.*) – Type of *Begonia bruneiana* subsp. *retakensis* Sands.
- **IUCN category: LC**. Known from four localities spread throughout Brunei.

Begonia burbidgei Stapf [§ Petermannia], Trans. Linn. Soc. London, Bot., II 4: 165 (1894); Ridley, J. Malayan Branch Roy. Asiat. Soc. 46: 256 (1906); Gibbs, J. Linn. Soc., Bot. 42: 84 (1914); Sands, Pl. Mt. Kinabalu 148 (2001). – Type: Borneo, Sabah, Mt. Kinabalu, *G.D. Haviland* 1188 (syn K); Borneo, Sabah, Mt. Kinabalu, 6°2′N, 116°30′30″E, *F.W.T. Burbidge s.n.* (syn not located).
- **BORNEO:** 4 ix 1913, *Native collector* 79 (P). **Sabah:** Mt. Kinabalu, *F.W.T. Burbidge s.n.* (syn not located) – Type of *Begonia burbidgei* Stapf; Mt. Kinabalu, Marai Parai, 18 ix 1958, *S.H. Collenette* A52 (BM); Mt. Kinabalu, Kambarangoh Relay Stn, 17 v 1967, *W.R. Price* 197 (K); Mt. Kinabalu, Kambarangoh Relay Stn, 19 v 1967, *W.R. Price* 246 (K); Mt. Kinabalu, Mesilau, 30 iii 1982, *J. Sinclair* 259 (E); Mt. Kinabalu, 16 xii 1960, *S.H. Collenette* 621 (K); Mt. Kinabalu, 22 viii 1961, *W.L. Chew, E.J.H. Corner, A. Stainton* 866 (K); Mt. Kinabalu, 22 viii 1961, *W.L. Chew, E.J.H. Corner, A. Stainton* 938 (K); Mt. Kinabalu, 16 vi 1961, *W.L. Chew, E.J.H. Corner, A. Stainton* 1043 (K, L); Mt. Kinabalu, *G.D. Haviland* 1188 (syn K) – Type of *Begonia burbidgei* Stapf; Pinosuk Plateau, 16 viii 1961, *W.L. Chew, E.J.H. Corner, A. Stainton* 1309 (K); Mt. Kinabalu, 31 iii 1980, *G.C.G. Argent* 1604 (E); Mt. Kinabalu, Carsons Camp – Ulu Berambung, 14 i 1969, *M. Hotta* 3806 (L); Mt. Kinabalu, Mesilau – Ulu Berambung, 21 i 1969, *Kokawa, Hotta* 4038 (L); ii 1910, *L.S. Gibbs* 4124 (BM); Mt. Kinabalu, Mesilau Cave, 3 iv 1964, *W.L. Chew, E.J.H. Corner* 4822 (K) [as *Begonia cf. burbidgei*]; Mt. Kinabalu, Mamut Ridge – Ulu Berambung, 19 ii 1969, *Kokawa, Hotta* 5828 (L); Mt. Kinabalu, Mamut Ridge, 22 ii 1969, *Kokawa, Hotta* 5984 (L); Mt. Kinabalu, Mesilau Cave, 30 xii 1983, *J.H. Beaman et al.* 7973 (L); Mt. Kinabalu, Mesilau Cave, 30 xii 1983, *J.H. Beaman et al.* 8137 (L); Mt. Kinabalu, Mesilau River, 26 iii 1984, *J.H. Beaman et al.* 9097 (L); Mt. Kinabalu, Mesilau Cave, 24 iii 1984, *J.H. Beaman et al.* 9097 (L); Mt. Kinabalu, Lumu Lumu, 15 vi 1957, *J. Sinclair* 9211 (E, K,

L); Mt. Kinabalu, Mesilau Cave – Janets Halt, 29 viii 1963, *H.P. Fuchs, S.H. Collenette* 21402 (K) [as *Begonia cf. burbidgei*]; Mt. Kinabalu, Ulu Liwago, 7 iii 1961, *W. Meijer* 24136 (K, L); Mt. Kinabalu, Silau Basin, 9 iv 1932, *J. & M.S. Clemens* 29076 (BM); Mt. Kinabalu, Mesilau Cave – Janets Halt, 29 viii 1963, *H.P. Fuchs, S.H. Collenette* 31402 (L); Mt. Kinabalu, Marai Parai, 23 iii 1933, *J. & M.S. Clemens* 32270 (A, BM); Mt. Kinabalu, Marai Parai, 29 iii 1933, *J. & M.S. Clemens* 32431 (BM, K, L); Colombon River Basin, 1933, *J. & M.S. Clemens* 33733 (BM); Mt. Kinabalu, Marai Parai, 25 iv 1933, *J. & M.S. Clemens* 35159 (BM); Mt. Kinabalu, Gurulau Spur, 12 xii 1932, *J. & M.S. Clemens* 50924 (A, BM, K, L).

- **IUCN category: LC**. One of the most frequently collected begonias from Kinabalu, locally common along trails (G. Argent, pers. comm.).

Begonia burkillii Dunn [§ Sphenanthera], Bull. Misc. Inform. Kew 1920: 110 (1920); Tebbitt, Edinburgh J. Bot. 60: 7 (2003). – Type: India, Abor Hills, *I.H. Burkill* 36121 (syn not located); India, Abor Hills, *I.H. Burkill* 36315 (syn not located); India, Abor Hills, *I.H. Burkill* 36910 (syn not located); India, Abor Hills, *I.H. Burkill* 37121 (syn not located); India, Abor Hills, *I.H. Burkill* 37139 (syn K *n.v.*, B *n.v.*); India, Abor Hills, *I.H. Burkill* 37375 (syn not located); India, Abor Hills, *I.H. Burkill* 37455 (syn K *n.v.*); India, Abor Hills, *I.H. Burkill* 37706 (syn not located).

- **BURMA:** Kachin Hills, *S.M. Toppin* 4276 (K); Kachin Hills, *S.M. Toppin* 4371 (K); Katha Distr., Kadu Hill, 22 ii 1910, *J.H. Lace* 5105 (B, E, K).
- **Notes:** Difficult to separate from *B. handellii* var. *prostrata* using herbarium material. Tebbitt (2003a) keeps them separate largely on fruit shape. Toppin's collections from the Kadu Hills are split between these two taxa.
- **IUCN category: LC**. Tebbitt (2003a) notes that this species (from northeast India and north Burma) is locally common along streams and has an altitudinal range of 200–1200 m.

Begonia burmensis L.B. Sm. & Wassh. [§ Lauchea], Phytologia 52: 445 (1983). – *Begonia macgregorii* W.W. Sm., Rec. Bot. Surv. India 6(4): 99 (1914); Smith & Wasshausen, Phytologia 52: 445 (1983). – Type: Burma, Keng Tung, vii 1909, *R.C. McGregor* 553 (holo K).

- **BURMA:** Keng Tung, vii 1909, *R.C. McGregor* 553 (holo K) – Type of *Begonia macgregorii* W.W. Sm.
- **Notes:** See notes under *B. Craib*.
- **IUCN category: DD**. Not evaluated due to its unclear distinction from the earlier described *B. demissa* Craib.

Begonia caespitosa Jack [§ Diploclinium], Malayan Misc. 2(7): 8 (1822); Candolle, Prodr. 15(1): 397 (1864). – *Diploclinium caespitosa* Miq., Fl. Ned. Ind. 1(1): 685 (1856); Candolle, Prodr. 15(1): 397 (1864). – Type: not located.

- **SUMATRA:** *T. Horsfield s.n.* (BM) [as *Begonia caespitosa* ?].
- **IUCN category: DD**. Insufficient specimens could be georeferenced with certainty.

Begonia calcarea Ridl. [§ Diploclinium], J. Straits Branch Roy. Asiat. Soc. 46: 260 (1906); Kiew & Geri, Gard. Bull. Singapore 55: 115 (2003); Kiew & Tan, Gard. Bull. Singapore 56: 73 (2004). – Type: Borneo, Sarawak, Mt. Braang, *G.D. Haviland s.n.* (holo not located).

- **BORNEO:** 1876, *J.E. Teijsmann* 11384 (Fl). **Sarawak:** Mt. Braang, *G.D. Haviland s.n.* (holo not located) – Type of *Begonia calcarea* Ridl.; Dulit, 14 viii 1932, *Native collector* 1255 (K).
- **Notes:** Unusual in § *Diploclinium* in having orange-red flowers. See also *B. sabahensis* Kiew.
- **IUCN category: EN B2ab(iii)**. A Bau limestone endemic, although there is a possible record further west (Dulit). It is known from only two limestone hills in Bau (out of 20 surveyed;

Kiew & Geri, 2003), and grows at the base of limestone outcrops which are more prone to agricultural disturbance.

Begonia calcicola Merr. [§ Diploclinium], Philipp. J. Sci. 6: 400 ('1911', 1912); Merrill, Enum. Philipp. Fl. Pl. 3: 120 (1923). – Type: Philippines, Luzon, Rizal Province, Montalban, i 1910, *E.D. Merrill* 7062 (syn B, BM, K, P).
- **PHILIPPINES: Luzon:** ix 1908, *E.B. Copeland s.n.* (para not located) – Type of *Begonia calcicola* Merr.; Rizal Province, Montalban, i 1910, *E.D. Merrill* 7062 (syn B, BM, K, P) – Type of *Begonia calcicola* Merr.; Rizal Province, Montalban, ix 1909, *A. Loher* 12032 (P); Laguna, Makiling, ix 1911, *M. Ramos* 13675 (B, BM, K); Laguna, Makiling, 24 ii 1913, *C.B. Robinson, W.H. Brown* 17330 (K, P); Rizal Province, Mt. Irid, xi 1926, *M. Ramos, G. Edano* 48413 (B, K); Rizal Province, San Andales, xii 1926, *G. Edano* 48780 (E). **Mindanao:** Davao Province, Mati, iii 1927 – iv 1927, *M. Ramos, G. Edano* 49597 (K).
- **IUCN category: VU D2.** Occurs in three areas: Mt. Sto. Thomas, Mt. Makiling and Mascap. The first two are protected areas, but in all cases the populations are small in both number of plants and number of localities (R. Rubite, pers. comm.).

Begonia calliantha Merr. & L.M. Perry [§ Petermannia], J. Arnold Arbor. 24: 47 (1943). – Type: New Guinea, Papua New Guinea, Mt. Tafa, Mavi, ix 1933, *L.J. Brass* 4986 (holo NY *n.v.*; iso B).
- **NEW GUINEA: Papua New Guinea:** Mt. Tafa, Mavi, ix 1933, *L.J. Brass* 4986 (holo NY *n.v.*; iso B) – Type of *Begonia calliantha* Merr. & L.M. Perry.
- **IUCN category: NT.** The Southeastern Papuan Rainforest eco-region still contains major wilderness areas, although the type locality is not within the nearby Mts. Albert Edward and Victoria protected area.

Begonia capituliformis Irmsch. [§ Petermannia], Bot. Jahrb. Syst. 50: 354 (1913). – Type: Sulawesi, Minahassa, Bojong, *O. Warburg* 15190 (holo B; iso B).
- **SULAWESI:** E of Tongoa, *Johansson, H. Nybom, S. Riebe* 125 (L) [as *Begonia capituliformis* ?]; Minahassa, G. Klabat, 29 vi 1956, *L.L. Forman* 314 (K); Minahassa, G. Klabat, 30 vi 1956, *L.L. Forman* 344A (K); Minahassa, G. Klabat, 30 vi 1956, *L.L. Forman* 344B (K); SW of Dongala, 11 v 1975, *W. Meijer* 10093a (L) [as *Begonia capituliformis* ?]; Minahassa, Bojong, *O. Warburg* 15190 (holo B; iso B) – Type of *Begonia capituliformis* Irmsch.
- **Notes:** Has a similar congested male inflorescence to *B. congesta* Ridl., but is distinct in having longer petioles and broadly lanceolate (not sub-spathulate) leaves.
- **IUCN category: LC.** The Sulawesi montane rain forest eco-region is relatively intact and the locality in the G. Klabat reserve is still largely forested from 400 to 2000 m.

Begonia cardiophora Irmsch. [§ Reichenheimia], Mitt. Inst. Allg. Bot. Hamburg 8: 104 (1929). – Type: Thailand, Punga, Pulau Tebun, 29 xi 1918, *M. Haniff, Md. Nur* 3580 (holo SING *n.v.*; iso K).
- **THAILAND:** Punga, Pulau Tebun, 29 xi 1918, *M. Haniff, Md. Nur* 3580 (holo SING *n.v.*; iso K) – Type of *Begonia cardiophora* Irmsch.
- **Notes:** Seems to be allied to *B. brandiseana* Kurz and *B. demissa* Craib.
- **IUCN category: DD.** Insufficient specimens could be georeferenced with certainty.

Begonia carnosa (Teijsm. & Binn.) Teijsm. & Binn. [§ Petermannia], Epim. Lugd. Bat. 4 (1863). – *Diploclinium carnosum* Teijsm. & Binn., Tijdschr. Ned.-Indië 25: 420 (1863). – Type: Sulawesi, Menado, Kapateran, *J.E. Teijsmann s.n.* (syn? K).

- **SULAWESI:** Menado, Kapateran, *J.E. Teijsmann s.n.* (syn? K) – Type of *Diploclinium carnosum* Teijsm. & Binn.
- **Notes:** There is some confusion about the type locality; the protologue states it is in 'districto Kapetaran, prov. Menado ins. Celebes', whereas the specimen label is annotated 'Ternate'.
- **IUCN category: DD.** Insufficient specimens could be georeferenced with certainty.

Begonia carnosula Ridl. [§ Parvibegonia], J. Fed. Malay States Mus. 4: 20 (1909); Ridley, Fl. Malay Penins. 1: 857 (1922); Irmscher, Mitt. Inst. Allg. Bot. Hamburg 8: 146 (1929); Kiew, Begonias Penins. Malaysia 70 (2005). – Type: Peninsular Malaysia, Pahang, Telom, 1908, *H.N. Ridley* 14124 (lecto SING *n.v.*).
- **PENINSULAR MALAYSIA:** Pahang, Telom, 1908, *H.N. Ridley* 14124 (lecto SING *n.v.*) – Type of *Begonia carnosula* Ridl.
- **Notes:** See notes under *B. integrifolia* Dalzell.
- **IUCN category: LC.** Locally common in the foothills of the Cameron Highlands, including near roadsides (Kiew, 2005).

Begonia casiguranensis Quisumb. & Merr. [§ Petermannia], Philipp. J. Sci. 37: 172 (1928). – Type: Philippines, Luzon, Tayabas, Casiguran, Cabulig River, v 1925 – vi 1925, *M. Ramos, G. Edano* 45277 (syn B, BM[2], BO, P).
- **PHILIPPINES: Luzon:** Tayabas, Casiguran, Cabulig River, v 1925 – vi 1925, *M. Ramos, G. Edano* 45277 (iso B, BM[2], BO, P) – Type of *Begonia casiguranensis* Quisumb. & Merr.; Tayabas, Casiguran, Cabulig River, 28 iii 1925, *M. Ramos, G. Edano* 45399 (para not located) – Type of *Begonia casiguranensis* Quisumb. & Merr.
- **IUCN category: DD.** Insufficient specimens could be georeferenced with certainty.

Begonia castilloi Merr. [§ Diploclinium], Philipp. J. Sci. 13: 38 (1918); Merrill, Enum. Philipp. Fl. Pl. 3: 120 (1923). – Type: Philippines, Luzon, Cagayan Province, iv 1915 – v 1915, *E. Castillo* 22723 (syn K, P).
- **PHILIPPINES: Luzon:** Cagayan Province, iv 1915 – v 1915, *E. Castillo* 22723 (syn K, P) – Type of *Begonia castilloi* Merr.
- **IUCN category: DD.** Insufficient specimens could be georeferenced with certainty.

Begonia cathayana Hemsl. [§ Platycentrum], Bot. Mag., n.s. 134: 8202 (1908); Tebbitt, Begonias 96 (2005). – Type: China, *Henry* 13516 (holo K *n.v.*; iso NY *n.v.*).
- **VIETNAM:** 13 viii 1926, *E. Poilane* 12917 (P); Tonkin, Sai Wong Mo Shan, 18 v 1940 – 5 vii 1940, *Tsang Wai Tak* 29976 (A, E, P).
- **Notes:** This species has orange-red flowers.
- **IUCN category: DD.** Insufficient specimens could be georeferenced with certainty.

Begonia cathcartii Hook. f. [§ Platycentrum], Ill. Himal. Pl. 13 (1855); Klotzsch, Abh. Kon. Akad. Wiss. Berlin 1854: 245 (1855); Candolle, Prodr. 15(1): 349 (1864); Clarke, Fl. Brit. Ind. 2: 646 (1879); Craib, Fl. Siam. 1: 772 (1931); Hara, Fl. E. Himalaya 1: 84 (1971). – *Platycentrum cathcartii* (Hook. f.) Klotzsch, Abh. Kon. Akad. Wiss. Berlin 1854: 245 (1855); Klotzsch, Begoniac. 125 (1855). – Type: not located.

    Begonia nemophila Kurz, J. Asiat. Soc. Bengal 46(2): 108 (1877); Clarke, Fl. Brit. Ind. 2: 646 (1879). – Type: not located.
- **BURMA:** Moulmein, *W. Lobb s.n.* (K); Moulmein, 1868, *C.S.P. Parish s.n.* (K); North Burma, Vernay-Cutting expedition, 26 xi 1938, *F.K. Ward* 41a (A); Burma, *W.S. Kurz* 1051 (K); Burma, 15 viii 1892,

W.S. Kurz 1933 (K) [as *Begonia aff. cathcartii*]; Burma, *W. Griffith* 2582 (A, B, K, P); 20 iv 1926, *F.K. Ward* 6640 (K).

THAILAND: Doi Har, 25 x 1910, *H.B.G. Garrett* 84 (ABD, K); Doi Angka, 8 xii 1926, *H.B.G. Garrett* 340 (ABD, B, K, P); Chiangmai, Doi Inthanon National Park, 25 i 1990, *W.S. Hoover* 725 (A); Chiangmai, Doi Inthanon National Park, 25 i 1990, *W.S. Hoover* 726 (A).

- **IUCN category: LC.** Occurs in Burma and Thailand (Doi Inthanon National Park).

Begonia caudata Merr. [§ Petermannia], Philipp. J. Sci. 13: 41 (1918); Merrill, Enum. Philipp. Fl. Pl. 3: 120 (1923). – Type: Philippines, Luzon, Apayao, Mt. Sulu, v 1917, *E. Fenix* 28414 (syn BM, K, P).
- **PHILIPPINES: Luzon:** Apayao, Mt. Sulu, v 1917, *E. Fenix* 28414 (syn BM, K, P) – Type of *Begonia caudata* Merr.
- **IUCN category: DD.** Insufficient specimens could be georeferenced with certainty.

Begonia cauliflora Sands [§ Petermannia], Kew Mag. 7: 68 (1990). – Type: Borneo, Sabah, Tenom, Crocker Range, 30 iii 1984, *M.J.S. Sands* 3711 (holo K *n.v.*).
- **BORNEO: Sabah:** Tenom, Crocker Range, 30 iii 1984, *M.J.S. Sands* 3711 (holo K *n.v.*) – Type of *Begonia cauliflora* Sands; Mt. Sapong, *H.F. Camber* 4093 (L); Nabawan, Syarikat Undan Sdn Bhd (logging plot?), 19 iii 1990, *T. Sawan* 128494 (L).
- **IUCN category: DD.** Insufficient specimens could be georeferenced with certainty.

Begonia cavaleriei H. Lév. [§ Diploclinium], Repert. Spec. Nov. Regni Veg. 7: 20 (1909). – Type: China, Kouy-Tcheou, Tou-Chan, *J. Cavalerie* 2592 (syn P); China, Kouy-Tcheou, Tin-Fan, 1899, *J. Layes s.n.* (syn not located).

Begonia cavaleriei var. cavaleriei
- **VIETNAM:** Son La Prov., Moc Chau Distr., Cho Long, *V.X. Phuong* 79 (HN *n.v.*); Son La Prov., Yen Chau Distr., Muong Lum, *D.K. Harder et al.* DKH576 (HN *n.v.*, LE *n.v.*, MO *n.v.*); Thanh Hoa Prov., Ba Thuoc Distr., Co Lung, *N.T. Hiep et al.* HAL948 (HN *n.v.*, LE *n.v.*, MO *n.v.*); Ha Giang Prov., Yen Minh Distr., Lao Va Chai, *N.T. Hiep* NTH3451 (HN *n.v.*, LE *n.v.*, MO *n.v.*); Ha Giang Prov., Quan Ba Distr., Can Ty, *D.K. Harder et al.* DKH4796 (HN *n.v.*, LE *n.v.*, MO *n.v.*).
- **IUCN category: LC.** Appears to cope with some level of disturbance, as it has been collected from amongst limestone rocks in agricultural fields.

Begonia celebica Irmsch. [§ Petermannia], Bot. Jahrb. Syst. 50: 343 (1913); Irmscher, Bot. Jahrb. Syst. 50: 573 (1914). – Type: Sulawesi, Mt. Poanaa, 22 ix 1902, *K.F. & P.B. Sarasin* 2069 (holo B).
- **SULAWESI:** Mt. Poanaa, 22 ix 1902, *K.F. & P.B. Sarasin* 2069 (holo B) – Type of *Begonia celebica* Irmsch.
- **IUCN category: DD.** Type locality cannot be georeferenced.

Begonia chiasmogyna M. Hughes [§ Petermannia], Edinburgh J. Bot. 63: 193 (2006). – Type: Sulawesi, Gorantalo, Gunung Boliohutu, 23 iv 2002, 0°53′N, 122°30′E, *M. Mendum, H.J. Atkins, M. Newman, Hendrian, A. Sofyan* 167 (holo E).
- **SULAWESI:** Gorantalo, Nr Gunung Gambuta, 9 iv 2002, *M. Mendum, H.J. Atkins, M. Newman, Hendrian, A. Sofyan* 46 (E, L); Gorantalo, Gunung Boliohutu, 22 iv 2002, *M. Mendum, H.J. Atkins, M. Newman, Hendrian, A. Sofyan* 146 (E); Gorantalo, Gunung Boliohutu, 23 iv 2002, *M. Mendum, H.J. Atkins, M. Newman, Hendrian, A. Sofyan* 167 (holo E) – Type of *Begonia chiasmogyna* M. Hughes.
- **IUCN category: VU D2** (Hughes, 2006).

Begonia chlorandra Sands [§ Petermannia], in Coode *et al.*, Checkl. Fl. Pl. Gymnosperms Brunei Darussalam App. 2: 432 (1997). – Type: Borneo, Brunei, Lamunin, Ladan Hills Forest Reserve, 30 iii 1991, *M.J.S. Sands* 5700 (holo K; iso BRUN *n.v.*).
- **BORNEO: Brunei:** Lamunin, Ladan Hills Forest Reserve, 30 iii 1991, *M.J.S. Sands* 5700 (holo K; iso BRUN *n.v.*) – Type of *Begonia chlorandra* Sands.
- **IUCN category: VU D2**. Although this species occurs in a Forest Reserve (Ladan Hills), the area is being logged (M. Sands, pers. comm.).

Begonia chlorocarpa Irmsch. ex Sands [§ Sphenanthera], Pl. Mt. Kinabalu 149 (2001). – Type: Borneo, Sabah, Mt. Kinabalu, Liwagu River, 3 ix 1961, *W.L. Chew, E.J.H. Corner, A. Stainton* 2693 (holo K; iso SAN *n.v.*).
- **BORNEO: Sabah:** Mt. Kinabalu, Penibukan, *Clemens s.n.* (para HBG *n.v.*) – Type of *Begonia chlorocarpa* Irmsch. ex Sands; Mt. Kinabalu, Boundary Rentis, 22 i 1976, *P.F. Stevens* 638 (A); Mt. Kinabalu, Liwagu River, 3 ix 1961, *W.L. Chew, E.J.H. Corner, A. Stainton* 2693 (holo K; iso SAN *n.v.*) – Type of *Begonia chlorocarpa* Irmsch. ex Sands; Mt. Kinabalu, Liwagu River, *M.J.S. Sands* 3893 (para K *n.v.*) – Type of *Begonia chlorocarpa* Irmsch. ex Sands; Mt. Kinabalu, Mt. Nungkek, *M.J.S. Sands* 3970 (para K *n.v.*) – Type of *Begonia chlorocarpa* Irmsch. ex Sands; Mt. Kinabalu, Kiau/ Tahubang River, *M.J.S. Sands* 4006 (para K *n.v.*) – Type of *Begonia chlorocarpa* Irmsch. ex Sands; Mt. Kinabalu, Tenompok F.R., 3 ii 1932, *J. & M.S. Clemens* 4931 (BM); Mt. Kinabalu, Tenompok F.R., 7 iii 1932, *Clemens* 28692 (para A, BM, BO *n.v.*, K, L *n.v.*) – Type of *Begonia chlorocarpa* Irmsch. ex Sands; Mt. Kinabalu, Kilembun Basin, 5 vii 1933, *Clemens* 33882 (A *n.v.*, BM, BO *n.v.*, HBG *n.v.*, L *n.v.*) – Type of *Begonia chlorocarpa* Irmsch. ex Sands; Mt. Kinabalu, Gurulau Spur, *Clemens* 50379 (para K) – Type of *Begonia chlorocarpa* Irmsch. ex Sands; Mt. Kinabalu, Tahubang River, 11 ix 1933, *Clemens* 51687 (para BM) – Type of *Begonia chlorocarpa* Irmsch. ex Sands; Mt. Kinabalu, Tenompok F.R., 8 xii 2003, *de Wilde, de Wilde-Duyfies* 142223 (L).
- **IUCN category: LC**. Reasonably commonly collected in the Kinabalu National Park, and possibly occurs at other sites in Borneo (Sands, 2001).

Begonia chloroneura P. Wilkie & Sands [§ Diploclinium], New Plantsman 6(3): 132 (1999). – Type: Philippines, Luzon, Isabela Province, Baranguay San Jose, 6 iii 1997, *P. Wilkie* 29015 (holo PNH *n.v.*; iso E[2], K *n.v.*).
- **PHILIPPINES: Luzon:** Isabela Province, Baranguay San Jose, 6 iii 1997, *P. Wilkie* 29015 (holo PNH *n.v.*; iso E[2], K *n.v.*) – Type of *Begonia chloroneura* P. Wilkie & Sands.
- **IUCN category: VU D2**. The population in the type locality has tolerated heavy logging in the immediate vicinity, due to the streamside habitat remaining intact. However, the level of threat would increase considerably and immediately if any of the streamside canopy cover were to be removed (P. Wilkie, pers. comm.).

Begonia chlorosticta Sands [§ Petermannia], Bot. Mag., n.s. 183(4): 134 (1982); Tebbitt, Begonias 98 (2005). – Type: Borneo, Sarawak, 5 viii 1967, *B.L. Burtt, T.C. Whitmore* 4795 (holo E *n.v.*).
- **BORNEO: Sarawak:** 5 viii 1967, *B.L. Burtt, T.C. Whitmore* 4795 (holo E *n.v.*) – Type of *Begonia chlorosticta* Sands.
- **IUCN category: DD**. Insufficient specimens could be georeferenced with certainty.

Begonia chongii Sands [§ Petermannia], Pl. Mt. Kinabalu 150 (2001). – Type: Borneo, Sabah, Mt. Kinabalu, Kiau River, 5 v 1984, 6°2′N, 116°31′E, *M.J.S. Sands* 4001 (holo K).

- **BORNEO: Sabah:** Mt. Kinabalu, Mt. Nungkek, *M.J.S. Sands* 3969 (para K *n.v.*) – Type of *Begonia chongii* Sands; Mt. Kinabalu, Kiau River, 5 v 1984, *M.J.S. Sands* 4001 (holo K) – Type of *Begonia chongii* Sands; Mt. Kinabalu, Penibukan, i 1933, *Clemens* 31037 (para BM, BO *n.v.*) – Type of *Begonia chongii* Sands; Mt. Kinabalu, Mt. Nungkek, 22 iv 1933, *Clemens* 32892 (para A *n.v.*, BM, K *n.v.*) – Type of *Begonia chongii* Sands; Mt. Kinabalu, Mt. Nungkek, *Clemens* 32982 (para BO *n.v.*) – Type of *Begonia chongii* Sands; Mt. Kinabalu, Penibukan, *Clemens* 50126A (para A, BM) – Type of *Begonia chongii* Sands.
- **IUCN category: VU D2.** Restricted to Penibukan forest on the west of Mt. Kinabalu.

Begonia ciliifera Merr. [§ Petermannia], Philipp. J. Sci. 6: 376 ('1911', 1912); Merrill, Enum. Philipp. Fl. Pl. 3: 121 (1923). – Type: Philippines, Mindanao, Zamboanga, Port Banga, *H.N. Whitford, W.I. Hutchinson* FB9318 (syn not located).
- **PHILIPPINES: Mindanao:** Zamboanga, xi 1911 – xii 1911, *E.D. Merrill* 8176 (BM); Zamboanga, Port Banga, *H.N. Whitford, W.I. Hutchinson* FB9318 (syn not located) – Type of *Begonia ciliifera* Merr.; Zamboanga, Malangas, x 1919 – xi 1919, *M. Ramos, G. Edano* 36991 (K, P); Zamboanga, Malangas, x 1919 – xi 1919, *M. Ramos, G. Edano* 37369 (B, P); Zamboanga, Malangas, x 1919 – xi 1919, *M. Ramos, G. Edano* 37373 (B).
- **IUCN category: EN B2ab(iii).** Known only from two localities, at low altitude sites which are more prone to disturbance. Its preference for canyons (Merrill, 1911) may afford it some protection.

Begonia cincinnifera Irmsch. [§ Petermannia], Webbia 9: 494 (1953). – Type: Borneo, Sarawak, Pangkalan ampat, xi 1866, *O. Beccari* PB2732 (holo FI).
- **BORNEO: Sarawak:** Pangkalan ampat, xi 1866, *O. Beccari* PB2732 (holo FI) – Type of *Begonia cincinnifera* Irmsch.
- **IUCN category: VU D2.** Known only from the type locality in the far west of Sarawak. Although the locality is not precise enough to georeference, the forests in this area are sparse and poorly protected.

Begonia cladotricha M. Hughes [§ Diploclinium], Edinburgh J. Bot. 64: 102 (2007). – Type: Laos, Khammouan, Mahaxi Distr., 8 xi 2005, *M. Newman, P.I. Thomas, K.A. Armstrong, K. Sengdala, V. Lamxay* LAO985 (holo E).
- **LAOS:** Khammouan, Mahaxi Distr., 8 xi 2005, *M. Newman, P.I. Thomas, K.A. Armstrong, K. Sengdala, V. Lamxay* LAO985 (holo E) – Type of *Begonia cladotricha* M. Hughes.
- **IUCN category: LC** (Hughes, 2007).

Begonia clemensiae Merr. & L.M. Perry [§ Petermannia], J. Arnold Arbor. 29: 161 (1948). – Type: New Guinea, Papua New Guinea, Boana, v 1940 – xi 1940, *J. & M.S. Clemens* 41672 (holo A).
- **NEW GUINEA: Papua New Guinea:** Morobe Distr., Wantoat, ii 1940, *J. & M.S. Clemens* 11159 (para not located) – Type of *Begonia clemensiae* Merr. & L.M. Perry; Matap, ii 1940 – iv 1940, *J. & M.S. Clemens* 41102 (para not located) – Type of *Begonia clemensiae* Merr. & L.M. Perry; Boana, v 1940 – xi 1940, *J. & M.S. Clemens* 41672 (holo A) – Type of *Begonia clemensiae* Merr. & L.M. Perry.
- **IUCN category: VU D2.** Known from only three locations.

Begonia cognata Irmsch. [§ Petermannia], Webbia 9: 477 (1953). – Type: Borneo, Sarawak, Colline del Bellaga, ix 1867, *O. Beccari* PB3776 (holo FI; mero B).

- **BORNEO: Sarawak:** Colline del Bellaga, ix 1867, *O. Beccari* PB3776 (holo FI; mero B) – Type of *Begonia cognata* Irmsch.
- **IUCN category: DD.** Insufficient specimens could be georeferenced with certainty.

Begonia collisiae Merr. [§ Diploclinium], Philipp. J. Sci. 14: 424 (1919); Merrill, Enum. Philipp. Fl. Pl. 3: 121 (1923). – Type: Philippines, Panay, Capiz, Libacao, 5 v 1918, *M. Ramos, G. Edano* BS31469 (syn not located).
- **PHILIPPINES: Panay:** Capiz, Libacao, 5 v 1918, *M. Ramos, G. Edano* BS31469 (syn not located) – Type of *Begonia collisiae* Merr.
- **IUCN category: DD.** Known only from the type locality, which cannot be georeferenced with certainty. There is a Watershed Forest Reserve in the province which may provide suitable habitat.

Begonia colorata Warb. in Perkins [§ Diploclinium], Fragm. Fl. Philipp. 51 (1904); Merrill, Philipp. J. Sci. 6: 386 ('1911', 1912); Merrill, Enum. Philipp. Fl. Pl. 3: 121 (1923). – Type: Philippines, Mindanao, Davao Province, Sibulan, vii 1888, *O. Warburg* 14633 (holo B).
- **PHILIPPINES: Mindanao:** Davao Province, Sibulan, vii 1888, *O. Warburg* 14633 (holo B) – Type of *Begonia colorata* Warb.
- **IUCN category: DD.** Insufficient specimens could be georeferenced with certainty.

Begonia congesta Ridl. [§ Petermannia], J. Straits Branch Roy. Asiat. Soc. 46: 253 (1906); Kiew & Geri, Gard. Bull. Singapore 55: 116 (2003). – Type: Borneo, Sarawak, Bau, *H.N. Ridley s.n.* (syn not located).
- **BORNEO: Sarawak:** Bau, *H.N. Ridley s.n.* (syn not located) – Type of *Begonia congesta* Ridl.; Mt. Dulit, 23 viii 1932, *P.W. Richards* 1041 (K); Mt. Dulit, 2 viii 1932, *P.M. Synge* 1083 (K); Mt. Dulit, 23 viii 1932, *P.W. Richards* 1083 (K); Bau, Bukit Rapor, 22 v 1962, *B.L. Burtt, P.J.B. Woods* 1881 (E[2]); Mt. Trekan, vii 1895, *G.D. Haviland, C. Hose* 2017 (BM); Bau, Seburan Mine, viii 1962, *P.J.B. Woods* 2890 (E); Bau, 15 ix 1955, *J.W.P. Purseglove* 4468 (L); Bau, 24 iv 1955, *W.M.A. Brooke* 9875 (BM, L); Bau, 13 xi 1955, *W.M.A. Brooke* 10822 (L); Bau, Bukit Jebong, 28 iv 1967, *Paul, Ilias* 25626 (L); Bau, Gunong Kawa, 11 viii 1978, *G. Rena* 38280 (E, L); Mulu National Park, Gunung Buda, 1 iii 1978, *P. Chai* 39468 (K); Bau, Bidi, 22 x 1979, *J. Namit* 42147 (E, L); Bau, Bukit Kapor, 9 vii 1996, *Mohiza et al.* 74455 (L); Baram, Gunong Api, 26 ix 1971, *J.A.R. Anderson* 310756 (E) [as *Begonia cf. congesta*].
- **Notes:** Irmscher's unpublished name *B. densinervis* seems to be the same as this taxon. One of the syntypes cited by Ridley in the protologue (*G.D. Haviland, C. Hose* 3224 *n.v.*) has since been assigned to *B. niahensis* K.G. Pearce.
- **IUCN category: LC.** Currently known from two areas each consisting of several sites in north and south Sarawak, one centred in Gunung Mulu National Park, the other in the Bau limestone district, of which one site is in the Gunung Gadin National Park.

Begonia conipila Irmsch. ex Kiew [§ Petermannia], Gard. Bull. Singapore 53: 287 (2001). – Type: Borneo, Sarawak, Gunung Mulu National Park, *H. Low s.n.* (lecto K).
- **BORNEO: Sarawak:** Gunung Mulu National Park, *H. Low s.n.* (lecto K) – Type of *Begonia conipila* Irmsch. ex Kiew; Gunung Mulu, Melinau River, 27 iv 1997, *R.M.A.P. Haegens, N. Klazenga* 512 (L); Baram, Sungei Melinau, 10 ii 1966, *W.L. Chew* 1055 (A, L); Gunung Mulu, Melinau River, 24 vi 1962, *B.L. Burtt, P.J.B. Woods* 2260 (E); Baram, Gunong Api, 7 vii 1961, *J.A.R. Anderson* 4287 (K, L); Mulu National Park, 2 iv 1978, *B.C. Stone* 13598 (L); Baram, Gunong Api, 31 viii 1970, *P. Chai* 30054 (L); Baram, Gunong Api, 26 ix 1971, *J.A.R. Anderson* 30732 (A, E, L); Gunung Mulu, 6 ii 1995, *A. Mohtar* 49606 (K, L); Baram, Sungei Melinau, vi 1996, *Yii Puan Ching, A. Talip* 58810 (L).

- **IUCN category: LC.** Endemic to the Melinau limestone in Mulu National Park. Although the area of occupancy is thus <500 km², this area is sufficiently well protected to make this taxon LC.

Begonia consanguinea Merr. [§ Petermannia], Sarawak Mus. J. 3: 531 (1928). – Type: not located.
- **BORNEO: Sarawak:** Gunung Skunyet, xi 1865, *O. Beccari* PB1052 (FI[2]); *O. Beccari* PB2904 (B, FI[2]); Colline del Bellaga, ix 1867, *O. Beccari* PB3795 (FI).
- **IUCN category: DD.** Insufficient specimens could be georeferenced with certainty.

Begonia contracta Warb. in Perkins **[§ Petermannia]**, Fragm. Fl. Philipp. 54 (1904); Merrill, Philipp. J. Sci. 6: 386 ('1911', 1912); Merrill, Enum. Philipp. Fl. Pl. 3: 121 (1923). – Type: Philippines, Luzon, Tayabas, Sampaloc, *O. Warburg* 13085 (holo B).
- **PHILIPPINES: Luzon:** Cagayan Province, Littoc, v 1917 – vi 1917, *M. Adduru* 149 (K, P); Tayabas, Lucban, v 1907, *A.D.E. Elmer* 7430 (E); Isabela Province, v 1909, *M. Ramos* 8008 (B); Tayabas, Sampaloc, *O. Warburg* 13085 (holo B) – Type of *Begonia contracta* Warb.; Tayabas, Kabibihan, ii 1911 – iii 1911, *M. Ramos* 13237 (B); Rizal, Mt. Kamunay, x 1911, *M. Ramos* 13781 (B, K, L) [as *Begonia contracta* ?]; Tayabas, Mauban, i 1913, *M. Ramos* 19482 (K, P); Tayabas, Mt. Binuang, v 1917, *M. Ramos, G. Edano* 28730 (K, P); Cagayan Province, Claveria, viii 1918, *M. Ramos* 33060 (BM) [as *Begonia contracta* ?]; Sorsogon, Mt. Bulusan, 17 v 1957, *G. Edano, H. Gutierrez* 38454 (K, L); Sorsogon, Mt. Bulusan, v 1957, *G. Edano, H. Gutierrez* 38480 (K, L); Sorsogon, Mt. Bulusan, 18 v 1957, *G. Edano, H. Gutierrez* 38501 (L); Isabela Province, San Mariano, ii 1926 – iii 1926, *M. Ramos, G. Edano* 47172 (BM). **Mindoro:** Pinamalayan, vi 1922, *M. Ramos* 40984 (BM). **Panay:** Capiz, Mt. Macosolon, iv 1918 – v 1918, *M. Ramos, G. Edano* 30751 (L); Capiz, Jamindan, iv 1918 – v 1918, *M. Ramos, G. Edano* 30918 (L); Capiz, Jamindan, iv 1918 – v 1918, *M. Ramos, G. Edano* 30919 (BM); Capiz, Jamindan, iv 1918 – v 1918, *M. Ramos, G. Edano* 30920 (B); Capiz, Jamindan, iv 1918 – v 1918, *M. Ramos, G. Edano* 31030 (P). **Negros:** Negros Occidental, Mt. Silay, v 1906, *H.N. Whitford* 1504 (P); Negros Occidental, Gimagaan River, v 1906, *H.N. Whitford* 1588 (P). **Bohol:** x 1923, *M. Ramos* 43096 (B, BM, P); viii 1923 – x 1923, *M. Ramos* 43149 (B, BM[2], K, P). **Leyte:** iii 1916, *C.A. Wenzel* 1687 (BM); So-ong Lake road, 14 iv 1957, *G.M. Frohne* 35118 (K, L); Tigbao, 31 viii 1957, *G.M. Frohne* 35719 (L); Cabalian, xii 1922, *M. Ramos* 41580 (K). **Samar:** Loquilocon, iv 1948 – v 1948, *M.D. Sulit* 6106 (L); Laquilacon, vi 1924, *R.C. McGregor* 43748 (K). **Mindanao:** Butuan, Tungao, 2 vi 1961, *D.R. Mendoza* 61-116 (L); Surigao, iv 1928, *C.A. Wenzel* 3232 (K); Zamboanga, Malangas, x 1919 – xi 1919, *M. Ramos, G. Edano* 36756 (L).
- **IUCN category: LC.** One of the most widespread species in the Philippines, but occurring only in low to medium altitude forests. Despite being widespread, there are no collections post 1960, and it is possible that this species is close to qualifying for the category VU A2.

Begonia copelandii Merr. [§ Diploclinium], Philipp. J. Sci. 6: 401 ('1911', 1912). – Type: Philippines, Mindanao, Davao Province, Todaya, iv 1904, *E.B. Copeland* 1255 (syn B *n.v.*).
- **PHILIPPINES: Luzon:** Cagayan Province, Penablanca, 1 v 1917 – 18 vi 1917, *M. Adduru* 259 (K, P). **Mindanao:** Davao Province, Todaya, iv 1904, *E.B. Copeland* 1255 (syn B *n.v.*) – Type of *Begonia copelandii* Merr.
- **IUCN category: DD.** I have not seen the type and cannot verify the specimens from Luzon. It may be that this species is only known from the type locality in Mindanao.

Begonia coriacea Hassk. [§ Reichenheimia], Cat. Hort. Bogor. Alt. 192 & 311 (1844); Hasskarl, Pl. Jav. Rar. 239 (1848); Hasskarl, Hort. Bogor. Descr. 328 (1858); Candolle, Prodr. 15(1): 390 (1864). – *Mitscherlichia coriacea* (Hassk.) Klotzsch, Abh. Kon. Akad. Wiss. Berlin 1854: 194 (1855); Klotzsch,

Begoniac. 74 (1855); Miquel, Fl. Ned. Ind. 1(1): 696 (1856); Candolle, Prodr. 15(1): 274 (1864); Golding, Phytologia 47: 294 (1981). – Type: not located.

Begonia peltata Hassk., in Hoeven & de Vriese, Tijdschr. Nat. Gescheid. 10: 133 (1843); Candolle, Prodr. 15(1): 390 (1864). – Type: not located.

Begonia hasskarlii Zoll. & Moritzi, Syst. Verzeich. 31 (1846); Hasskarl, Pl. Jav. Rar. 240 (1848); Candolle, Prodr. 15(1): 390 (1864); Koorders, Exkurs.-Fl. Java 2: 644 (1912). – Type: not located.

Begonia hernandifolia Hook., Bot. Mag. 78: 4676 (1852); Seemann, Bot. Voy. Herald 128 (1854); Seemann, Bot. Voy. Herald 254 (1854); Klotzsch, Abh. Kon. Akad. Wiss. Berlin 1854: 194 (1855); Klotzsch, Begoniac. 74 (1855); Candolle, Prodr. 15(1): 390 (1864); Golding, Phytologia 47: 293 (1981). – Type: not located.

Begonia umbilicata hort. ex Planch., Fl. Serres Jard. 1(8): 810 (1853). – Type: not located.

Mitscherlichia junghuhniana Miq., Fl. Ned. Ind. 1(1): 696 (1856); Miquel, Pl. Jungh. 4: 418 ('1855', 1857). – Type: not located.

Begonia junghuhniana Miq., Pl. Jungh. 4: 418 ('1855', 1857); Candolle, Prodr. 15(1): 390 (1864); Koorders, Exkurs.-Fl. Java 2: 644 (1912); Backer & Brink, G. v.d., Fl. Java 1: 309 (1964). – Type: not located.

Begonia junghuhniana forma *acutifolia* Miq. ex Koord., Exkurs.-Fl. Java 2: 644 (1912); Backer & Brink, G. v.d., Fl. Java 1: 309 (1964). – Type: not located.

- **SUMATRA:** Bencoolen, 22 v 1931, *L.v.d. Pijl* 362 (BO) [as *Begonia* aff. *coriacea*]; Bencoolen, Bukit Daoen, 27 ii 1931, *C.N.A. de Voogd* 1055 (BO) [as *Begonia* aff. *coriacea*]; Palembang, Aer Telanai cleft, 2 xi 1929, *C.G.G.J.v. Steenis* 3934 (BO) [as *Begonia* aff. *coriacea*].
  JAVA: *unknown s.n.* (B); *F.W. Junghuhn s.n.* (B); 10 xi 1929, *P. Groenhart* 95 (U) [as *Begonia coriacea* ?]; Buitenzorg, 9 x 1843, *H. Zollinger* 1613 (A, B, BO, FI, L, P); Manding River, 2 x 1844 – 4 x 1844, *H. Zollinger* 2431 (BM, P); Besoeki, Sadeng, 13 vi 1935, *C.G.G.J.v. Steenis* 7366 (BO); Gunung Ardjoeno, 17 x 1899, *S.H. Koorders* 38205B (BO).
- **IUCN category: NT**. Although it appears to be widespread in Java and also occurs in south Sumatra, the vast majority of collections of this species are historic.

Begonia corneri Kiew [§ Reichenheimia], Bot. Jahrb. Syst. 113: 271 (1991); Kiew, Begonias Penins. Malaysia 202 (2005). – Type: Peninsular Malaysia, Trengganu, Kemaman, Sungai Nipa, 20 xi 1935, *E.J.H. Corner* SFN30748 (holo SING *n.v.*).

- **PENINSULAR MALAYSIA:** Trengganu, Kemaman, Sungai Nipa, 20 xi 1935, *E.J.H. Corner* SFN30748 (holo SING *n.v.*) – Type of *Begonia corneri* Kiew.
- **IUCN category: CR B2ab(iii)**. Currently known from a single locality. A rare and extremely local species (Kiew, 2005).

Begonia coronensis Merr. [§ Diploclinium], Philipp. J. Sci. 26: 480 (1925). – Type: Philippines, Palawan, Coron Island, ix 1922, *M. Ramos* 41164 (syn BO, L, P).

- **PHILIPPINES: Palawan:** ix 1910, *E.D. Merrill* 7245 (B, BM[3], K[2], P); Coron Island, ix 1922, *M. Ramos* 41164 (syn BO, L, P) – Type of *Begonia coronensis* Merr.
- **Notes:** Allied to *B. woodii* Merr., from which it appears to differ mainly in the larger size of its vegetative and floral parts.
- **IUCN category: LC**. The forests on Coron Island have a tradition of sustainable use by the indigenous Tagbanua people; the dissected limestone habitat is also likely to afford protection for this species.

Begonia crenata Dryand. [§ Parvibegonia], Trans. Linn. Soc. 1: 162 (1791). – Type: not located.

Saueria crenata Hassk., Verslagen Meded. Afd. Natuurk. Kon. Akad. Wetensch. 4: 139 (1855). – Type: not located.

Begonia minima Bedd., Madras J. Lit. Sci. 1: 48 (1864); Beddome, Icon. Pl. Ind. Orient 1(2): 23 (1874); Clarke, Fl. Brit. Ind. 2: 651 (1879). – Type: not located.

- **BURMA:** Upper Cheidwin Distr., Kindat, 24 viii 1908, *J.H. Lace s.n.* (E) [as *Begonia cf. crenata*]; Tenasserim, *Helfer* 2584 (K) [as *Begonia cf. crenata*].
- **IUCN category: DD**. Insufficient specimens could be georeferenced with certainty.

Begonia crispipila Elmer [§ Petermannia], Leafl. Philipp. Bot. 2: 737 (1910); Merrill, Philipp. J. Sci. 6: 384 ('1911', 1912); Merrill, Enum. Philipp. Fl. Pl. 3: 121 (1923). – Type: Philippines, Luzon, Benguet, Baguio, Irisan Creek, iii 1907, *A.D.E. Elmer* 8687 (iso BM, E, K, L).

- **PHILIPPINES: Luzon:** Benguet, Sablan(g), iv 1904, *A.D.E. Elmer* 6149 (K); Benguet, Baguio, Irisan Creek, iii 1907, *A.D.E. Elmer* 8687 (iso BM, E, K, L) – Type of *Begonia crispipila* Elmer; Benguet, Sablan(g), xi 1910 – xii 1910, *E. Fenix* 12602 (BM, K); Apayao, Mt. Sulu, v 1917, *E. Fenix* 28373 (P); Benguet, Baguio, v 1925, *M. Ramos, G. Edano* 45035 (P).
- **IUCN category: VU B2ab(iii)**. Known only from Mt. Sulu in Apayao and the creeks surrounding Baguio. The forest has been heavily degraded in these areas, and the possibility of this species being able to tolerate disturbance of the canopy over its gravelly streamside habitat is unlikely.

Begonia cucphuongensis H.Q. Nguyen & Tebbitt [not placed to section], Gard. Bull. Singapore 57: 248 (2005). – Type: Vietnam, Cuc Phuong National Park, 19 vi 1999, *M.A. Jaramillo* 493 (holo HN *n.v.*; iso CPNP *n.v.*, F *n.v.*, MO *n.v.*).

- **VIETNAM:** Cuc Phuong National Park, 19 vi 1999, *M.A. Jaramillo* 493 (holo HN *n.v.*; iso CPNP *n.v.*, F *n.v.*, MO *n.v.*) – Type of *Begonia cucphuongensis* H.Q. Nguyen & Tebbitt; Thanh Hoa Prov., Ba Thuoc Distr., Pu Luong, *N.T. Hiep et al.* 937 (para CPNP *n.v.*, F *n.v.*, HN *n.v.*) – Type of *Begonia cucphuongensis* H.Q. Nguyen & Tebbitt; Cuc Phuong National Park, 17 viii 2001, *N.M. Cuong* 1457 (L) [as *Begonia cf. cucphuongensis*]; Cuc Phuong National Park, Bong, 20 vii 2000, *P.K. Loc et al.* 10323 (para CPNP *n.v.*, F *n.v.*, HN *n.v.*) – Type of *Begonia cucphuongensis* H.Q. Nguyen & Tebbitt.
- **IUCN category: LC**. Despite being known from only two localities, the entire range of the species is within protected areas.

Begonia cumingiana (Klotzsch) A. DC. [§ Petermannia], Prodr. 15(1): 320 (1864); Fernández-Villar, Noviss. App. 98 (1880); Vidal, Phan. Cuming. Philipp. 116 (1885); Vidal, Rev. Pl. Vasc. Filip. 143 (1886); Merrill, Philipp. J. Sci. 6: 388 ('1911', 1912); Merrill, Enum. Philipp. Fl. Pl. 3: 121 (1923). – *Petermannia cumingiana* Klotzsch ex Klotzsch, Abh. Kon. Akad. Wiss. Berlin 1854: 195 (1855). – *Diploclinium cumingianum* (Klotzsch) Miq., Fl. Ned. Ind. 1(1): 691 (1856); Candolle, Prodr. 15(1): 320 (1864). – Type: Philippines, Luzon, Albay, *H. Cuming* 856 (syn B, BM, E, FI, K, P).

- **PHILIPPINES: Luzon:** Albay, *H. Cuming* 856 (syn B, BM, E, FI, K, P) – Type of *Petermannia cumingiana* Klotzsch ex Klotzsch; Albay – Camarines, vi 1908, *H.M. Curran* 12260 (B, K, P); Sorsogon, vii 1915 – viii 1915, *M. Ramos* 23483 (P); Cagayan Province, Bagio Cave, 22 iii 1981, *M.S. Allen* 150007 (L).
- **Notes:** The collection by Allen (150007) has very short internodes compared with the type, and may represent a different taxon. The specimen bears a curious ethnobotanical notation, stating that the leaves are burned and the resulting smoke is used in a ritual to remove bad luck from hunting dogs.
- **IUCN category: VU D2**. The distribution of this species can be narrowed only to province (Albay and Catanduanes). However, it is known only from three historical collections in the threatened Luzon rain forest eco-region.

Begonia cumingii A. Gray [§ Petermannia], U.S. Expl. Exped. Phan. 15: 658 (1854); Merrill, Philipp. J. Sci. 3: 84 (1908); Merrill, Philipp. J. Sci. 6: 384 ('1911', 1912); Merrill, Enum. Philipp. Fl. Pl. 3: 121 (1923). – Type: Philippines, Luzon, Tayabas, Mt. Banahao, 1811, *H. Cuming* 1897 (syn FI, K).
   Begonia philippinensis A. DC., Prodr. 15(1): 320 (1864); Fernández-Villar, Noviss. App. 98 (1880); Vidal, Phan. Cuming. Philipp. 116 (1885); Vidal, Rev. Pl. Vasc. Filip. 143 (1886). – Type: Philippines, Luzon, Tayabas, Mt. Banahao, 1811, *H. Cuming* 1897 (syn FI, K); Philippines, Luzon, *W. Lobb s.n.* (syn K).
- **PHILIPPINES: Luzon:** *W. Lobb s.n.* (syn K) – Type of *Begonia philippinensis* A. DC.; Tayabas, Mt. Banahao, 22 vi 1904, *W. Klemme* 885 (B); Benguet, Mt. Santo Tomas, 29 xi 1904, *R.S. Williams* 1532 (B); Tayabas, Mt. Banahao, 1811, *H. Cuming* 1897 (syn FI, K) – Type of *Begonia cumingii* A. Gray; Tayabas, Mt. Banahao, 1811, *H. Cuming* 1897 (syn FI, K) – Type of *Begonia philippinensis* A. DC.; Tayabas, Mt. Banahao, 20 xii 1913, *C.F. Baker* 2466 (K); Laguna, iii 1908, *C.B. Robinson* 6068 (B, P); Laguna, Makiling, ii 1902, *E.D. Merrill* 6305 (B, K, P); Laguna, Makiling, ii 1907, *E.D. Merrill* 6305 (B, K, P); Benguet, Baguio, iii 1907, *A.D.E. Elmer* 8521 (B, BM, E); Camarines, Alanao, 14 vi 1958, *J. Sinclair, G. Edano* 9554 (E, L); Laguna, Makiling, ix 1911, *M. Ramos* 13697 (E); Tayabas, Mt. Banahao, i 1913, *M. Ramos* 19572 (BM, P); Tayabas, Mt. Binuang, v 1917, *M. Ramos, G. Edano* 28790 (BM, K, P); Mountain Province, Bontoc Road, 27 iv 1959, *M.L. Dr. Steiner* 41584 (L); Mountain Province, Banaue, Bayinan, 9 xii 1962, *H. Conklin, Buwaya* 78637 (L); Laguna, Mt. Banajao, i 1913, *M. Ramos* 195783 (L).
- **IUCN category: LC**. Known from at least five localities in Luzon, including the Mt. Maquiling protected area, and Mt. Banahao, where it is stated to be common (Merrill, 1923).

Begonia cuneatifolia Irmsch. [§ Petermannia], Bot. Jahrb. Syst. 50: 370 (1913). – Type: Sulawesi, Minahassa, Tomohon, 19 vi 1894, *K.F. & P.B. Sarasin* 494 (holo B; iso K).
- **SULAWESI:** Gorantalo, Gunung Boliohutu, 24 iv 2002, *M. Mendum, H.J. Atkins, M. Newman, Hendrian, A. Sofyan* 189 (E); Minahassa, Tomohon, 19 vi 1894, *K.F. & P.B. Sarasin* 494 (holo B; iso K) – Type of *Begonia cuneatifolia* Irmsch.
- **IUCN category: VU D2**. Known only from two localities, neither of which are protected.

Begonia cyanescens Sands [§ Petermannia], in Coode *et al.*, Checkl. Fl. Pl. Gymnosperms Brunei Darussalam App. 2: 433 (1997). – Type: Borneo, Brunei, Temburong, 24 iii 1991, 4°30′N, 115°8′E, *M.J.S. Sands* 5577 (holo K; iso BRUN *n.v.*).
- **BORNEO: Sarawak:** Baram, Sg. Chipidi, 8 viii 1974, *P. Chai* 34692 (L). **Brunei:** Batu Apoi Forest Reserve, 7 xi 1991, *C. Hansen* 1515 (L); Batu Apoi Forest Reserve, 20 xi 1991, *C. Hansen* 1588 (L); Temburong, 24 iii 1991, *M.J.S. Sands* 5577 (holo K; iso BRUN *n.v.*) – Type of *Begonia cyanescens* Sands; Temburong – Belalang, *M. Jacobs* 5609 (K, L).
- **IUCN category: LC**. Occurs in the protected Ulu Temburong National Park, and one location in Sarawak. Forest cover is high in Brunei, and clearance rates are currently low.

Begonia decora Stapf [§ Platycentrum], Gard. Chron., III 12: 621 (1892); Ridley, Fl. Malay Penins. 1 (1922); Irmscher, Mitt. Inst. Allg. Bot. Hamburg 8: 121 (1929); Craib, Fl. Siam. 1: 772 (1931); Henderson, Malayan Wild Flowers – Dicotyledons 164 (1959); Kiew, Gard. Bull. Singapore 51: 117 (1999); Kiew, Begonias Penins. Malaysia 133 (2005). – Type: 10 ix 1891, *J.G. Veitch s.n.* (holo K *n.v.*).
   Begonia praeclara King, J. Asiat. Soc. Bengal 71: 66 (1902); Ridley, J. Fed. Malay States Mus. 4: 21 (1909); Ridley, Fl. Malay Penins. 1: 863 (1922); Kiew, Begonias Penins. Malaysia 133 (2005). – Type: Peninsular Malaysia, Perak, Gunung Batu Puteh, 1890, *L. Wray Jr.* 349 (lecto K *n.v.*; isolecto BM *n.v.*).
- **PENINSULAR MALAYSIA:** Perak, Gunung Kerbau, vi 1913, *H.C. Robinson s.n.* (BM); Pahang, Cameron Highlands, Gunung Brinchang, 26 vii 2002, *S. Neale, G. Bramley* 8 (E); Perak, Gunong

Batu Puteh, 1890, *L. Wray Jr.* 349 (lecto K *n.v.*; isolecto BM *n.v.*) – Type of *Begonia praeclara* King; Pahang, Cameron Highlands, Gunung Berumban, 21 vi 1984, *R. Kiew* 410 (L); Perak, Gunung Batu Puteh, *L. Wray Jr.* 427 (L); Pahang, Cameron Highlands, Robinsons Falls, 20 iii 1992, *Klackenberg, Lundin* 700 (L); Pahang, Cameron Highlands, 3 x 1963, *W.L. Chew* 777 (L); Pahang, Cameron Highlands, 6 x 1963, *W.L. Chew* 848 (L); Perak, Tapah Hills, Sungei Woh, 13 vii 1966, *unknown* 1360 (L); Perak, Gunong Batu Puteh, viii 1885, *Kings Collector* 8077 (B); Pahang, Telom, xi 1900, *Anon.* 14005 (B, BM).

- **IUCN category: LC.** Found in three protected areas (Ulu Muda, Endau Rompin and Cameron Highlands).

Begonia delicatula Parish ex C.B. Clarke in Hook. f. **[§ Apterobegonia]**, Fl. Brit. Ind. 2: 652 (1879). – Type: Burma, Moulmein, 1862, *C.S.P. Parish s.n.* (holo K).

- **BURMA:** Moulmein, 1862, *C.S.P. Parish s.n.* (holo K) – Type of *Begonia delicatula* Parish ex C.B. Clarke.
- **IUCN category: DD.** Insufficient specimens could be georeferenced with certainty.

Begonia demissa Craib [not placed to section], Bull. Misc. Inform. Kew 1930: 409 (1930); Craib, Fl. Siam. 1: 773 (1931). – Type: Thailand, Rachaburi, Kanburi, Baw Re, 21 vii 1926, *N. Put* 218 (holo ABD; iso BM, K).

- **BURMA:** Tenasserim Division, Tavoy District, Paungdaw Power Station, viii 1961, *J. Keenan et al.* 742 (A, E, K); Tenasserim Division, Tavoy District, Paungdaw Power Station, viii 1961, *J. Keenan et al.* 818 (A, E, K); Tenasserim Division, Tavoy District, Paungdaw Power Station, 6 ix 1961, *J. Keenan et al.* 1428 (A, E, K).

  THAILAND: Phitsanulok Prov., Thungsalaeng Luang, Kaeng So Pha waterfall, 22 x 1984, *A. McAllan* 2 (K); Rachaburi, Kanburi, Baw Re, 21 vii 1926, *N. Put* 218 (holo ABD; iso BM, K) – Type of *Begonia demissa* Craib; Kanchanaburi, Tung Yai Naresuan Wildlife Sanct., 19 viii 2001, *M. van de Bult* 458 (A, L); Uthai Thani Prov., Huai Kha Kaeng Wildlife Sanct., 27 vii 2002, *M. van de Bult* 594 (L); Phitsanulok Prov., So Pah Waterfall, 22 vii 1966, *K. Larsen et al.* 719 (P); SangKlaburi Distr., Teeng Yai Naresuan Wildlife Reserve, Pawng, 15 viii 1993, *J.F. Maxwell* 93-900 (A, L); Tak Prov., Po Tip Tawng Cave, 20 viii 1994, *J.F. Maxwell* 94-913 (A, L); Payap, Doi Chiengdao, 21 viii 1935, *H.B.G. Garrett* 989 (K); Chiangmai, Chiang Dao, Pa Blawag Cave, 5 viii 1989, *J.F. Maxwell* 89-993 (A); Payap, Doi Chiengdao, 30 viii 1935, *H.B.G. Garrett* 1004 (A, ABD, K, L, P); Lampang Prov., Chae Son Nat. Park, 26 vii 1996, *J.F. Maxwell* 96-1025 (A, L); Kin Sayok, 9 vii 1946, *A.J.G.H. Kostermans* 1041A (A); Payap, Doi Chiengdao, 17 viii 1963, *T. Smitinand, H. Sleumer* 1057 (L); Payap, Doi Chiengdao, 14 viii 1949, *H.B.G. Garrett* 1261 (K, L, P); Payap, Doi Chiengdao, 20 vi 1968, *C.F. van Beusekom, C. Phengkhlai* 1339 (L); Kin Sayok, 27 vii 1946, *A.J.G.H. Kostermans* 1373 (A); Kanchanaburi, Dongyai, 14 viii 1971, *C. Pengkali, B. Sangkhachand* 2937 (L); Teen Tok, 3 viii 1968, *K. Larsen et al.* 3093 (L); Kanchanaburi, Kritee – Huay ban Kao, 12 vii 1973, *R. Geesink, C. Phengkhlai* 6228 (K, L, P); Payap, Doi Chiengdao, 11 ix 1967, *M. Tagawa et al.* 9814 (L); Tung Salaeng Luang Nat. Park, 30 ix 1967, *M. Tagawa et al.* 11231 (L, P); Phitsanulok Prov., Tung Salaeng Luang Nat. Park, Kang So Pa Waterfall, So Pah Waterfall, 11 x 1979, *T. Shimizu et al.* 18356 (L, P).

- **Notes:** Placed in § *Parvibegonia* by Doorenbos *et al.* (1998), but differs markedly in having three stigmas and an asymmetric androecium reminiscent of a tiny bunch of bananas (see illustration on page 34). It may be best placed in § *Lauchea* with *B. adenopoda* Lem. and *B. burmensis* L.B. Sm. & Wassh., and indeed is probably synonymous with the latter. See notes under *B. brandisiana* Kurz.
- **IUCN category: LC.** Widespread and occurring in several national parks.

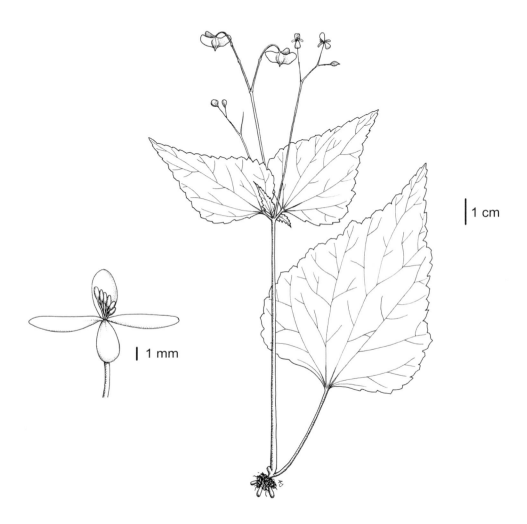

1 cm

1 mm

Begonia demissa

Begonia densiretis Irmsch. [§ Petermannia], Webbia 9: 490 (1953). – Type: Borneo, Sarawak, Busso, 1865, *O. Beccari* PB216 (holo FI).
- **BORNEO: Sarawak:** Busso, 1865, *O. Beccari* PB216 (holo FI) – Type of *Begonia densiretis* Irmsch.
- **IUCN category: DD**. Insufficient specimens could be georeferenced with certainty.

Begonia diffusiflora Merr. & L.M. Perry [§ Petermannia], J. Arnold Arbor. 24: 46 (1943). – Type: New Guinea, Papua New Guinea, Palmer River, vii 1936, *L.J. Brass* 7318 (holo A *n.v.*).
- **NEW GUINEA: Papua New Guinea:** Fly River, v 1936, *L.J. Brass* 6713 (para not located) – Type of *Begonia diffusiflora* Merr. & L.M. Perry; Fly River, v 1936, *L.J. Brass* 7010 (para not located) – Type of *Begonia diffusiflora* Merr. & L.M. Perry; Palmer River, vii 1936, *L.J. Brass* 7318 (holo A *n.v.*) – Type of *Begonia diffusiflora* Merr. & L.M. Perry.
- **IUCN category: NT**. Although no specific threat could be identified to the forests in these localities, this species grows at relatively low elevations (80–100 m) and is stated to be 'uncommon' and 'a rare species' according to specimens cited in the protologue.

Begonia discrepans Irmsch. [§ Platycentrum], Bot. Jahrb. Syst. 76: 100 (1953). – *Begonia tenuicaulis* Irmsch., Mitt. Inst. Allg. Bot. Hamburg 10: 543 (1939). – Type: not located.
- **BURMA:** North Triangle, 12 vii 1953, *F.K. Ward* 21141 (A, K) [as *Begonia aff. tenuicaulis*].
- **IUCN category: DD**. Insufficient specimens could be georeferenced with certainty.

Begonia discreta Craib [§ Diploclinium], Bull. Misc. Inform. Kew 1930: 410 (1930); Craib, Fl. Siam. 1: 773 (1931). – Type: Thailand, Chiangmai, Doi Sootep, 25 x 1925, 18°50′N, 98°54′E, *A.F.G. Kerr* 3442 (holo ABD; iso BM, K, NY).
- **THAILAND:** Chiangmai, Doi Sootep, 25 x 1925, *A.F.G. Kerr* 3442 (holo ABD; iso BM, K, NY) – Type of *Begonia discreta* Craib.
- **IUCN category: VU D2**. Appears to be endemic to the Doi Suthep National Park. Whilst this is protected, the park is heavily used as a recreation area.

Begonia divaricata Irmsch. [§ Petermannia], Webbia 9: 473 (1953). – Type: Sumatra, Mt. Singalan, 1878, *O. Beccari* CB4505 (syn B[2], FI); Sumatra, Mt. Singalan, 1878, *O. Beccari* CB4506 (syn B, FI).
- **SUMATRA:** Mt. Singalan, 1878, *O. Beccari* PS125 (holo FI; iso FI) – Type of *Begonia divaricata* forma *minor* Irmsch.; Mt. Singalan, 1878, *O. Beccari* PS125 (holo FI; iso FI) – Type of *Begonia divaricata* forma *minor* Irmsch.; Mt. Singalan, 1878, *O. Beccari* CB4505 (syn B[2], FI) – Type of *Begonia divaricata* Irmsch.; Mt. Singalan, 1878, *O. Beccari* CB4506 (syn B, FI) – Type of *Begonia divaricata* Irmsch.

Begonia divaricata forma divaricata (Irmsch.), Webbia 9: 473 (1953). – *Begonia divaricata* forma *typica* Irmsch., Webbia 9: 473 (1953). – Type: not located.
- **SUMATRA:** Jambi Prov., Gunung Tujuh, 26 vii 2006, *D. Girmansyah, A. Poulsen, I. Hatta, R. Neivita* 785 (BO, E[2]).

Begonia divaricata forma minor Irmsch., Webbia 9: 473 (1953). – Type: Sumatra, Mt. Singalan, 1878, *O. Beccari* PS125 (holo FI; iso FI).
- **Notes:** Tentatively assigned to § *Bracteibegonia* (Doorenbos *et al.*, 1998), but drawings by Irmscher show this to have inflorescences with female flowers basal to the male, typical for § *Petermannia*.

- **IUCN category: LC**. The G. Singgalang Protection Forest is in quite good condition. The species is also found on Gunung Tujuh in the Kerinci-Seblat National Park (D. Girmansyah, pers. comm.).

Begonia diwolii Kiew [§ Diploclinium], Gard. Bull. Singapore 53: 254 (2001). – Type: Borneo, Sabah, Lahad Datu District, Segama River, Tempadong, *R. Kiew* 4767 (holo SAN *n.v.*; iso BRUN *n.v.*, K *n.v.*, SAR *n.v.*, SING *n.v.*).
- **BORNEO: Sabah:** Lahad Datu District, Segama River, Tempadong, *R. Kiew* 4767 (holo SAN; iso BRUN *n.v.*, K *n.v.*, SAR *n.v.*, SING *n.v.*) – Type of *Begonia diwolii* Kiew.
- **IUCN category: VU D2**. Known only from limestone hills in the vicinity of the type locality, which are not protected.

Begonia djamuensis Irmsch. [§ Petermannia], Bot. Jahrb. Syst. 50: 364 (1913). – Type: New Guinea, Papua New Guinea, Djamu, 20 ii 1908, *F.R.R. Schlechter* 17310 (holo B; iso P).
- **NEW GUINEA: Papua New Guinea:** Djamu, 20 ii 1908, *F.R.R. Schlechter* 17310 (holo B; iso P) – Type of *Begonia djamuensis* Irmsch.
- **IUCN category: DD**. Insufficient specimens could be georeferenced with certainty.

Begonia dolichotricha Merr. [§ Petermannia], Philipp. J. Sci. 17: 292 (1920); Merrill, Enum. Philipp. Fl. Pl. 3: 122 (1923). – Type: Philippines, Luzon, Camarines, Mt. Bagacay, xi 1918 – xii 1918, *M. Ramos, G. Edano* 33926 (syn B, BM, P).
- **PHILIPPINES: Luzon:** Camarines, Mt. Bagacay, xi 1918 – xii 1918, *M. Ramos, G. Edano* 33855 (para K) – Type of *Begonia dolichotricha* Merr.; Camarines, Mt. Bagacay, xi 1918 – xii 1918, *M. Ramos, G. Edano* 33926 (syn B, BM, P) – Type of *Begonia dolichotricha* Merr.; Tayabas, Mt. Alzapan, v 1925 – vi 1925, *M. Ramos, G. Edano* 45673? (B, K, P); Tayabas, Mt. Alzapan, v 1925 – vi 1925, *M. Ramos, G. Edano* 45773 (BO, K).
- **IUCN category: VU D2**. Known only from the type locality.

Begonia dosedlae Gilli [§ Petermannia], Ann. Naturhist. Mus. Wien 83: 421 (1979). – Type: New Guinea, Papua New Guinea, Mt. Hagen, 20 vi 1971, *Dosedla* D139 (holo W *n.v.*).
- **NEW GUINEA: Papua New Guinea:** Mt. Hagen, 20 vi 1971, *Dosedla* D139 (holo W *n.v.*) – Type of *Begonia dosedlae* Gilli.
- **IUCN category: DD**. I have not seen any specimens to check if this is distinct from *B. naumoniensis*.

Begonia dux C.B. Clarke in Hook. f. **[§ Platycentrum]**, Fl. Brit. Ind. 2: 637 (1879); Tebbitt & Dickson, Brittonia 52(1): 116 (2000). – Type: Burma, Moulmein, Moolee, *C.S.P. Parish s.n.* (holo K).
- **BURMA:** Moulmein, Moolee, *C.S.P. Parish s.n.* (holo K) – Type of *Begonia dux* C.B. Clarke; Moulmein, Moolee, xii 1880, *R.H. Beddome s.n.* (K); Mooley, *R.H. Beddome* 3196 (BM *n.v.*); Mooley, *R.H. Beddome* 3197 (BM *n.v.*).
- **IUCN category: DD**. Insufficient specimens could be georeferenced with certainty.

Begonia eberhardtii Gagnep. [§ Petermannia], Bull. Mus. Hist. Nat. (Paris) 25: 198 (1919); Gagnepain, in Lecomte (ed.), Fl. Indo-Chine 2: 1116 (1921). – Type: Vietnam, Annam, Thua-thien Province, *P.A. Eberhardt* 3106 (holo P).
- **VIETNAM:** Annam, Thua-thien Province, *P.A. Eberhardt* 3106 (holo P) – Type of *Begonia eberhardtii* Gagnep.
- **IUCN category: DD**. Insufficient specimens could be georeferenced with certainty.

Begonia edanoi Merr. [§ Petermannia], Philipp. J. Sci. 13: 314 (1918); Merrill, Enum. Philipp. Fl. Pl. 3: 122 (1923). – Type: Philippines, Luzon, Rizal Province, Mt. Susong-Dalanga, viii 1917, 14°37′59′′N, 121°19′E, *M. Ramos, G. Edano* 29374 (syn BM, K, L, P).

- **PHILIPPINES: Luzon:** Rizal Province, Mt. Irid, xi 1926, *M. Ramos, G. Edano* 18532 (B); Rizal Province, Mt. Susong-Dalanga, viii 1917, *M. Ramos, G. Edano* 29374 (syn BM, K, L, P) – Type of *Begonia edanoi* Merr.; Rizal Province, Mt. Irid, xi 1926, *M. Ramos, G. Edano* 48532 (BM, K).
- **IUCN category: LC**. Endemic to the Marikina Watershed Forest Reserve.

Begonia eiromischa Ridl. [§ Ridleyella], J. Straits Branch Roy. Asiat. Soc. 75: 36 (1917); Ridley, Fl. Malay Penins. 1: 860 (1922); Kiew, Begonias Penins. Malaysia 35 (2005). – Type: Peninsular Malaysia, Penang, Pulau Betong, xi 1898, *C. Curtis* 1028 (lecto SING *n.v.*; isolecto SING *n.v.*).

- **PENINSULAR MALAYSIA:** Penang, Pulau Betong, xi 1898, *C. Curtis* 1028 (lecto SING *n.v.*; isolecto SING *n.v.*) – Type of *Begonia eiromischa* Ridl.
- **Notes:** Now considered extinct (Kiew, 2005).
- **IUCN category: EX**. Described from a single locality which has since been destroyed by farming (Kiew, 2005).

Begonia elatostematoides Merr. [§ Petermannia], Philipp. J. Sci. 7: 309 (1912); Merrill, Enum. Philipp. Fl. Pl. 3: 122 (1923). – Type: Philippines, Mindanao, Zamboanga, Sax River Mountains, xi 1911 – xii 1911, *E.D. Merrill* 8232 (syn BM, BO, E, L, P).

- **PHILIPPINES: Mindanao:** Zamboanga, Sax River Mountains, xi 1911 – xii 1911, *E.D. Merrill* 8232 (syn BM, BO, E, L, P) – Type of *Begonia elatostematoides* Merr.
- **IUCN category: DD**. Insufficient specimens could be georeferenced with certainty.

Begonia elatostemma Ridl. [§ Petermannia], J. Straits Branch Roy. Asiat. Soc. 46: 255 (1906). – Type: Borneo, Sarawak, Rejang, *G.D. Haviland* 2946 (holo not located).

- **BORNEO: Sarawak:** Rejang, *G.D. Haviland* 2946 (holo not located) – Type of *Begonia elatostemma* Ridl.
- **IUCN category: DD**. Insufficient specimens could be georeferenced with certainty.

Begonia eliasii Warb. [§ Petermannia], Bot. Jahrb. Syst. 13: 387 (1891); Warburg, Fl. Deutsch. Schutzgeb. Südsee 457 (1901). – Type: New Guinea, Papua New Guinea, Kaiser-Wilhelmsland, Station Finschhafen, *O. Warburg* 20480 (holo B).

- **NEW GUINEA: Papua New Guinea:** Kaiser-Wilhelmsland, Station Finschhafen, *O. Warburg* 20480 (holo B) – Type of *Begonia eliasii* Warb.
- **IUCN category: DD**. Insufficient specimens could be georeferenced with certainty.

Begonia elisabethae Kiew [§ Parvibegonia], Begonias Penins. Malaysia 98 (2005). – Type: Peninsular Malaysia, Langkawi, x 1996, *E. Eber-Chan* 124 (holo SING *n.v.*).

- **THAILAND:** Surat Thani, 13 viii 1975, *D. Prapat* 9 (L); Klong Chilat, 13 vii 1992, *K. Larsen et al.* 43343 (P).
  PENINSULAR MALAYSIA: Kedah, Bukit Weng, *C.M. Weber s.n.* (para SING *n.v.*) – Type of *Begonia elisabethae* Kiew; Langkawi, x 1996, *E. Eber-Chan* 124 (holo SING *n.v.*) – Type of *Begonia elisabethae* Kiew.
- **Notes:** See notes under *B. vagans* Craib.
- **IUCN category: LC**. Found on both granite and limestone in southern Thailand and in northern Peninsular Malaysia.

Begonia elmeri Merr. [§ Diploclinium], Philipp. J. Sci. 13: 39 (1918). – *Begonia peltata* Elmer. Later homonym, Leafl. Philipp. Bot. 7: 2556 (1915); Merrill, Philipp. J. Sci. 13: 39 (1918); Merrill, Enum. Philipp. Fl. Pl. 3: 122 (1923). – Type: Agusan, Mt. Urdaneta, *A.D.E. Elmer* 14183 (syn E, L).

- **PHILIPPINES: Mindanao:** Agusan, Mt. Urdaneta, *A.D.E. Elmer* 14183 (syn E, L) – Type of *Begonia peltata* Elmer. Later homonym.
- **IUCN category: DD**. Insufficient specimens could be georeferenced with certainty.

Begonia erosa Blume [§ Parvibegonia], Enum. Pl. Javae 1: 96 (1827); Candolle, Prodr. 15(1): 276 (1864); Koorders, Exkurs.-Fl. Java 2: 647 (1912). – *Platycentrum erosum* (Blume) Miq., Fl. Ned. Ind. 1(1): 694 (1856); Klotzsch, Bot. Zeitung 15: 182 (1857); Candolle, Prodr. 15(1): 276 (1864). – *Casparya erosa* (Blume) A. DC., Prodr. 15(1): 276 (1864). – Type: not located.
    Sphenanthera erosa Hassk. ex Klotzsch, Bot. Zeitung 15: 182 (1857); Candolle, Prodr. 15(1): 276 (1864). – Type: not located.

- **JAVA:** 1846, *W. Lobb s.n.* (K); 1846, *W. Lobb s.n.* (K); *W.H. de Vriese s.n.* (K); *C.L. von Blume* 1664 (B); 22 ii 1845, *H. Zollinger* 2720 (P); Banjoemas, Noesa Kambangan, 2 ii 1896, *S.H. Koorders* 21926B (K).
- **Notes:** See notes under *B. integrifolia*.
- **IUCN category: DD**. Boundary between this and similar species not clear.

Begonia erythrogyna Sands [§ Petermannia], Kew Mag. 7: 81 (1990); Sands, Pl. Mt. Kinabalu 153 (2001). – Type: Borneo, Sabah, Ranau District, Poring Hot Springs, Poring Hot Springs – Gua Kechapi, 18 iv 1984, *M.J.S. Sands* 3291 (holo K *n.v.*).

- **BORNEO: Sabah:** Ranau District, Poring Hot Springs – Langanan Falls, 18 iv 1984, *M.J.S. Sands* 3291 (holo K *n.v.*) – Type of *Begonia erythrogyna* Sands.
- **IUCN category: VU D2**. Endemic to the eastern side of Mt. Kinabalu on the limits of the park boundary.

Begonia esculenta Merr. [§ Petermannia], Philipp. J. Sci. 6: 389 ('1911', 1912); Merrill, Enum. Philipp. Fl. Pl. 3: 122 (1923). – Type: Philippines, Luzon, Tayabas, Mt. Binuang, viii 1909, *C.B. Robinson* BS9449 (syn not located).

- **PHILIPPINES: Luzon:** Tayabas, Mt. Binuang, viii 1909, *C.B. Robinson* BS9449 (syn not located) – Type of *Begonia esculenta* Merr.; Ilocos Norte Province, Mt. Palimlim, viii 1918, *M. Ramos* 33347 (B, K, P).
- **IUCN category: VU D2**. Known only from two localities.

Begonia eutricha Sands [§ Petermannia], in Coode *et al.*, Checkl. Fl. Pl. Gymnosperms Brunei Darussalam App. 2: 434 (1997). – Type: Borneo, Brunei, Kuala Belalong Field Centre, *J. Dransfield* 6708 (holo K; iso BRUN *n.v.*).

- **BORNEO: Brunei:** Batu Apoi Forest Reserve, 6 xi 1991, *C. Hansen* 1504 (L); Kuala Belalong Field Centre, *J. Dransfield* 6708 (holo K; iso BRUN *n.v.*) – Type of *Begonia eutricha* Sands; Kuala Belalong Field Centre, 22 ii 1991, *G.C.G. Argent, D. Mitchell* 9194 (E[3]).
- **IUCN category: LC**. Endemic to the Ulu Temburong National Park in Brunei.

Begonia everettii Merr. [§ Petermannia], Philipp. J. Sci. 6: 390 ('1911', 1912); Merrill, Enum. Philipp. Fl. Pl. 3: 122 (1923). – Type: Philippines, Negros, Sicaba, Daluapan River, *A.H. Everett* FB5587 (syn PNH).

- **PHILIPPINES: Luzon:** Tayabas, *W.H. Kobbe* 6717 (para not located) – Type of *Begonia everettii* Merr.; Ifugao, Mt. Mahaling, 28 x 1994, *R.S. Majaducon* 8601 (L). **Panay:** Antique Province, v

1918 – viii 1918, *R.C. McGregor* 32501 (L). **Negros:** Sicaba, Daluapan River, *A.H. Everett* FB5587 (syn PNH) – Type of *Begonia everettii* Merr.
- **IUCN category: DD**. Insufficient specimens could be georeferenced with certainty.

Begonia fasciculata Jack [§ Petermannia], Malayan Misc. 2(7): 12 (1822); Candolle, Prodr. 15(1): 322 (1864). – *Petermannia fasciculata* (Jack) Klotzsch, Monatsber. Kon. Preuss. Akad. Wiss. Berlin 1854: 124 (1854); Klotzsch, Abh. Kon. Akad. Wiss. Berlin 1854: 195 (1855); Klotzsch, Begoniac. 75 (1855); Miquel, Fl. Ned. Ind. 1(1): 690 (1856); Candolle, Prodr. 15(1): 322 (1864). – *Diploclinium fasciculatum* (Jack) Miq., Fl. Ned. Ind. 1(1): 690 (1856). – Type: not located.
- **SUMATRA:** *Fide* Jack (1822).
- **IUCN category: DD**. Insufficient specimens could be georeferenced with certainty.

Begonia fasciculiflora Merr. [§ Petermannia], Philipp. J. Sci. 6: 376 ('1911', 1912). – Type: Philippines, Mindanao, Zamboanga, Port Banga, *H.N. Whitford, W.I. Hutchinson* FB9316 (syn not located).
- **PHILIPPINES: Mindanao:** Zamboanga, Port Banga, xii 1907 – i 1913, *H.N. Whitford, W.I. Hutchinson* FB9248 (para not located) – Type of *Begonia fasciculiflora* Merr.; Zamboanga, Port Banga, *H.N. Whitford, W.I. Hutchinson* FB9316 (syn not located) – Type of *Begonia fasciculiflora* Merr.; Zamboanga, Malangas, x 1919 – xi 1919, *M. Ramos, G. Edano* 36792 (B, K, P); Zamboanga, Malangas, x 1919 – xi 1919, *M. Ramos, G. Edano* 37228 (BM, BO).
- **IUCN category: VU D2**. Known from two localities only.

Begonia fenicis Merr. [§ Diploclinium], Philipp. J. Sci. 3: 421 (1908); Merrill, Philipp. J. Sci. 6: 401 ('1911', 1912); Merrill, Enum. Philipp. Fl. Pl. 3: 122 (1923); Tebbitt, Begonias 225 (2005). – Type: Philippines, Batan Islands, Santo Domingo de Basco, 27 v 1907, *E.A. Mearns* 3207 (syn not located); Philippines, Batan Islands, Santo Domingo de Basco, 30 v 1907, *E. Fenix* 3619 (syn not located); Philippines, Babuyan Islands, Babuyan, 17 vi 1907, *E. Fenix* 3893 (syn not located).
- **PHILIPPINES: Palawan:** Balabac Island, 16 x 1906, *E.D. Merrill* 5375 (B, P). **Batan Islands:** Santo Domingo de Basco, 27 v 1907, *E.A. Mearns* 3207 (syn not located) – Type of *Begonia fenicis* Merr.; Santo Domingo de Basco, 30 v 1907, *E. Fenix* 3619 (syn not located) – Type of *Begonia fenicis* Merr. **Babuyan Islands:** Babuyan, 17 vi 1907, *E. Fenix* 3893 (syn not located) – Type of *Begonia fenicis* Merr. **Samar:** Lanang, 3 x 1906, *E.D. Merrill* 5237 (B, P).
- **Notes:** The records from Palawan and Samar may be referable to other species in § *Diploclinium*. This species extends north into Taiwan.
- **IUCN category: LC**. Also occurs in Taiwan.

Begonia festiva Craib [§ Diploclinium], Bull. Misc. Inform. Kew 1930: 411 (1930); Craib, Fl. Siam. 1: 773 (1931). – Type: Thailand, Kao Nawng, 8°50′N, 99°30′E, *A.F.G. Kerr s.n.* (holo ABD).
- **THAILAND:** Kao Nawng, *A.F.G. Kerr s.n.* (holo ABD) – Type of *Begonia festiva* Craib.
- **IUCN category: VU D2**. Collected from the Tai Rom Yen National Park, which is adjacent to Khao Luang National Park. However, the species is known only from the type despite the area being reasonably well collected.

Begonia fibrosa C.B. Clarke in Hook. f. **[§ Reichenheimia]**, Fl. Brit. Ind. 2: 652 (1879); Craib, Fl. Siam. 1: 773 (1931). – Type: Burma, Moulmein, 1862, *C.S.P. Parish s.n.* (holo K).
- **BURMA:** Moulmein, 1862, *C.S.P. Parish s.n.* (holo K) – Type of *Begonia fibrosa* C.B. Clarke. **THAILAND:** Chiengrai, Doi Tain Jup, 14 ix 1924, *H.B.G. Garrett* 199 (ABD[2], B, BM, L).
- **IUCN category: DD**. Insufficient specimens could be georeferenced with certainty.

Begonia filibracteosa Irmsch. [§ Petermannia], Bot. Jahrb. Syst. 50: 361 (1913); Irmscher, Bot. Jahrb. Syst. 50: 565 (1914). – Type: New Guinea, Papua New Guinea, Jaduna, 8 iii 1908, *F.R.R. Schlechter* 17380 (holo B).
- **NEW GUINEA: Papua New Guinea:** Jaduna, 8 iii 1908, *F.R.R. Schlechter* 17380 (holo B) – Type of *Begonia filibracteosa* Irmsch.
- **IUCN category: DD**. Insufficient specimens could be georeferenced with certainty.

Begonia flacca Irmsch. [§ Petermannia], Webbia 9: 486 (1953). – Type: Sulawesi, Kendari, Peninsula SE of Kendari, 1874, *O. Beccari* CB4501 & CB4501A (syn FI).
- **SULAWESI:** Kendari, Peninsula SE of Kendari, 1874, *O. Beccari* CB4501 & CB4501A (syn FI) – Type of *Begonia flacca* Irmsch.
- **IUCN category: VU D2**. The locality of the type is rather vague, but is possibly in the forest covered by the Tanjung Peropa Game Reserve.

Begonia flaccidissima Kurz [§ Parvibegonia], J. Asiat. Soc. Bengal 41: 308 (1872). – Type: Burma, Tenasserim, *unknown s.n.* (holo not located).
- **BURMA:** Tenasserim, *unknown s.n.* (holo not located) – Type of *Begonia flaccidissima* Kurz.
- **IUCN category: DD**. Insufficient specimens could be georeferenced with certainty.

Begonia flaviflora H. Hara [§ Platycentrum], J. Jap. Bot. 45: 91 (1970). – Type: India, Darjeeling, Senchal, 5 vii 1969, *H. Hara et al.* 69218 (holo TI *n.v.*).

Begonia flaviflora var. flaviflora
- **BURMA:** Mali Hka, 31 vii 1937, *F.K. Ward* 12859 (BM, E).

Begonia flaviflora var. vivida (Irmsch.) Golding & Kareg., Phytologia 54(7): 496 (1984). – *Begonia laciniata* subsp. *flaviflora* Irmsch., Mitt. Inst. Allg. Bot. Hamburg 10: 529 (1939); Golding & Karegeannes, Phytologia 54(7): 496 (1984). – Type: not located.
- **BURMA:** Sadon – Eansi Gorge, ix 1912, *G. Forrest* 9100 (B, E); Tzi-Tzo-ti, vii 1925, *G. Forrest* 27146 (B[2], E); Hpimaw, *G. Forrest* 29922 (B, E).
- **IUCN category: LC**. This species occurs in the vicinity of the Gaoligongshan Nature Reserve (south portion), as well as in the relatively undisturbed forests of north Burma.

Begonia flexicaulis Ridl. [§ Petermannia], Trans. Linn. Soc. London, Bot., II 9: 59 (1916). – Type: New Guinea, Papua, Camp I – Camp III, *C.B. Kloss s.n.* (holo BM).
- **NEW GUINEA: Papua:** Camp I – Camp III, *C.B. Kloss s.n.* (holo BM) – Type of *Begonia flexicaulis* Ridl.
- **IUCN category: DD**. Insufficient specimens could be georeferenced with certainty.

Begonia flexula Ridl. [§ Petermannia], J. Malayan Branch Roy. Asiat. Soc. 1(87): 63 (1923). – Type: Sumatra, Bukit Khsang, Sibolangit, 4 viii 1921, 3°18'3"N, 98°24'3"E, *A.J.G.H. Kostermans, Md. Nur* 7444 (holo K).
- **SUMATRA:** Bukit Khsang, Sibolangit, 4 viii 1921, *A.J.G.H. Kostermans, Md. Nur* 7444 (holo K) – Type of *Begonia flexula* Ridl.
- **IUCN category: CR B2ab(iii)**. The lowland forests in the Sibolangit vicinity are quite heavily degraded.

Begonia forbesii King [§ Reichenheimia], J. Asiat. Soc. Bengal, Pt. 2, Nat. Hist. 71: 58 (1902); Irmscher, Mitt. Inst. Allg. Bot. Hamburg 8: 159 (1929); Kiew, Begonias Penins. Malaysia 206

(2005). – Type: Peninsular Malaysia, Perak, Sungai Larut, 1888, *L. Wray Jr.* 2476 (lecto K *n.v.*); Sumatra, *Forbes* 2666 (syn not located).

- **PENINSULAR MALAYSIA:** Perak, Sungai Larut, 1888, *L. Wray Jr.* 2476 (lecto K *n.v.*) – Type of *Begonia forbesii* King; Perak, Sungai Larut, *R. Kiew* RK4909 (KEP *n.v.*, SING *n.v.*).
- **Notes:** The specimen from Sumatra (*Forbes* 2666) cannot be located; however, this record is considered doubtful (Kiew, 2005).
- **IUCN category: CR B2ab(iii).** Known only from a single population which is in danger of having its habitat destroyed through the encroachment of farmland (Kiew, 2005).

Begonia forrestii Irmsch. [§ Platycentrum], Mitt. Inst. Allg. Bot. Hamburg 10:548 (1939). – Type: not located.

- **BURMA:** Tara Hka, 24 viii 1939, *R. Kaulback* 359 (BM); Kachin, Sumprabum sub-division, Tsuptaung – Kanang, 27 xii 1961, *J. Keenan et al.* 3010 (E); Kachin, Sumprabum sub-division, Ndum-Zup – Hpuginkhu, 30 xii 1961, *J. Keenan et al.* 3095 (E); Kachin, Sumprabum sub-division, Ning W'Krok – Kanang, 21 i 1962, *J. Keenan* 3370 (E); Kachin, Sumprabum sub-division, Magi – Sumprabum, 14 ii 1962, *J. Keenan* 3400 (E); Kachin, Sumprabum sub-division, Kanang – Ning W'Krok, 16 ii 1962, *J. Keenan* 3434 (E); Mali Hka, 16 vii 1937, *F.K. Ward* 12827 (BM); North Triangle, Hkinlum, 2 v 1953, *F.K. Ward* 20770 (BM).
- **IUCN category: LC.** Although this species has an apparent area of occupancy of <2000 km² and hence qualifies for VU D2, there are four collections spread through the relatively stable Kachin Hills area.

Begonia foxworthyi Burkill ex Ridl. [§ Reichenheimia], Fl. Malay Penins. 5(suppl.): 311 (1925); Irmscher, Mitt. Inst. Allg. Bot. Hamburg 8: 100 (1929); Kiew, Begonias Penins. Malaysia 227 (2005). – Type: Peninsular Malaysia, Kelantan, Kuala Rek, *M. Haniff, Md. Nur* 10199 (holo SING *n.v.*; iso BM, K[2]).

- **PENINSULAR MALAYSIA:** Pahang, Sungei Teku, 23 vii 1936, *K.H. Salleh* 1781 (L) [as *Begonia cf. foxworthyi*]; Kelantan, Gunung Renayang, 8 v 1990, *R. Kiew, S. Anthonysamy* 2866 (L); Kelantan, Elephant Cave, 15 v 1990, *R. Kiew, S. Anthonysamy* 2946 (L); Kelantan, Kampung Sta, 15 v 1990, *R. Kiew, S. Anthonysamy* 2961 (L); Kelantan, Kuala Rek, *M. Haniff, Md. Nur* 10199 (holo SING *n.v.*; iso BM, K[2]) – Type of *Begonia foxworthyi* Burkill ex Ridl.; Kelantan, Gua Musang, 13 viii 1929, *M.R. Henderson* SFN22710 (K); Kelantan, Bukit Batu Papan, Sungai Lebir, 4 vii 1935, *M.R. Henderson* 29509 (P); Kelantan, Bukit Batu Papan, Sungai Lebir, 4 vii 1935, *M.R. Henderson* SFN29509 (not located); Kelantan, Bukit Batu Papan, Sungai Lebir, 4 vii 1935, *M.R. Henderson* SFN29509 (BO); Kelantan, Gua Teja, Sungai Betis, 15 vii 1935, *M.R. Henderson* 29602 (BM); Kelantan, Gua Teja, Sungai Betis, 15 vii 1935, *M.R. Henderson* SFN29672 (K); Pahang, Sungai Teku (near), SFN31781 (K) [as *Begonia cf. foxworthyi*].
- **IUCN category: LC.** Known from over seven sites. Is not restricted to limestone.

Begonia fraseri Kiew [§ Platycentrum], Sandakania 6: 64 (1995); Kiew, Begonias Penins. Malaysia 184 (2005). – Type: Peninsular Malaysia, Frasers Hill, 17 vii 1995, *R. Kiew* 3831 (holo KEP ex UPM *n.v.*).

- **PENINSULAR MALAYSIA:** Pahang, Gunung Benom, 2 viii 1925, *Native collector s.n.* (SING *n.v.*); Frasers Hill, *R. Kiew* 3831 (holo KEP ex UPM *n.v.*) – Type of *Begonia fraseri* Kiew; Frasers Hill, 23 iv 1955, *J.W.P. Purseglove* 4315 (K); Frasers Hill, 16 ix 1922 – 30 ix 1922, *I.H. Burkill, R.E. Holttum* SFN8248 (K); Frasers Hill, 28 viii 1923, *Md. Nur* 11104 (BM, K); Frasers Hill, 12 viii 1937, *E.J.H. Corner* SFN33176 (K, L).

■ **IUCN category: VU D2.** Known from two areas. Although these are not currently under threat the species is not common and is known from a few small populations.

Begonia fruticella Ridl. [§ Petermannia], Trans. Linn. Soc. London, Bot., II 9: 60 (1916). – Type: New Guinea, Papua, Camp VIb, *C.B. Kloss s.n.* (syn BM); New Guinea, Papua, Camp I – Camp III, *C.B. Kloss s.n.* (syn BM); New Guinea, Papua, Camp 4 – Camp 6, *C.B. Kloss s.n.* (syn BM).
■ **NEW GUINEA: Papua:** Camp VIb, *C.B. Kloss s.n.* (syn BM) – Type of *Begonia fruticella* Ridl.; Camp 4 – Camp 6, *C.B. Kloss s.n.* (syn BM) – Type of *Begonia fruticella* Ridl.; Camp I – Camp III, *C.B. Kloss s.n.* (syn BM) – Type of *Begonia fruticella* Ridl.
■ **IUCN category: DD.** Insufficient specimens could be georeferenced with certainty.

Begonia fulvo-villosa Warb. [§ Symbegonia], Bot. Jahrb. Syst. 13: 386 (1891); Irmscher, Bot. Jahrb. Syst. 50: 381 (1913); Forrest & Hollingsworth, Plant Syst. Evol. 241: 208 (2003). – *Symbegonia fulvo-villosa* (Warb.) Warb., Nat. Pflanzenfam. 3(6A): 149 (1894); Warburg, Fl. Deutsch. Schutzgeb. Südsee 459 (1901); Forrest & Hollingsworth, Plant Syst. Evol. 241: 208 (2003). – Type: New Guinea, Papua New Guinea, Morobe Distr., Sattelberg, *O. Warburg* 20476 (holo B).
■ **NEW GUINEA: Papua New Guinea:** Morobe Distr., Sattelberg, 11 xii 1893, *L. Karnbach* 70 (B); Kaiser-Wilhelmsland, 3 i 1889, *F.C. Hellwig* 237 (B[2], BO, K); Morobe Distr., Sattelberg, *K. Weinland* 327 (B); Morobe Distr., Sattelberg, 23 vi 1899, *E. Nyman* 511 (B); Kaiser-Wilhelmsland, 8 ix 1889, *F.C. Hellwig* 517 (B); Southern Highlands Distr., Mendi Valley, Above Kiburu, 1 vii 1961, *R. Schodde* 1379 (K, L); Southern Highlands Distr., Ebenda, 21 vii 1961, *R. Schodde* 1547 (L) [as *Begonia aff. fulvo-villosa*]; Morobe Distr., Aseki Valley, 25 iv 1966, *R. Schodde* 5041 (L *n.v.*); Eastern Highlands Distr., Al River, 7 ix 1953, *J.S. Womersley* 5334 (K); Eastern Highlands Distr., Aiyura Range, 7 vii 1954, *J.S. Womersley* 6060 (L); Mt. Bosavi, 27 x 1973, *M. Jacobs* 8880 (L); Chimbu Province, Crater Mt. Wildlife Management area, 17 vii 1998, *W. Takeuchi* 12154 (L); Eastern Highlands Distr., Warrapuri River, 2 ix 1963, *P.v. Royen* 18117 (BO, K, L); Western Highland Province, Mazmal, 1 ix 1963, *A. Millar, N.G.F. van Royen* 18573 (BO, K); Morobe Distr., Sattelberg, *O. Warburg* 20476 (holo B) – Type of *Begonia fulvo-villosa* Warb.; Eastern Highlands Distr., 4 miles N of Okapa, 28 ix 1959, *L.J. Brass* 31778 (K, L); Morobe Distr., Wau Subprov., Webiak Creek, 6 ii 1968, *A. Kairo* NGF35803 (BO, K, L); Mt. Hagen – Mur Mur Pass, 1 x 1963, *J. Vandenberg, P. Katik, A. Kairo* NGF39909 (BO, K); Morobe Distr., Mt. Shungol, 14 xi 1970, *P.F. Stevens* 50475 (BO, L); Morobe Distr., Mt. Dilmargi, 16 xii 1972, *P.F. Stevens* LAE58032 (L).
■ **IUCN category: LC.** Reasonably widespread across the Central Range in Papua New Guinea.

Begonia furfuracea Hook. f. [§ Tetraphila], Fl. Trop. Afr. 2: 571 (1871). – Type: Africa, Equatorial Guinea, Bioko, *Mann* 304 (holo K; iso K, P).
■ **THAILAND:** Kanchanaburi, 7 vii 1973, *R. Geesink, C. Phengkhlai* 6144 (L) [as *Begonia cf. furfuracea*]; Kanchanaburi, 26 vi 1974, *K. Larsen et al.* 33821 (K) [as *Begonia cf. furfuracea*].
  LAOS: Laos, 6 iv 1932, *A.F.G. Kerr* 20938 (BM, K) [as *Begonia cf. furfuracea*].
■ **Notes:** Although the determinations are tentative due to the lack of female flowers on the specimens, to find a species belonging to the previously solely African section *Tetraphila* in Thailand and Laos is remarkable enough to justify the publication of this record. There is no reason to suspect, given the location of the collections and the local distribution of this species in Africa, that this is anything other than a natural extension to the known range of § *Tetraphila* (Hans de Wilde, pers. comm.).
■ **IUCN category: DD.** Insufficient specimens could be georeferenced with certainty.

Begonia fuscisetosa Sands [§ Petermannia], in Coode *et al.*, Checkl. Fl. Pl. Gymnosperms Brunei Darussalam App. 2: 433 (1997); Sands, Pl. Mt. Kinabalu 153 (2001). – Type: Borneo, Brunei, Temburong, 25 iii 1991, 4°33´N, 115°9´E, *M.J.S. Sands* 5606 (holo K; iso BRUN *n.v.*).

- **BORNEO: Sarawak:** Kalabit Highlands, Mt. Murud, Belapan River – Dapo River, 1 iv 1970, *H.P. Nooteboom, P. Chai* 1841a (B, L); Kalabit Highlands, Batu Lawi, 29 iv 1970, *H.P. Nooteboom, P. Chai* 2302 (B); Lambir National Park, 21 ix 1978, *B.L. Burtt* 11571 (E); Bario, Ulu Baram, 27 vi 1964, *J.A.R. Anderson* 20132 (L); Kapit Distr., Balleh River, Sungei Balang, *J.A.R. Anderson, I. Paie* 28877 (E, L); Limbang, Bt. Pagon, Sg. Sipayan, vii 1989, *D. Awa, B. Lee* 47518 (L); Bario, Bt. Lawi, 16 ii 1988, *D. Awa, B. Lee* 50955 (L); Batang Balui, Ulu Penuan, Bukit Tasu, vii 1989, *Ching* 53681 (L); Kanit Div., Sg. Kui, 26 iv 1991, *Yii Puan Ching et al.* 63032 (L). **Sabah:** Mt. Kinabalu, 23 iii 1982, *J. Sinclair* 174 (E); Nungkok Mt., 8 iii 1954, *Darnton* 494 (BM); Mt. Kinabalu, Liwagu River, 8 iii 1954, *Darnton* 564 (BM); Mt. Kinabalu, Silam-Silam Trail, 23 i 1976, *P.F. Stevens* 665 (L); Mt. Kinabalu, 6 vii 1989, *P.L. Swanborn* 898 (L); Mt. Kinabalu, Ulu Liwago – Ulu Mesilau, 3 ix 1961, *W.L. Chew, E.J.H. Corner, A. Stainton* 1436 (K); Mt. Kinabalu, Sungai Payau, 22 iii 1964, *M. Hotta* 1533 (L); Mt. Kinabalu, Ulu Langanani, 5 viii 1961, *W.L. Chew, E.J.H. Corner, A. Stainton* 1634 (K); Mt. Kinabalu, Mesilau River, 7 iv 1964, *W.L. Chew, E.J.H. Corner* 4855 (K, L); Mt. Kinabalu, Mamut Ridge, 20 ii 1696, *Kokawa, Hotta* 5908 (L); Mt. Kinabalu, Mesilau River – Mesilau Cave, 20 xii 1983, *J.H. Beaman* 7971 (L); Mt. Kinabalu, Tenompok F.R., 6 xi 1959, *W. Meijer* 20321 (L); Mt. Kinabalu, Tenompok F.R., 6 xi 1959, *W. Meijer* 20332 (K, L); Mt. Kinabalu, Tenompok F.R., 15 ii 1932, *J. & M.S. Clemens* 28139 (BM, K); Mt. Kinabalu, Upper Kinabalu, 19 v 1932, *J. & M.S. Clemens* 29736 (BM, K); Mt. Kinabalu, Kina Taki River (upper), 16 i 1933, *J. & M.S. Clemens* 31089 (BM); Mt. Kinabalu, Marai Parai, 23 iii 1933 – 27 iii 1933, *J. & M.S. Clemens* 32321 (BM, L); Mt. Kinabalu, Penataran River, 28 vii 1933, *J. & M.S. Clemens* 34287 (BM, K, L); Mt. Kinabalu, Keembambang, 1933, *J. & M.S. Clemens* 34339 (BM, L); Mt. Kinabalu, Mesilau, 27 vii 1963, *W. Meijer* 38072 (K); Mt. Kinabalu, Penibukan, 1933, *J. & M.S. Clemens* 40322 (BM, K); Mt. Kinabalu, Gurulau Spur, 23 xi 1933, *J. & M.S. Clemens* 50427 (BM[2], K, L). **Brunei:** Temburong – Belalang, ix 1930, *M. Jacobs* 5587 (B, K, L); Temburong, 25 iii 1991, *M.J.S. Sands* 5606 (holo K; iso BRUN *n.v.*) – Type of *Begonia fuscisetosa* Sands; Temburong – Belalang, 3 x 1958, *M. Jacobs* 5629 (B, K, L); Kuala Belalong Field Centre, 20 ii 1991, *G.C.G. Argent et al.* 9145 (E). **Kalimantan:** Long Bawan – Panado, 12 vii 1981, *R. Geesink* 9025 (L); Papadi – Pamilau, 7 viii 1981, *R. Geesink* 9274 (L); Bulungan, Gunung Malim, 14 ix 1990, *M. Kato et al.* 23818 (A, BM, L); Putussibau, 8 i 1992, *H. Okada et al.* 32247 (L).
- **Notes:** Perhaps the most widespread species in Borneo, easily recognised by the distinctive dark-brown bristles on the upper leaf surface, each often arising from a dark reddish spot (M. Sands, pers. comm.).
- **IUCN category: LC.** Widespread and occurs at a range of altitudes.

Begonia garrettii Craib [§ Diploclinium], Bull. Misc. Inform. Kew 1930: 411 (1930); Craib, Fl. Siam. 1: 773 (1931). – Type: Thailand, Doi Angka, 22 ix 1927, *H.B.G. Garrett* 462 (holo ABD; iso ABD, K).

- **THAILAND:** Doi Angka, 22 ix 1927, *H.B.G. Garrett* 462 (holo ABD; iso ABD, K) – Type of *Begonia garrettii* Craib.
- **IUCN category: DD.** Insufficient specimens could be georeferenced with certainty.

Begonia gemella Warb. ex L.B. Sm. & Wassh. [§ Petermannia], Phytologia 52(7): 443 (1983). – Type: Sulawesi, Minahassa, *S.H. Koorders* 16243B (lecto K; isolecto B).

    Begonia gemella Warb. ex Koord. *nom. nud.*, Natuurw. Tijdschr. Ned.-Indië 63: 91 (1904); Koorders-Schumacher, Suppl. Fl. Celebes 3: 46 (1922); Smith & Wasshausen, Phytologia 52(7): 443 (1983). – Type: not located.

- **SULAWESI:** Batui River, 17 x 1989, *M. Coode* 5985 (L) [as *Begonia cf. gemella*]; Minahassa, *S.H. Koorders* 16243B (lecto K; isolecto B) – Type of *Begonia gemella* Warb. ex L.B. Sm. & Wassh.
- **IUCN category: DD.** Insufficient specimens could be georeferenced with certainty.

Begonia geoffrayi Gagnep. [§ Parvibegonia], Bull. Mus. Hist. Nat. (Paris) 25: 199 (1919); Gagnepain, in Lecomte (ed.), Fl. Indo-Chine 2: 1105 (1921). – Type: Cambodia, Kampot, Kamchay Mts., 30 vii 1904, *Geoffray* 416 (holo P).
- **CAMBODIA:** Kampot, Kamchay Mts., 30 vii 1904, *Geoffray* 416 (holo P) – Type of *Begonia geoffrayi* Gagnep.
- **Notes:** Tentatively placed in § *Diploclinium* by Doorenbos *et al.* (1998), but examination of the type and beautiful sketches by Gagnepain show this species to be in § *Parvibegonia*, probably allied to the *B. integrifolia* complex.
- **IUCN category: DD.** Insufficient specimens could be georeferenced with certainty.

Begonia gibbsiae Irmsch. ex Sands [§ Petermannia], Pl. Mt. Kinabalu 154 (2001). – Type: Borneo, Sabah, Mt. Kinabalu, Melangkap Kappa, 6 iii 1984, 6°10′N, 116°32′E, *J.H. Beaman* 8804 (holo K).
- **BORNEO: Sabah:** Mt. Kinabalu, Kaung, *Darnton* 499 (para BM) – Type of *Begonia gibbsiae* Irmsch. ex Sands; Mt. Kinabalu, Mt. Nungkek, *M.J.S. Sands* 3975 (para K *n.v.*) – Type of *Begonia gibbsiae* Irmsch. ex Sands; Mt. Kinabalu, Kaung, *L.S. Gibbs* 4313 (para BM) – Type of *Begonia gibbsiae* Irmsch. ex Sands; Mt. Kinabalu, Melangkap Kappa, 6 iii 1984, *J.H. Beaman* 8804 (holo K) – Type of *Begonia gibbsiae* Irmsch. ex Sands; Mt. Kinabalu, Kaung, *Clemens* 9861 (para BO *n.v.*) – Type of *Begonia gibbsiae* Irmsch. ex Sands; Mt. Kinabalu, Dallas, *Clemens* 26274 (para BM, K *n.v.*) – Type of *Begonia gibbsiae* Irmsch. ex Sands; Mt. Kinabalu, Dallas, *Clemens* 30003 (para K *n.v.*) – Type of *Begonia gibbsiae* Irmsch. ex Sands; Mt. Kinabalu, Bungol, *Clemens* 51321 (para BM) – Type of *Begonia gibbsiae* Irmsch. ex Sands.
- **IUCN category: VU D2.** Restricted to hill forest and lower montane forest on the west side of Mt. Kinabalu (Sands, 2001).

Begonia gilgiana Irmsch. [§ Petermannia], Bot. Jahrb. Syst. 50: 340 (1913). – Type: New Guinea, Papua New Guinea, West Sepik Province, Kaiserin-Augusta-Flusses, *C.L. Ledermann* 7070 (holo not located).
- **NEW GUINEA: Papua New Guinea:** West Sepik Province, Kaiserin-Augusta-Flusses, *C.L. Ledermann* 7070 (holo not located) – Type of *Begonia gilgiana* Irmsch.; West Sepik Province, Kaiserin-Augusta-Flusses, 19 iv 1912, *C.L. Ledermann* 7077 (B[2]); West Sepik Province, Kaiserin-Augusta-Flusses, 3 vi 1912, *C.L. Ledermann* 7496 (B[2]).
- **IUCN category: DD.** No locality could be georeferenced with certainty.

Begonia gitingensis Elmer [§ Diploclinium], Leafl. Philipp. Bot. 2: 738 (1910); Merrill, Philipp. J. Sci. 6: 396 ('1911', 1912); Merrill, Enum. Philipp. Fl. Pl. 3: 122 (1923). – Type: Philippines, Sibuyan, Mt. Giting-Giting, iv 1910, *A.D.E. Elmer* 12368 (iso BM, E, FI, K, U).
- **PHILIPPINES: Sibuyan:** Mt. Giting-Giting, iv 1910, *A.D.E. Elmer* 12368 (iso BM, E, FI, K, U) – Type of *Begonia gitingensis* Elmer.
- **IUCN category: VU D2.** Known only from Mt. Giting-Giting on Sibuyan, which is still largely covered by forest. However, the low altitude of this species (500 m) takes it into forests which have been logged to some extent (Proctor & Argent, 1998).

Begonia glabricaulis Irmsch. [§ Petermannia], Bot. Jahrb. Syst. 50: 371 (1913). – Type: New Guinea, Papua New Guinea, West Sepik Province, Kaiserin-Augusta-Flusses, *C.L. Ledermann* 6611 (syn B[2]); New Guinea, Papua New Guinea, West Sepik Province, Kaiserin-Augusta-Flusses, *C.L. Ledermann* 6886 (syn B[2]); New Guinea, Papua New Guinea, West Sepik Province, Kaiserin-Augusta-Flusses, *C.L. Ledermann* 7754 (syn B[2]).

Begonia glabricaulis var. glabricaulis
- **NEW GUINEA: Papua New Guinea:** West Sepik Province, Kaiserin-Augusta-Flusses, *C.L. Ledermann* 6611 (syn B[2]) – Type of *Begonia glabricaulis* Irmsch.; West Sepik Province, Kaiserin-Augusta-Flusses, *C.L. Ledermann* 6886 (syn B[2]) – Type of *Begonia glabricaulis* Irmsch.; West Sepik Province, Kaiserin-Augusta-Flusses, *C.L. Ledermann* 7754 (syn B[2]) – Type of *Begonia glabricaulis* Irmsch.; West Sepik Province, Kaiserin-Augusta-Flusses, *C.L. Ledermann* 12248a (B[3]).

Begonia glabricaulis var. brachyphylla Irmsch., Bot. Jahrb. Syst. 50: 373 (1913). – Type: New Guinea, Papua New Guinea, West Sepik Province, Kaiserin-Augusta-Flusses, *C.L. Ledermann* 6619 (holo B; iso B).
- **NEW GUINEA: Papua New Guinea:** West Sepik Province, Kaiserin-Augusta-Flusses, *C.L. Ledermann* 6619 (holo B; iso B) – Type of *Begonia glabricaulis* var. *brachyphylla* Irmsch.
- **IUCN category: DD**. Insufficient specimens could be georeferenced with certainty.

Begonia goegoensis N.E. Br. [§ Reichenheimia], Gard. Chron., II 71 (1882); Fotsch, Begonien 38 (1933); Graf, Exotica 3: 306 (1963); Tebbitt, Begonias 134 (2005). – Type: Sumatra, Goegoe, *C. Curtis s.n.* (holo K).
- **SUMATRA:** Goegoe, *C. Curtis s.n.* (holo K) – Type of *Begonia goegoensis* N.E. Br.; Pajakumbuh, Mt. Sago, 15 iv 1956, *W. Meijer* 60 (L).
- **IUCN category: DD**. Insufficient specimens could be georeferenced with certainty.

Begonia gomantongensis Kiew [§ Petermannia], Gard. Bull. Singapore 50: 164 (1998); Kiew, Gard. Bull. Singapore 53: 259 (2001). – Type: Borneo, Sabah, Kinabatangan, Bukit Dulung Lambu, Gomantong Cave, *J. Awing* SAN47257 (holo SAN *n.v.*).
- **BORNEO: Sabah:** Kinabatangan, Bukit Dulung Lambu, 29 x 1996, *R. Kiew, S.P. Lim* BDL3 (para SAN *n.v.*, SING *n.v.*) – Type of *Begonia gomantongensis* Kiew; Kinabatangan, Bukit Dulung Lambu, 6 viii 1996, *S.P. Lim, Ubaldus* LSP785 (para SAN *n.v.*, SING *n.v.*) – Type of *Begonia gomantongensis* Kiew; Kinabatangan, Bukit Dulung Lambu, Gomantong Cave, 11 x 1964, *J. Awing* SAN47257 (holo SAN *n.v.*) – Type of *Begonia gomantongensis* Kiew.
- **IUCN category: EN B2ab(iii)**. Endemic to Bukit Dulong Lambu, where it grows at the base of limestone cliffs. Given that this is not a protected area, and the pressures on limestone habitats, this species is considered to be endangered.

Begonia goniotis C.B. Clarke in Hook. f. **[§ Platycentrum]**, Fl. Brit. Ind. 2: 648 (1879). – Type: Burma, *W. Griffith* 2579 (holo K; iso P).
- **BURMA:** *W. Griffith* 2579 (holo K; iso P) – Type of *Begonia goniotis* C.B. Clarke.
- **Notes:** This species has distinctive oblong-lanceolate leaves.
- **IUCN category: DD**. Insufficient specimens could be georeferenced with certainty.

Begonia gracilipes Merr. [§ Petermannia], Philipp. J. Sci. 6: 405 ('1911', 1912); Merrill, Enum. Philipp. Fl. Pl. 3: 123 (1923). – Type: Philippines, Luzon, Cagayan Province, Claveria, iii 1909, *M. Ramos* 7395 (syn B, K).

- **PHILIPPINES: Luzon:** Cagayan Province, Claveria, iii 1909, *M. Ramos* 7395 (syn B, K) – Type of *Begonia gracilipes* Merr.; Catanduanes, 14 xi 1917 – 11 xii 1917, *M. Ramos* 30561 (K).
- **Notes:** See notes under *B. aequata* A. Gray.
- **IUCN category: DD**. Taxonomically uncertain.

Begonia grandipetala Irmsch. [§ Petermannia], Bot. Jahrb. Syst. 50: 377 (1913). – Type: Sulawesi, Bowonglangi, 24 iv 1902, *K.F. & P.B. Sarasin* 2154 (holo B).
- **SULAWESI:** Bowonglangi, 24 iv 1902, *K.F. & P.B. Sarasin* 2154 (holo B) – Type of *Begonia grandipetala* Irmsch.
- **IUCN category: DD**. Insufficient specimens could be georeferenced with certainty.

Begonia grantiana Craib [§ Parvibegonia], Gard. Chron., III 83: 66 (1928); Craib, Fl. Siam. 1: 773 (1931). – Type: Thailand, Chantaburi, Koh Chang Island, Klawng Prao, 30 ix 1924, 12°3´N, 102°19´E, *A.F.G. Kerr* 9266 (holo ABD; iso BM, K).
- **THAILAND:** Chantaburi, Koh Chang Island, Klawng Prao, 30 ix 1924, *A.F.G. Kerr* 9266 (holo ABD; iso BM, K) – Type of *Begonia grantiana* Craib.
- **IUCN category: LC**. Endemic to Koh Chang, and hence in a protected area with a considerable proportion of undisturbed forest.

Begonia grata Geddes ex Craib [§ Parvibegonia], Gard. Chron., III 83: 66 (1928); Craib, Fl. Siam. 1: 774 (1931). – Type: Thailand, Chantaburi, Kao Soi Dao, *A.F.G. Kerr s.n.* (holo ABD; iso ABD).
- **THAILAND:** Chantaburi, Kao Soi Dao, *A.F.G. Kerr s.n.* (holo ABD; iso ABD) – Type of *Begonia grata* Geddes ex Craib.
- **IUCN category: DD**. Insufficient specimens could be georeferenced with certainty.

Begonia griffithiana (A. DC.) Warb. [§ Monopteron], Nat. Pflanzenfam. 3(6A): 142 (1894). – *Mezierea griffithiana* A. DC., Ann. Sci. Nat. Bot., IV 11: 144 (1859); Clarke, Fl. Brit. Ind. 2: 644 (1879); Warburg, Nat. Pflanzenfam. 3(6A): 142 (1894). – Type: Bhutan, *Griffith* 2504 (syn not located).
  Begonia episcopalis C.B. Clarke *nom. superfl.* in C.B. Clarke, Fl. Brit. Ind. 2: 644 (1879). – Type: Bhutan, *Griffith* 2504 (syn not located).
- **BURMA:** Khasi Mts., *fide* C.B. Clarke (1879) [as *Begonia episcopalis* C.B. Clarke *nom. superfl.*]; Mali Hka, 19 vii 1937, *F.K. Ward* 12823 (BM).
- **IUCN category: DD**. Insufficient specimens could be georeferenced with certainty.

Begonia gueritziana Gibbs [§ Diploclinium], J. Linn. Soc., Bot. 42: 82 (1914); Kiew, Gard. Bull. Singapore 53: 260 (2001); Sands, Pl. Mt. Kinabalu 155 (2001). – Type: Borneo, Sabah, Tenom, Kayoh Hills, i 1910, *L.S. Gibbs* 2892 (holo BM).
- **BORNEO: Sarawak:** Kapit Distr., Balleh Ridge, Bukit Tibang, 14 vii 1969, *Ilias* 28751 (L) [as *Begonia aff. gueritziana*]. **Sabah:** 1895, *C.M.G. Creagh s.n.* (K); Mt. Kinabalu, Kaung, 4 iii 1954, *Darnton* 288 (BM, L); Bt. Doji, 24 x 1968, *Kokawa, Hotta* 380 (L); Bt. Doji, 24 x 1968, *Kokawa, Hotta* 419 (L); Kinabatangan, Bukit Dulung Lambu, Gomantong Cave, 31 x 1968, *Kokawa, Hotta* 577 (L); Mt. Kinabalu, 26 vii 1961, *W.L. Chew, E.J.H. Corner, A. Stainton* 1187 (K, L); Tenom, Kayoh Hills, i 1910, *L.S. Gibbs* 2892 (holo BM) – Type of *Begonia gueritziana* Gibbs; Ranau District, Poring Hot Springs, 1 ii 1969, *Kokawa, Hotta* 4771 (L); Kinabatangan, Bukit Dulung Lambu, Gomantong Cave, 23 iii 1996, *M.M.J.v. Balgooy* 7277 (L); Kalang Waterfall, 11 x 1986, *J.M. Huisman, J.J. Vermeulen, E.F. de Vogel, P.C.v. Welzen* 8246 (L); Mt. Kinabalu, Penataran River, 11 iii 1984, *J.H. Beaman* 8859 (A, L); Ranau District, Kg. Takuthan, *J.H. Beaman* 9644 (L); Kinabatangan, Supu F.R., 16 x 1938, *Puasa* 10115 (K); Kinabatangan, Bukit Dulung Lambu, Gomantong Cave, 25

ix 1968 – 26 ix 1968, *K. Ogata* 11015 (L); Sandakan District, Gomantung, 14 ii 1960, *W. Meijer* 20754 (K); Mt. Kinabalu, Dallas, 23 viii 1931, *J. & M.S. Clemens* 26139 (BM, K); Keningau, Laing Cave, 4 viii 1965, *J.K. Lajangah* 44562 (L); Lahad Datu District, Ulu Segama, 19 vii 1970, *A. Talip* 70979 (K) [as *Begonia gueritziana* ?]; Tamhunan, Kg. Mansaralong, 8 vii 1978, *A.N. Abas* 85913 (L); Lamag, Sg. Pin, 28 xi 1978, *S. Dewol, T. Harun* 89912 (L); Ranau District, Kg. Miruru, 5 vi 1979, *G. Aban* 90246 (L); Ranau District, Kg. Sagindai – Kg. Nabutan, 16 vii 1979, *Petrus, G. Aban* 90624 (L); Ranau District, Kg. Paus, 23 ii 1982, *Amin et al.* 94680 (L); Keningau, Ulu Puntih, 21 ii 1989, *F. Krispinum* 125408 (K, L); Tenom, Ulu Sungei Kalang, 16 ix 1991, *unknown* 133057 (E[2], L); Tenom, Ulu Kalang, 16 ix 1991, *Maikin et al.* 133653 (L). **Kalimantan:** Kutei, G. Beratus, 17 vii 1952, *W. Meijer* 815 (L); G. Meratus, 27 iii 1995, *P.J.A. Kessler et al.* 931 (L); Pajungan Distr., Kayan Mentarang Reserve, 14 vii 1992, *McDonald, Ismail* 3587 (L).

- **Notes:** This species was originally placed in § *Reichenheimia* by Gibbs, and subsequently moved to § *Platycentrum* on account of it having a 2-locular ovary (Kiew, 2001). However, *B. gueritziana* is very closely allied to other (as yet undescribed) species from Palawan which also have 2-locular ovaries. Molecular data (C. Coyle, pers. comm., unpublished data) show these Palawan species to be nested within Philippine members of § *Diploclinium. Begonia guertiziana* also has female flowers with four tepals (as opposed to five in § *Platycentrum*) and anthers without an extended connective.
- **IUCN category: LC**. Widespread within Borneo, including some localities in national parks.

Begonia hahiepiana H.Q. Nguyen & Tebbitt [§ Sphenanthera], Novon 16(3): 374 (2006). – Type: Vietnam, Phu Tho Province, Xuan Son Nat. Park, 30 xi 2000, *V.X. Phuong, N.K. Khoi, N.Q. Binh, H.Q. Nguyen* 3911 (holo HN; iso MO).

- **VIETNAM:** Nui Phu Tho Province, Xuan Son Nat. Park, 30 xi 2000, *V.X. Phuong, N.K. Khoi, N.Q. Binh, H.Q. Nguyen* 3911 (holo HN; iso MO) – Type of *Begonia hahiepiana* H.Q. Nguyen & Tebbitt; Nui Phu Tho Province, Xuan Son Nat. Park, *N.T. Hiep et al.* 6112 (para HN) – Type of *Begonia hahiepiana* H.Q. Nguyen & Tebbitt.
- **IUCN category: VU D1,2** (Nguyen & Tebbitt, 2006).

Begonia halconensis Merr. [§ Petermannia], Philipp. J. Sci. 6: 385 ('1911', 1912); Merrill, Enum. Philipp. Fl. Pl. 3: 123 (1923); Tebbitt, Begonias 186 (2005). – Type: Philippines, Mindoro, Mt. Halcon, xi 1906, *E.D. Merrill* 5515 (syn B, P).

- **PHILIPPINES: Mindoro:** Mt. Halcon, xi 1906, *E.D. Merrill* 5515 (syn B, P) – Type of *Begonia halconensis* Merr.; Mt. Halcon, xi 1906, *E.D. Merrill* 5607 (para not located) – Type of *Begonia halconensis* Merr.; Mt. Halcon, 15 iii 1997, *P. Wilkie et al.* 29107 (E[2]).
- **Notes:** Currently in § *Petermannia*, but its turbinate fruit, spherical male buds and axillary inflorescences are reminiscent of § *Sphenanthera*. This affinity is congruent with the phylogeny in Tebbitt *et al.* (2006).
- **IUCN category: CR B2ab(iii)**. Recorded from altitudes of 1100–1800 m, which effectively restricts its distribution to the slopes of Mt. Halcon. This area is not protected and is heavily degraded.

Begonia handelii Irmsch. [§ Sphenanthera], Anz. Akad. Wiss. Wien, Math.-Naturwiss. Kl. 58: 24 (1921); Irmscher, Mitt. Inst. Allg. Bot. Hamburg 6(3): 348 (1927); Handel-Mazzetti, Symb. Sin. 3: 385 (1931); Chun & Chun, Sunyatsenia 4: 22 (1939); Yu, Bull. Fan. Mem. Inst. Biol. n.s. 1: 115 (1948); Hô, Ill. Fl. Vietnam 1(2): 733 (1991); Tebbitt, Edinburgh J. Bot. 60: 1 (2003); Tebbitt, Begonias 208 (2005). – Type: Vietnam, Tonkin, 1 xi 1914, *H.R.E. Handel-Mazzetti* 12 (holo WU *n.v.*; iso B, E *n.v.*).

Begonia lecomtei Gagnep., **syn. nov.**, Bull. Mus. Hist. Nat. (Paris) 25: 276 (1919); Gagnepain, in Lecomte (ed.), Fl. Indo-Chine 2: 1112 (1921). – Type: Vietnam, Tonkin, Yen-bay, 28 x 1911, *P.H. Lecomte, Finet* 392 (lecto P, here designated); Vietnam, Annam, Quang-tri Province, *P.A. Eberhardt* 2070 (syn P); Vietnam, Laoke, Chapa, Chapa – Bhamo, 29 x 1911, *P.H. Lecomte, Finet* 464 (syn P).

Begonia handelii var. handelii

- **VIETNAM:** Tonkin, 1 xi 1914, *H.R.E. Handel-Mazzetti* 12 (holo WU *n.v.*; iso B, E *n.v.*) – Type of *Begonia handelii* Irmsch.; Cuc Phuong National Park, 14 xi 2000, *N.T. Hiep* 4244 (L); Cuc Phuong National Park, 19 vii 2000, *unknown* PKL10389 (not located); Tonkin, Yen-bay, 28 x 1911, *P.H. Lecomte, Finet* 392 (lecto P) – Type of *Begonia lecomtei* Gagnep.; Laoke, Chapa – Muong-xen, 29 x 1911, *P.H. Lecomte, Finet* 464 (syn P) – Type of *Begonia lecomtei* Gagnep.; Annam, Quang-tri Province, *P.A. Eberhardt* 2070 (syn P) – Type of *Begonia lecomtei* Gagnep.; *P.A. Eberhardt* 4690 (P); Tonkin, Mt. Bavi, 25 i 1941, *P.A. Petelot* 7105 (B[3]); Muong-xen – Chapa, 5 xii 1913, *A.J.B. Chevalier* 29365 (B, P).
- **LAOS:** Luang Namtha, Viengphoukha, 3 iii 2006, *E.C.S. Lundh & L.J. Ahnby* 129 (E).

Begonia handelii var. prostrata (Irmsch.) Tebbitt, Edinburgh J. Bot. 60: 6 (2003). – *Begonia prostrata* Irmsch., Mitt. Inst. Allg. Bot. Hamburg 10: 516 (1939); Tebbitt, Edinburgh J. Bot. 60: 6 (2003). – Type: China, Yunnan, *Henry* 11628 (holo LE *n.v.*; iso E *n.v.*, K *n.v.*).

- **BURMA:** Southern Shan States, King Jung, xii 1909, *R.W. MacGregor* 1239 (E); Keng Tung, Valley of the Meh Len, Muang Hypak, 27 i 1922, *J.F.C. Rock* 2072 (E); Kachin Hills, *S.M. Toppin* 4137 (E, K).
  THAILAND: Fang – Chiengrai, 28 ii 1958, *T. Sorensen et al.* 1804 (E *n.v.*, GB *n.v.*, L); Fang – Chiengrai, ii 1928, *T. Sorensen et al.* 1804 (E *n.v.*, GB *n.v.*, L); Doi Pae Poe, 14 iii 1968, *B. Hansen, T. Smitinand* 12905 (E *n.v.*, L).
  LAOS: Tatom, Chieng kwang (Xiang Khoang), 1 iv 1932, *A.F.G. Kerr* 21772 (K *n.v.*); 5 iv 1932, *A.F.G. Kerr* 21778 (K) [as *Begonia cf. handelii* var. *prostrata*].
  VIETNAM: Son La Prov., Moc Chau Distr., Suan nha, *H.T. Dung* 197 (HN *n.v.*); Van Son, 4 i 1964, *Sino-Vietnam expedition* 954 (KUN *n.v.*); Vinh Yen Province, Tam Dao, 4 ii 1962, *Sino-Vietnam expedition* 1982 (KUN *n.v.*); Vinh Yen Province, Tam Dao, 8 ii 1965, *Sino-Vietnam expedition* 2070 (KUN *n.v.*); *unknown* 3454 (HN *n.v.*); Vinh Yen Province, Tam Dao, *unknown* 4570 (HN *n.v.*); Lan-Tsang Hsien, v 1936, *C.W. Wang* 76618 (KUN *n.v.*).
- **Notes:** Of the other Asian species with 4-locular wingless fruits (*B. aborensis*, *B. acetosella*, *B. silletensis*, *B. roxburghii*, *B. tessaricarpa*, *B. burkillii*), this species (especially var. *prostrata*) is most easily confused with *B. burkillii*. *Begonia lecomtei* Gagnep. is included here as a synonym (M.C. Tebbitt, pers. comm.).
- **IUCN category: LC.** This species is widespread in the Indo-China region.

Begonia harmandii Gagnep. [§ Reichenheimia], Bull. Mus. Hist. Nat. (Paris) 25: 200 (1919); Gagnepain, in Lecomte (ed.), Fl. Indo-Chine 2: 1099 (1921). – Type: Vietnam, Nui-Cam, *F.F.J. Harmand* 543 (holo P; iso P).

- **VIETNAM:** Nui-Cam, *F.F.J. Harmand* 543 (holo P; iso P) – Type of *Begonia harmandii* Gagnep.
- **IUCN category: DD.** Insufficient specimens could be georeferenced with certainty.

Begonia hasskarliana (Miq.) A. DC. [§ Diploclinium], Prodr. 15(1): 329 (1864). – *Diploclinium hasskarlianum* Miq., Fl. Ned. Ind. 1(1): 1091 (1856); Miquel, Fl. Ned. Ind., Suppl. 1: 332 (1861); Candolle, Prodr. 15(1): 329 (1864). – Type: not located.

- **SUMATRA:** *Fide* de Candolle (1864). No specimens located.
- **IUCN category: DD.** Insufficient specimens could be georeferenced with certainty.

Begonia hatacoa Buch.-Ham. ex D. Don [§ Platycentrum], Prodr. Fl. Nepal 223 (1825); Hara, J. Jap. Bot. 47: 143 (1972); Hara, Fl. E. Himalaya 3: 85 (1975). – Type: *Buchanan-Hamilton s.n.* (not located).

Begonia rubrovenia Hook., Bot. Mag. 79: 4689 (1853); Candolle, Prodr. 15(1): 347 (1864); Gagnepain, in Lecomte (ed.), Fl. Indo-Chine 2: 1102 (1921); Craib, Fl. Siam. 1: 778 (1931). – *Platycentrum rubrovenium* (Hook.) Klotzsch, Abh. Kon. Akad. Wiss. Berlin 1854: 244, pl. 11B (1855). – Type: India, *Wallich* 36798 (holo K *n.v.*).

- **BURMA:** Mala Hka, 28 xii 1930, *F.K. Ward* 9067 (BM) [as *Begonia rubrovenia* ?].
  THAILAND: Chantaburi, Kao Soi Dao, 12 xii 1924, *A.F.G. Kerr* 9623 (ABD, BM, K, P).
  VIETNAM: Annam, *M. Krempf s.n.* (P); Tonkin, Phuong-Lam, 2 v 1888, *B. Balansa* 3763 (K, P) [as *Begonia aff. rubrovenia*]; Tonkin, Muong Thon, x 1937, *P.A. Petelot* 7113 (B[2]); Annam, 9 i 1935, *E. Poilane* 23784 (P).
- **IUCN category: DD**. Insufficient specimens could be georeferenced with certainty.

Begonia havilandii Ridl. [§ Diploclinium], J. Straits Branch Roy. Asiat. Soc. 46: 258 (1906). – Type: Borneo, Sarawak, Pengkulu Ampat, *G.D. Haviland* 279 (holo not located).

- **BORNEO: Sarawak:** Pengkulu Ampat, *G.D. Haviland* 279 (holo not located) – Type of *Begonia havilandii* Ridl.; *O. Beccari* CB4473 (FI).
- **IUCN category: DD**. Insufficient specimens could be georeferenced with certainty.

Begonia heliostrophe Kiew [§ Petermannia], Gard. Bull. Singapore 53: 262 (2001). – Type: Borneo, Sabah, Kinabatangan, Batu Batangan, *R. Kiew, S.P. Lim* RK4293 (holo SAN *n.v.*; iso BRUN *n.v.*, K *n.v.*, L *n.v.*, SAR *n.v.*, SING *n.v.*).

- **BORNEO: Sabah:** Kinabatangan, Batu Batangan, *R. Kiew, S.P. Lim* RK4293 (holo SAN *n.v.*; iso BRUN *n.v.*, K *n.v.*, L *n.v.*, SAR *n.v.*, SING *n.v.*) – Type of *Begonia heliostrophe* Kiew.
- **IUCN category: DD**. Insufficient specimens could be georeferenced with certainty.

Begonia hemsleyana Hook. f. [§ Platycentrum], Bot. Mag., n.s. 125: 7685 (1899); Tebbitt, Begonias 144 (2005). – Type: not located.

- **BURMA:** N. Shan State, Gokteik Gorge, 10 vii 1909, *J.H. Lace* 4861 (E, K).
- **IUCN category: DD**. Insufficient specimens could be georeferenced with certainty.

Begonia hernandioides Merr. [§ Diploclinium], Philipp. J. Sci. 6: 392 ('1911', 1912); Merrill, Enum. Philipp. Fl. Pl. 3: 123 (1923). – Type: Philippines, Luzon, Cagayan Province, Claveria, iii 1909, *M. Ramos* BS7393 (syn not located).

- **PHILIPPINES: Luzon:** Cagayan Province, Claveria, iii 1909, *M. Ramos* BS7387 (para not located) – Type of *Begonia hernandioides* Merr.; Cagayan Province, Claveria, iii 1909, *M. Ramos* BS7393 (syn not located) – Type of *Begonia hernandioides* Merr.; Ilocos Norte Province, Bangui, xi 1923, *R.C. McGregor* 43571 (B, K).
- **IUCN category: DD**. Insufficient specimens could be georeferenced with certainty.

Begonia herveyana King [§ Platycentrum], J. Asiat. Soc. Bengal, Pt. 2, Nat. Hist. 71: 63 (1902); Ridley, J. Linn. Soc., Bot. 38: 310 (1908); Ridley, Fl. Malay Penins. 1: 861 (1922); Irmscher, Mitt. Inst. Allg. Bot. Hamburg 8: 131 (1929); Kiew, Begonias Penins. Malaysia 272 (2005). – Type: Peninsular Malaysia, Malacca, *D.F.A. Hervey s.n.* (lecto K *n.v.*); Peninsular Malaysia, Malacca, *R. Derry s.n.* (syn not located).

Begonia herveyana var. herveyana
■ **PENINSULAR MALAYSIA:** Malacca, *D.F.A. Hervey s.n.* (lecto K *n.v.*) – Type of *Begonia herveyana* King; Malacca, *R. Derry s.n.* (syn not located) – Type of *Begonia herveyana* King.

Begonia herveyana var. barnesii Irmsch., Mitt. Inst. Allg. Bot. Hamburg 8: 131 (1929); Kiew, Begonias Penins. Malaysia 293 (2005). – Type: Peninsular Malaysia, Perak, Kluang Terbang, *Barnes s.n.* (holo SING *n.v.*).
■ **PENINSULAR MALAYSIA:** Perak, Kluang Terbang, *Barnes s.n.* (holo SING *n.v.*) – Type of *Begonia herveyana* var. *barnesii* Irmsch.
■ **Notes:** Irmscher's var. *barnesii* is considered to represent a new species (Kiew, 2005).
■ **IUCN category: EN B2ab(iii).** Known only from two sites where it is extremely local in its distribution. One of the sites is threatened by logging (Kiew, 2005).

Begonia heteroclinis Miq. ex Koord. [§ Petermannia], Meded. Lands Plantentuin 19: 484 (1898); Koorders, Natuurw. Tijdschr. Ned.-Indië 63: 89 (1904). – Type: not located.
■ **SULAWESI:** 1859 – 1860, *W.H. de Vriese, J.E. Teijsmann s.n.* (L); Manado, 1859 – 1860, *J.E. Teijsmann s.n.* (B); Pangkadjene, 12 i 1937, *P.J. Eyma* 301 (BO) [as *Begonia* aff. *heteroclinis*].
■ **Notes:** This species had been assigned to § *Sphenanthera* in error, as it has dehiscent fruit with distinct wings. I have re-assigned it here to § *Petermannia*, as it has (i) bifid placentae, (ii) male flowers with two tepals and (iii) two female flowers at the base of a larger cymose male inflorescence. The specimen collected by de Vriese and Teijsmann in L (annotated by Miquel) is a possible syntype, although the protologue refers only to the 'original exemplar von Miquel in Herb. Hort. Bog.', without collector information.
■ **IUCN category: DD.** Insufficient specimens could be georeferenced with certainty.

Begonia hexaptera Sands [§ Petermannia], in Coode *et al.*, Checkl. Fl. Pl. Gymnosperms Brunei Darussalam App. 2: 434 (1997). – Type: Borneo, Brunei, Belait, Batu Melintang, 25 vii 1993, 4°7′N, 114°43′E, *M.J.S. Sands* 5940 (holo K; iso BRUN *n.v.*).
■ **BORNEO: Brunei:** Belait, Batu Melintang, Melilas, 25 vii 1993, *M.J.S. Sands* 5940 (holo K; iso BRUN *n.v.*) – Type of *Begonia hexaptera* Sands.
■ **IUCN category: VU D2.** Known only from one locality in Brunei.

Begonia hirsuticaulis Irmsch. [§ Petermannia], Bot. Jahrb. Syst. 50: 346 (1913); Irmscher, Bot. Jahrb. Syst. 50: 566 (1914). – Type: New Guinea, Papua New Guinea, 3 iii 1912, *C.L. Ledermann* 6531 (syn B); New Guinea, Papua New Guinea, *C.L. Ledermann* 6613 (syn B); New Guinea, Papua New Guinea, *C.L. Ledermann* 6666 (syn B[2]).
■ **NEW GUINEA: Papua New Guinea:** Northeast New Guinea, 3 iii 1912, *C.L. Ledermann* 6531 (syn B) – Type of *Begonia hirsuticaulis* Irmsch.; Northeast New Guinea, *C.L. Ledermann* 6613 (syn B) – Type of *Begonia hirsuticaulis* Irmsch.; Northeast New Guinea, *C.L. Ledermann* 6666 (syn B[2]) – Type of *Begonia hirsuticaulis* Irmsch.; Northeast New Guinea, West Sepik Province, Kaiserin-Augusta-Flusses, 6 i 1913, *C.L. Ledermann* 10458 (B[3]); Northeast New Guinea, West Sepik Province, Kaiserin-Augusta-Flusses, *C.L. Ledermann* 10459 (B[5]) [as *Begonia* cf. *hirsuticaulis*]; Northeast New Guinea, West Sepik Province, Kaiserin-Augusta-Flusses, 28 iii 1913, *C.L. Ledermann* 11552 (B) [as *Begonia* cf. *hirsuticaulis*].
■ **IUCN category: DD.** Insufficient specimens could be georeferenced with certainty.

Begonia hispidissima Zipp. ex Koord. [§ Petermannia], Meded. Lands Plantentuin 19: 485 (1898); Koorders-Schumacher, Suppl. Fl. Celebes 3: 46 (1922); Smith & Wasshausen, Phytologia

52: 444 (1983). – Type: Sulawesi, Minahassa, 10 iv 1895, *S.H. Koorders* 16241B (lecto K; isolecto B, BO, L).

- **SULAWESI:** B. Watoewila, 25 iii 1929, *G.K. Kjellberg* 1031 (BO); Menado, Donggala, 9 xi 1930, *O. Posthumus* 2562 (BO); 1929, *G.K. Kjellberg* 2661 (BO); Mt. Nokilaki, 20 iv 1975, *W. Meijer* 9614A (L) [as *Begonia cf. hispidissima*]; Minahassa, 10 iv 1895, *S.H. Koorders* 16241B (lecto K; isolecto B, BO, L) – Type of *Begonia hispidissima* Zipp. ex Koord.
- **Notes:** The collections I have seen in B and L, collected and annotated by Zippelius, are from New Guinea, and obviously a different taxon to the lectotype. The collection by Koorders designated as the lectotype (Smith & Wasshausen, 1983) is annotated by Warburg as '*Begonia hispidissima* Warb. *n. sp.*', and it would seem that this is the taxon that Koorders validated.
- **IUCN category: DD**. The limits between this species and *B. masarangensis* Irmsch. need clarifying.

Begonia holosericea (Teijsm. & Binn.) Teijsm. & Binn. [not placed to section], Epim. Lugd. Bat. 5 (1863). – *Diploclinium holosericeum* Teijsm. & Binn., Tijdschr. Ned.-Indië 25: 421 (1863). – Type: Ternate Island, *J.E. Teijsmann s.n.* (BM).
- **MOLUCCAS: Halmahera:** Ternate Island, *J.E. Teijsmann s.n.* (BM) – Type of *Diploclinium holosericeum* Teijsm. & Binn.
- **Notes:** Tentatively assigned to § *Petermannia* (Doorenbos *et al.*, 1998).
- **IUCN category: DD**. Insufficient specimens could be georeferenced with certainty.

Begonia holttumii Irmsch. [§ Petermannia], Mitt. Inst. Allg. Bot. Hamburg 8: 113 (1929); Henderson, Malayan Wild Flowers – Dicotyledons 167 (1959); Kiew, Malayan Nat. J. 47: 312 (1994); Kiew, Begonias Penins. Malaysia 112 (2005). – Type: Peninsular Malaysia, Penang, Batu Etam, x 1889, *C. Curtis* 1262 (lecto SING *n.v.*; isolecto KEP *n.v.*).
   Begonia isoptera auct. non Dryand. ex Sm.: Ridley, Mat. Fl. Malay. Penins. 586 (1902), *pro parte*.
- **PENINSULAR MALAYSIA:** Pahang, Kemanshul Div., Sabai Estate, 10 vi 1958, *M. Shah* 197 (L); Penang, Batu Etam, x 1889, *C. Curtis* 1262 (lecto SING *n.v.*; isolecto KEP *n.v.*) – Type of *Begonia holttumii* Irmsch.; Negri Sembilan, 7 ix 1937, *C.W. Franck* 1409 (P); Malacca, 1894, *unknown* 1703 (BM, P); Perak, *B. Scortechini* 2000 (P); Perak, ix 1885, *Kings Collector* 2260 (P); Perak, Kledang Saiong, 25 iv 1987, *R. Kiew* 2564 (L); Perak, vi 1883, *Kings Collector* 4428 (ABD, E); Penang, xii 1895, *H.N. Ridley* 7094 (BM); Perak, ix 1885, *Kings Collector* 8262 (L, P); Negeri Sembilan, Perhentian Tinggi, xii 1898, *H.N. Ridley* 10028 (BM); Trengganu, Kemaman, 2 xi 1935, *E.J.H. Corner* 30206 (BM, L); Johore, Sungai Segun, 2 ii 1936, *E.J.H. Corner* 30750 (BM, L); Johore, Gunong Pulai, 15 iv 1953, *J. Sinclair* 39555 (E, L, P); Trengganu, Sungei Kemia, 30 viii 1996, *Chua et al.* 40563 (L); Negeri Sembilan, 9 x 1989, *B. Everett* 104918 (L).
   SUMATRA: North Sumatra, Bohorok, Bukit Lawang, 19 ii 1973, *Soedarsono* 289 (K); Sumatra East coast, 1914 – 1917, *H. Surbeck* 453 (L); Asahan, Silo Maradja, vii 1928 – viii 1928, *Rahmat Si Boeea* 857 (A, E); *H.S. Yates* 1138 (BM[2]); Batak Lands, Tapanoeli, 20 vi 1933, *Rahmat Si Boeea* 4626 (K); Asahan, Hoeta Bagasan, 7 ix 1934 – 4 ii 1935, *Rahmat Si Boeea* 6555 (A).
- **Notes:** Many of the specimens from Sumatra were previously placed under the dustbin taxon *B. isoptera*, but here I consider them to belong to *B. holttumii*, hence extending the known distribution of this species.
- **IUCN category: LC**. Relatively widespread in Peninsular Malaysia.

Begonia horsfieldii Miq. ex A. DC. [§ Petermannia], Prodr. 15(1): 397 (1864). – Type: Sumatra, *T. Horsfield s.n.* (holo BM).

Diploclinium horsfieldii Miq., Fl.Ned.Ind.1(1):691 (1856); Candolle, Prodr.15(1):397 (1864). – Type: Sumatra, *T. Horsfield s.n.* (holo BM).
- **SUMATRA:** *T. Horsfield s.n.* (holo BM) – Type of *Begonia horsfieldii* Miq. ex A. DC.
- **IUCN category: DD**. Insufficient specimens could be georeferenced with certainty.

Begonia hullettii Ridl. [§ Petermannia], J. Straits Branch Roy. Asiat. Soc. 46: 255 (1906). – Type: Borneo, Sarawak, Matang, *H.N. Ridley* 11776 (holo K).
- **BORNEO: Sarawak:** Matang, v 1886, *O. Beccari* PB1573 (Fl[3]); Gunung Matang, Sungei China, 14 vii 1962, *B.L. Burtt, P.J.B. Woods* 2500 (E) [as *Begonia cf. hullettii*]; *O. Beccari* PB3784 (K) [as *Begonia hullettii* ?]; Matang, *H.N. Ridley* 11776 (holo K) – Type of *Begonia hullettii* Ridl.
- **IUCN category: VU D2**. Known only from Matang. The collection localities are too vague to determine whether the species occurs within the Matang National Park.

Begonia humboldtiana Gibbs [§ Petermannia], Fl. Arfak Mts. 215 (1917). – Type: New Guinea, Papua, Humboldt Bay, i 1914, *L.S. Gibbs* 6253 (holo BM; iso K, L).
- **NEW GUINEA: Papua:** Humboldt Bay, i 1914, *L.S. Gibbs* 6253 (holo BM; iso K, L) – Type of *Begonia humboldtiana* Gibbs; Vogelkopf, Aifat River Valley, Sururem – Son Village, 2 xi 1961, *P.v. Royen, H. Sleumer* 7503 (L); Vogelkopf, Aifat River Valley, Sururem – Son Village, 2 xi 1961, *P.v. Royen, H. Sleumer* 7507 (A, L); Vogelkopf, Bamfot Village, 2 xi 1961, *P.v. Royen, H. Sleumer* 7644 (L).
- **IUCN category: VU D2**. A reasonable amount of forest at the altitude given in the protologue still exists around Humboldt Bay, but is under pressure from development.

Begonia humericola Sands [§ Petermannia], Pl. Mt. Kinabalu 156 (2001). – Type: Borneo, Sabah, Mt. Kinabalu, Singh's Plateau, 10 vi 1961, 6°5′N, 116°38′E, *W.L. Chew, E.J.H. Corner, A. Stainton* RSNB 1008 (holo K; iso BO *n.v.*, L *n.v.*, SAN *n.v.*).
- **BORNEO: Sabah:** Mt. Kinabalu, Singh's Plateau, 10 vi 1961, *W.L. Chew, E.J.H. Corner, A. Stainton* RSNB 1008 (holo K; iso BO *n.v.*, L *n.v.*, SAN *n.v.*) – Type of *Begonia humericola* Sands; Mt. Kinabalu, Eastern shoulder, 31 viii 1961, RSNB 1571 (para A *n.v.*, BO *n.v.*, K, L *n.v.*, SAN *n.v.*) – Type of *Begonia humericola* Sands; Mt. Kinabalu, Eastern shoulder, *Kokawa, Hotta* 4998 (para L *n.v.*) – Type of *Begonia humericola* Sands; Mt. Kinabalu, Eastern shoulder, *Kokawa, Hotta* 5019 (para L *n.v.*) – Type of *Begonia humericola* Sands; Mt. Kinabalu, Eastern shoulder, *Kokawa, Hotta* 5036 (para L *n.v.*) – Type of *Begonia humericola* Sands; Mt. Kinabalu, Lugas Hill, *J.H. Beaman* 10539 (para K *n.v.*) – Type of *Begonia humericola* Sands; Tambunan, Trusmadi, 10 x 1962, *O. Mikil* 31868 (K).
- **IUCN category: VU D2**. Fairly localised in lower montane forest on the eastern and southern sides of Mt. Kinabalu (Sands, 2001).

Begonia humilicaulis Irmsch. [§ Petermannia], Bot. Jahrb. Syst. 50: 356 (1913). – Type: Sulawesi, Angabe, *A.B. Meyer s.n.* (holo B).
- **SULAWESI:** Angabe, *A.B. Meyer s.n.* (holo B) – Type of *Begonia humilicaulis* Irmsch.; Menado, Maraowa, 5 viii 1937, *P.J. Eyma* 1569 (K, L, U); Menado, Maraowa, 5 viii 1937, *P.J. Eyma* 1570 (L, U).
- **IUCN category: DD**. Insufficient specimens could be georeferenced with certainty.

Begonia hymenophylla Gagnep. [§ Reichenheimia], Bull. Mus. Hist. Nat. (Paris) 25:200 (1919); Gagnepain, in Lecomte (ed.), Fl. Indo-Chine 2: 1099 (1921). – Type: Laos, Mt. Bassac, 1866 – 1888, *C. Thorel* 2358 (holo P; iso P[2]).

- **LAOS:** Mt. Bassac, 1866 – 1888, *C. Thorel* 2358 (holo P; iso P[2]) – Type of *Begonia hymenophylla* Gagnep.
  CAMBODIA: Dangrek Mts., 4 xi 1927, *E. Poilane* 14001 (P).
- **Notes:** A miniscule species, and a new record for Cambodia.
- **IUCN category: VU D2.** The altitude range of this species is not known, and hence whether it is likely to occur in the protected area north of the site in the Dangkrek Mts. is unsure. The vegetation on Mt. Bassac is largely intact, but not formally protected.

Begonia hymenophylloides Kingdon-Ward ex L.B. Sm. & Wassh. [not placed to section], Phytologia 54(7): 467 (1984); Smith & Wasshausen, Phytologia 55: 112 (1984). – Type: Burma, Nam Tisang-Mali divide, 24 viii 1926, *F.K. Ward* 7334 (holo K; iso K).
  Begonia hymenophylloides Kingdon-Ward *nom. nud.*, Gard. Chron., III 104: 474 (1938); Smith & Wasshausen, Phytologia 54(7): 467 (1984); Smith & Wasshausen, Phytologia 55: 112 (1984). – Type: not located.

- **BURMA:** Nam Tisang-Mali divide, 24 viii 1926, *F.K. Ward* 7334 (holo K; iso K) – Type of *Begonia hymenophylloides* Kingdon-Ward ex L.B. Sm. & Wassh.; Mali Hka, 31 vii 1937, *F.K. Ward* 12858 (BM); Nam Tisang, 9 xii 1937, *F.K. Ward* 13555 (BM) [as *Begonia* cf. *hymenophylloides*].
- **Notes:** The type specimen is annotated 'B. dioica in extreme form'.
- **IUCN category: DD**. Taxonomically uncertain.

Begonia ignorata Irmsch. [§ Reichenheimia], Mitt. Inst. Allg. Bot. Hamburg 8: 97 (1929); Kiew, Begonias Penins. Malaysia 43 (2005). – Type: Peninsular Malaysia, Pahang, Kota Gelanggi, *H.N. Ridley* 2442 (holo SING *n.v.*).
  Begonia hasskarlii var. *hirsuta* Ridl., Fl. Malay Penins. 1: 860 (1922); Irmscher, Mitt. Inst. Allg. Bot. Hamburg 8: 97 (1929). – Type: Peninsular Malaysia, Pahang, Kota Gelanggi, *H.N. Ridley* 2442 (lecto SING *n.v.*).

- **PENINSULAR MALAYSIA:** Pahang, Gunong Senyum, vi 1917, *J.H. Evans s.n.* (BM); Pahang, Kota Gelanggi, *H.N. Ridley* 2442 (holo SING *n.v.*; lecto SING *n.v.*) – Type of *Begonia hasskarlii* var. *hirsuta* Ridl.; Pahang, Kota Gelanggi, *H.N. Ridley* 2442 (holo SING *n.v.*; lecto SING *n.v.*) – Type of *Begonia ignorata* Irmsch.; Pahang, Gunung Senyum, 31 vii 1929, *M.R. Henderson* 22327 (BM).
  JAVA: 22 ii 1914, *C.A.B. Backer* 11775 (B).
- **Notes:** There is a record from Java (*C.A.B. Backer* 11775 (B)) determined by Irmscher, but it is likely to represent a different taxon.
- **IUCN category: LC.** Described as 'locally common' (Kiew, 2005), and occurs in three provinces of Peninsular Malaysia.

Begonia imbricata Sands [§ Petermannia], Kew Mag. 7: 64 (1990); Sands, Pl. Mt. Kinabalu 157 (2001). – Type: Borneo, Sabah, Mt. Kinabalu, Mamut copper mine, 3 v 1984, 6°1'59"N, 116°39'59"E, *M.J.S. Sands* 3955 (holo K).

- **BORNEO: Sabah:** Mt. Kinabalu, Ulu Liwago – Ulu Mesilau, 3 ix 1961, *W.L. Chew, E.J.H. Corner, A. Stainton* 2685 (K); Mt. Kinabalu, Mamut copper mine, 3 v 1984, *M.J.S. Sands* 3955 (holo K) – Type of *Begonia imbricata* Sands; Ranau District, 9 vi 1987, *Amin* 117879 (L); Ranau District, Bukit Hampuan, 11 xii 1987, *Amin et al.* 121458 (K, L); Ranau District, Mosilou, 6 ix 1988, *Amin et al.* 123544 (K).
- **IUCN category: VU D2.** Although known only from the type locality, it is described as 'locally abundant' (Sands, 2001).

Begonia imperfecta Irmsch. [§ Petermannia], Bot. Jahrb. Syst. 50: 367 (1913). – Type: Sulawesi, Bada, *K.F. & P.B. Sarasin* 2126 (holo B).
■ **SULAWESI:** Bada, *K.F. & P.B. Sarasin* 2126 (holo B) – Type of *Begonia imperfecta* Irmsch.
■ **IUCN category: DD.** Insufficient specimens could be georeferenced with certainty.

Begonia incerta Craib [§ Diploclinium], Bull. Misc. Inform. Kew 1911: 57 (1911); Gagnepain, in Lecomte (ed.), Fl. Indo-Chine 2: 1116 (1921); Craib, Fl. Siam. 1: 774 (1931). – Type: Thailand, Meh Ping Rapids, 15 xii 1908, *A.F.G. Kerr* 508 (holo K; iso BM).
■ **THAILAND:** Meh Ping Rapids, 15 xii 1908, *A.F.G. Kerr* 508 (holo K; iso BM) – Type of *Begonia incerta* Craib; Meh Ping Rapids, 17 xii 1913, *A.F.G. Kerr* 3043 (BM, K); Meh Ping Rapids, 23 xi 1910, *A.F.G. Kerr* 4368 (K); Meh Ping Rapids, 23 xi 1920, *A.F.G. Kerr* 4638 (ABD, BM).
■ **IUCN category: DD.** Insufficient specimens could be georeferenced with certainty.

Begonia incisa A. DC. [§ Petermannia], Ann. Sci. Nat. Bot., IV 11: 129 (1859); Candolle, Prodr. 15(1): 321 (1864); Fernández-Villar, Noviss. App. 99 (1880); Merrill, Philipp. J. Sci. 2: 285 (1907); Merrill, Philipp. J. Sci. 6: 382 ('1911', 1912); Merrill, Enum. Philipp. Fl. Pl. 3: 123 (1923). – Type: not located.
■ **PHILIPPINES:** Philippines, *unknown s.n.* (B). **Luzon:** Camarines, xii 1913, *M. Ramos* 1579 (B, BM[2], BO, L, P); Tayabas, Atimonan, iii 1905, *E.D. Merrill* 3994 (B, BO, K, L, P); Zambales, xi 1907 – xii 1907, *M. Ramos* 4993 (P); Pampanga, Mt. Pinatubo, v 1927, *A.D.E. Elmer* 22132 (B, BM, K, L, P); Tayabas, iii 1917, *G. Edano* 26961 (P). **Mindoro:** Mt. Halcon, ii 1922, *M. Ramos, G. Edano* 40592 (B, L); Mt. Halcon, iii 1922, *M. Ramos, G. Edano* 40663 (P). **Panay:** Capiz, Jamindan, iv 1918 – v 1918, *M. Ramos, G. Edano* 31352 (K, P). **Negros:** Negros Occidental, Cuernos Mts., v 1908, *A.D.E. Elmer* 10033 (BM, BO, E). **Bohol:** *M. Ramos* 43304 (B, BM). **Mindanao:** Bukidnon subprovince, Tangculan, *M. Ramos, G. Edano* 39103 (B, K).
■ **IUCN category: NT.** Despite this species occurring throughout the Philippines, I have not come across any collections made after the 1920s.

Begonia incondita Craib [§ Diploclinium], Bull. Misc. Inform. Kew 1930: 412 (1930); Craib, Fl. Siam. 1: 774 (1931). – Type: Thailand, Nakawn Sritamarat, Kao Luang, 29 iv 1920, 8°30′N, 99°40′E, *A.F.G. Kerr* 15457 (holo ABD; iso BM, K).
■ **THAILAND:** Nakawn Sritamarat, Kao Luang, 29 iv 1920, *A.F.G. Kerr* 15457 (holo ABD; iso BM, K) – Type of *Begonia incondita* Craib.
■ **IUCN category: LC.** The type locality is in the Khao Luang National Park.

Begonia inostegia Stapf [§ Petermannia], Trans. Linn. Soc. London, Bot., II 4: 166 (1894); Stapf, Icon. Pl. 24: 2309 (1895); Ridley, J. Malayan Branch Roy. Asiat. Soc. 46: 253 (1906); Sands, Pl. Mt. Kinabalu 157 (2001). – Type: Borneo, Sabah, Mt. Kinabalu, 1 ii 1893, *G.D. Haviland* 1190 (holo K).
■ **BORNEO: Sabah:** Mt. Kinabalu, Boundary Rentis, 22 i 1976, *P.F. Stevens* 645 (L); Mt. Kinabalu, 1 ii 1893, *G.D. Haviland* 1190 (holo K) – Type of *Begonia inostegia* Stapf; Mt. Kinabalu, 4 i 1969, *Kokawa, Hotta* 3091 (L); Mt. Kinabalu, 4 i 1969, *Kokawa, Hotta* 3092 (L); Mt. Kinabalu, Tenompok F.R., 4 ii 1932, *J. & M.S. Clemens* 3504 (BO); 2 ii 1932, *J. & M.S. Clemens* 28214 (B, BM, K, L); Colombon River Basin, Ulu Sadikan, 21 vi 1933, *J. & M.S. Clemens* 34009 (A, BM, BO); Colombon River Basin, Ulu Sadikan, 5 v 1933, *J. & M.S. Clemens* 34009 (A, BM, BO).
■ **IUCN category: LC.** The type locality is in the Kinabalu National Park, where it is found at a number of localities.

Begonia insularum Irmsch. [§ Petermannia], Bot. Jahrb. Syst. 50: 353 (1913). – Type: Sulawesi, Sangir Island, *O. Warburg* 16107 (holo B).

■ **SULAWESI:** Sangir Island, *O. Warburg* 16107 (holo B) – Type of *Begonia insularum* Irmsch.

■ **IUCN category: DD**. The locality on Sangir Island is not defined in the protologue, nor is there sufficient information available on the state of forest cover on the island.

Begonia integrifolia Dalzell [§ Parvibegonia], Hooker's J. Bot. Kew Gard. Misc. 3: 230 (1851); Candolle, Prodr. 15(1): 351 (1864); Gagnepain, in Lecomte (ed.), Fl. Indo-Chine 2: 1113 (1921); Craib, Fl. Siam. 1: 775 (1931); Kiew, Begonias Penins. Malaysia 74 (2005). – Type: India, Bombay, *N.A. Dalzell s.n.* (lecto K *n.v.*).

Begonia guttata Wall. ex A. DC., Prodr. 15(1): 352 (1864); Clarke, Fl. Brit. Ind. 2: 648 (1879); Irmscher, Mitt. Inst. Allg. Bot. Hamburg 8: 153 (1929). – *Begonia integrifolia* var. *guttata* (A. DC.) Gagnep., in Lecomte, Fl. Indo-Chine 2: 1114 (1921). – Type: Peninsular Malaysia, Penang, *N. Wallich* 3671A (holo K; iso P).

Begonia debilis King, J. Asiat. Soc. Bengal, Pt. 2, Nat. Hist. 71: 60 (1902). – Type: Peninsular Malaysia, Perak, *Kings Collector* 8289 (lecto K *n.v.*).

Begonia curtisii Ridl., J. Straits Branch Roy. Asiat. Soc. 59: 106 (1911); Irmscher, Mitt. Inst. Allg. Bot. Hamburg 8: 149 (1929); Craib, Fl. Siam. 1: 772 (1931). – Type: Thailand, Kasum, xi 1896, *C. Curtis* 3234 (holo SING; iso K).

Begonia leucantha Ridl., J. Straits Branch Roy. Asiat. Soc. 57: 49 (1911); Ridley, Fl. Malay Penins. 1: 857 (1922); Irmscher, Mitt. Inst. Allg. Bot. Hamburg 8: 157 (1929); Kiew, Begonias Penins. Malaysia 74 (2005). – Type: Peninsular Malaysia, Perak, Temengoh, vii 1909, *H.N. Ridley* 14731 (lecto SING *n.v.*; isolecto BM, K *n.v.*).

Begonia haniffii Burkill, J. Straits Branch Roy. Asiat. Soc. 79: 103 (1918); Ridley, Fl. Malay Penins. 1: 856 (1922); Irmscher, Mitt. Inst. Allg. Bot. Hamburg 8: 151 (1929); Craib, Fl. Siam. 1: 774 (1931); Kiew, Begonias Penins. Malaysia 74 (2005). – Type: Peninsular Malaysia, Langkawi, 15 vii 1916, *M. Haniff* 1494 (lecto SING *n.v.*; iso K).

Begonia guttata var. *angopensis* Irmsch., Mitt. Inst. Allg. Bot. Hamburg 8: 154 (1929). – Type: Thailand, 14 xii 1918, *M. Haniff, Md. Nur* 4041 (holo SING *n.v.*).

Begonia debilis var. *punicea* Craib, Fl. Siam. 1: 772 (1931). – Type: Thailand, Ayuthia, Saraburi, Muak Lek, 30 viii 1924, 14°35′N, 100°55′E, *A.F.G. Kerr* 9075 (holo ABD; iso B, BM, K).

Begonia guttata forma *elongata* Irmsch., Mitt. Inst. Allg. Bot. Hamburg 8: 154 (1929); Kiew, Begonias Penins. Malaysia 74 (2005). – Type: Peninsular Malaysia, Penang, *C. Curtis s.n.* (lecto SING *n.v.*).

■ **BURMA:** Moulmein, *C.S.P. Parish s.n.* (K); Moulmein, 1859, *C.S.P. Parish s.n.* (K).

THAILAND: Doi Sootep National Park, Palaht Temple, 21 viii 1993, *A. Phuskan* (not located); Tarutao Island, 12 x 1979, *C. Congdon* 11 (A[2]); Khao Yai National Park, 19 vi 2000, *S. Chongko* 38 (L); Doi Sootep National Park, Palaht Temple, 21 viii 1993, *A. Phuakam* 46 (A); Nakawn Sritamarat, Kao Chem, Tung Song, 20 vii 1929, *N. Rabil Bunnag* 102 (BM); Satun Distr., Toong Ngui Subdistr., Nam Rah Village, 19 viii 1984, *J.F. Maxwell* 84-106 (A) [as *Begonia cf. integrifolia*]; Tan Te Waterfall, 9 xii 1974, *T. Santisuk* 341 (L); Chonburi, Siricha Distr., Kow Kieo, 6 vi 1976, *J.F. Maxwell* 76-385 (L[2]); Satun Distr., Talaybun National Park, 13 vii 2000, *D.J. Middleton et al.* 415 (A); Sriracha, 16 xi 1926, *N. Put* 488 (ABD, BM); See Bahn Pote Distr., Kao Boo-Kao Yah National Park, Riang Tong Falls, 30 vii 1986, *J.F. Maxwell* 86-521 (A, L); Saraburi Province, Sahm Lahn Forest, 29 vi 1974, *J.F. Maxwell* 74-636 (L); Chantaburi, Bong Nam Rawn Distr., Soi Dao Mt., 5 vii 1974, *J.F. Maxwell* 74-661 (L); Chiangmai, Doi Sootep, 17 vii 1909, *A.F.G. Kerr* 730 (B, BM, K); Trang Province, Khao Chong, 11 viii 1975, *J.F. Maxwell* 75-738 (L); Tarutao Island, 31 vii 1980, *C. Congdon* 820 (A); Satun Distr., Talaybun National Park, *P. Wilkin* 821 (K); Lampang Prov., Doi Kuhn Dahn National Park, 30 vii 1994, *J.F. Maxwell* 94-838 (A, L); Satun Distr., Talaybun National Park, 6 ix 1985, *J.F. Maxwell* 85-845 (L) [as *Begonia cf. integrifolia*]; Pattani, Koke Po Distr., Sai Kow Falls, 11 ix 1985,

J.F. Maxwell 85-853 (A, L); Chiangmai, Ban Pah Dahng, 11 viii 1990, *J.F. Maxwell* 90-865 (A, L); Lansagah Distr., Khao Luang National Park, Gahrome Falls, 15 ix 1985, *J.F. Maxwell* 85-889 (A, L); Chiangmai, Mawk Fa Falls, 18 viii 1990, *J.F. Maxwell* 90-894 (A, L); Doi Sootep National Park, 22 viii 1990, *J.F. Maxwell* 90-897 (A, L); Chiangmai, Doi Chiang Dao Animal Sanctuary, 25 viii 1990, *J.F. Maxwell* 90-910 (A, L); Chantaburi, 6 vi 1927, *N. Put* 917 (BM[2], K); Khon Chalat, 7 viii 1923, *D.J. Collins* 928 (K); Chonburi, Siricha Distr., Kow Kieo, 30 viii 1975, *J.F. Maxwell* 75-935 (L); Chumpawn, 7 ix 1927, *N. Put* 981 (BM); Chumpawn, Kao Sabap, 7 ix 1927, *N. Put* 982 (ABD, BM, K); Payap, Doi Chiengdao, 26 viii 1935, *H.B.G. Garrett* 995 (ABD, K[2], L, P); Payap, Doi Chiengdao, 26 viii 1935, *A.F.G. Kerr* 995 (ABD, K[2], L, P); Lampang Prov., Chae Son Nat. Park, 26 vii 1996, *J.F. Maxwell* 96-1016 (L); Kaeng Krachan National Park, 14 viii 2002, *D.J. Middleton et al.* 1041 (A); Chumpawn, 10 ix 1927, *N. Put* 1046 (ABD, BM, K); Champasak Prov., 13 x 1997, *J.F. Maxwell* 97-1150 (L); Surat, Khao Lang Tao, 26 ix 1963, *T. Smitinand, H. Sleumer* 1291 (BO) [as *Begonia aff. integrifolia*]; Huay Yang National Park, Bua Sawan Waterfall, 26 viii 2002, *D.J. Middleton et al.* 1357 (A); Kanchanaburi, Sangkhlaburi, 28 vii 1968, *B. Sangkhachand* 1424 (L); Sakaerat Forest Reserve, Huai Krae Stream, 1 xi 1969, *C.F. van Beusekom, C. Charoenpol* 1973 (L, P); Phu Langka N.P., Tad Kham Falls, 25 viii 2001, *R. Pooma, W.J.J.O. de Wilde, B.E.E. Duyfjes* 2632 (L); Kasum, xi 1896, *C. Curtis* 3234 (holo SING; iso K) – Type of *Begonia curtisii* Ridl.; Khao Yai National Park, 10 viii 1968, *K. Larsen et al.* 3249 (P); Nang Rong Falls, 13 viii 1968, *K. Larsen et al.* 3364 (L, P); Trang Province, Khao Chong, 9 x 1970, *K. Larsen et al.* 3494 (L, P) [as *Begonia aff. integrifolia*]; Yala Province, 21 x 1970, *C. Charoenpol et al.* 4111 (L); Yala Province, Kue Long waterfalls, 22 x 1970, *K. Larsen et al.* 4157 (P) [as *Begonia aff. integrifolia*]; Pangnga, Kan Bow Koranee cascade, 9 v 1973, *R. Geesink et al.* 5306 (L *n.v.*) [as *Begonia cf. integrifolia*]; Trang Province, 15 vi 1974, *R. Geesink et al.* 7259 (L, P); Banang Sta, 26 vii 1923, *A.F.G. Kerr* 7341 (ABD, BM); Telok Wau, 20 xi 1921, *M. Haniff, Md. Nur* 7495 (BM); Pattani, 18 viii 1925, *A.F.G. Kerr* 7609 (ABD); Banang Sta, 28 vii 1923, *A.F.G. Kerr* 7891 (BM); Ayuthia, Saraburi, Muak Lek, 30 viii 1924, *A.F.G. Kerr* 9075 (holo ABD; iso B, BM, K) – Type of *Begonia debilis* var. *punicea* Craib; Nakawn Sritamarat, Kao Luang, *A.F.G. Kerr* 10823 (BM); Ban Kawn Kep, 5 viii 1927, *A.F.G. Kerr* 13163 (ABD, BM, K); Surat Thani, 16 vii 1972, *K. Larsen et al.* 31051 (B) [as *Begonia cf. integrifolia*]; Nam Tok Taka Mao, 23 vii 1972, *K. Larsen et al.* 31263 (B, L) [as *Begonia cf. integrifolia*]; Nam Tok Taka Mao, 25 viii 1972, *K. Larsen et al.* 32018 (B, L) [as *Begonia cf. integrifolia*]; Krabi Prov., Than Bok Koroni Nat. Res., 15 vii 1992, *K. Larsen et al.* 43445 (P).

LAOS: Bassac, Khon Island, 1866 – 1888, *C. Thorel* 2226 (P).

VIETNAM: Annam, 20 x 1940, *E. Poilane s.n.* (P) [as *Begonia cf. integrifolia*]; Cochinchine, ix 1865, *J.B.L. Pierre s.n.* (BM, P); Cochinchine, vii 1866, *J.B.L. Pierre s.n.* (P); Me-Kong, 1866 – 1888, *C. Thorel* 2240 (P); Annam, 13 x 1923, *E. Poilane* 8192 (P).

PENINSULAR MALAYSIA: Penang, *C. Curtis s.n.* (lecto SING *n.v.*) – Type of *Begonia guttata* forma *elongata* Irmsch.; Langkawi, 9 vii 1891, *C. Curtis s.n.* (P); Perak, *B. Scortechini* 571 (L, P); Perak, Bukit Bujang Melaka, 19 v 1987, *S. Anthonysamy* 824 (L); Langkawi, 22 viii 1988, *S. Anthonysamy* 960 (L); Langkawi, 22 viii 1988, *S. Anthonysamy* 978 (L); Langkawi, 15 vii 1916, *M. Haniff* 1494 (iso K; lecto SING *n.v.*) – Type of *Begonia haniffii* Burkill; Penang, *N. Wallich* 3671A (P) – Type of *Begonia guttata* Wall. ex A. DC.; Perlis, Mata Ayer F.R., 1 vii 1993, *R. Kiew* 3698 (L[2]); 14 xii 1918, *M. Haniff, Md. Nur* 4041 (holo SING *n.v.*) – Type of *Begonia guttata* var. *angopensis* Irmsch.; *A.C. Maingay* 6759 (L); Perak, *Kings Collector* 8289 (holo not located) – Type of *Begonia debilis* King; Perak, Temengoh, vii 1909, *H.N. Ridley* 14731 (lecto SING *n.v.*; isolecto BM, K *n.v.*) – Type of *Begonia leucantha* Ridl.; Kelantan, Kuala Aring, 1992, *B.H. Kiew* 25 (L[3]).

■ **Notes:** The most widespread of a number of similar species which together make up a large portion of § *Parvibegonia*, namely *B. carnosula* Ridl., *B. erosa* Blume, *B. geoffrayi* Gagnep., *B. grantiana* Geddes ex Craib, *B. paleacea* Kurz, *B. phoeniogramma* Ridl., *B. rimarum* Craib, *B.*

rupicola Miq., *B. socia* Craib, *B. tenuifolia* Dryand., *B. variabilis* Ridl. and *B. zollingeriana* A. DC. It is possible that some of these could be placed within the variability of *B. integrifolia* as it is currently circumscribed.

■ **IUCN category: LC**. Widespread; described as 'common on the west coast' in Peninsular Malaysia (Kiew, 2005).

Begonia intermixta Irmsch. [§ Reichenheimia], Mitt. Inst. Allg. Bot. Hamburg 8: 101 (1929). – Type: Thailand, Pulau Panji, 16 xii 1918, *M. Haniff, Md. Nur* 4076 (syn SING *n.v.*); Thailand, Kasum, xi 1896, *C. Curtis s.n.* (syn K, SING *n.v.*).

■ **THAILAND:** Kasum, xi 1896, *C. Curtis s.n.* (syn K, SING *n.v.*) – Type of *Begonia intermixta* Irmsch.; Pulau Panji, 16 xii 1918, *M. Haniff, Md. Nur* 4076 (syn SING *n.v.*) – Type of *Begonia intermixta* Irmsch.

■ **IUCN category: DD**. Insufficient specimens could be georeferenced with certainty.

Begonia inversa Irmsch. [§ Reichenheimia], Webbia 9: 505 (1953). – Type: Sumatra, Padang, Sungei Bulu, ix 1878, *O. Beccari* PS903 (syn BM, FI[2], K, L).

Begonia inversa forma inversa

■ **SUMATRA:** Padang, Sungei Bulu, ix 1878, *O. Beccari* PS903 (syn BM, FI[2], K, L) – Type of *Begonia inversa* Irmsch.; Padang, Sungei Bulu, 1878, *O. Beccari* CB4517 (FI).

Begonia inversa forma nana Irmsch., Webbia 9: 507 (1953). – Type: Sumatra, Padang, Sungei Bulu, ix 1878, *O. Beccari* CB4517 (syn FI).

■ **SUMATRA:** Padang, Sungei Bulu, ix 1878, *O. Beccari* CB4517 (syn FI) – Type of *Begonia inversa* forma *nana* Irmsch.

■ **Notes:** Placed in § *Diploclinium* in Doorenbos *et al.* (1998), but has entire placentae and so belongs in § *Reichenheimia* and appears to be allied to the other Sumatran species in that section.

■ **IUCN category: DD**. Insufficient specimens could be georeferenced with certainty.

Begonia ionophylla Irmsch. [§ Diploclinium], Bot. Jahrb. Syst. 50: 378 (1913). – Type: Sumatra, Poeloe Bras, xii 1889, *V. Lehmann* 69 (holo B).

■ **SUMATRA:** Poeloe Bras, xii 1889, *V. Lehmann* 69 (holo B) – Type of *Begonia ionophylla* Irmsch.

■ **IUCN category: DD**. Insufficient specimens could be georeferenced with certainty.

Begonia iridescens Dunn [§ Platycentrum], Bull. Misc. Inform. Kew 1920: 110 (1920). – Type: India, Abor Hills, *I.H. Burkill* 36673 (syn K).

■ **BURMA:** Nam Tisang-Mali divide, 24 viii 1926, *F.K. Ward* 7336 (K) [as *Begonia aff. iridescens*].

■ **IUCN category: DD**. Insufficient specimens could be georeferenced with certainty.

Begonia isabelensis Quisumb. & Merr. [§ Diploclinium], Philipp. J. Sci. 37: 173 (1928). – Type: Philippines, Luzon, Isabela Province, San Mariano, ii 1926 – iii 1926, *M. Ramos, G. Edano* 47207 (iso BM, K, P).

■ **PHILIPPINES: Luzon:** Isabela Province, San Mariano, ii 1926 – iii 1926, *M. Ramos, G. Edano* 47207 (syn BM, K, P) – Type of *Begonia isabelensis* Quisumb. & Merr.

■ **IUCN category: DD**. Insufficient specimens could be georeferenced with certainty.

Begonia isoptera Dryand. ex Sm. [§ Petermannia], Pl. Icon. pl. 43 (1790); Dryander, Trans. Linn. Soc. 1: 160 (1791); Candolle, Prodr. 15(1): 320 (1864); Merrill, Philipp. J. Sci. 6: 406 ('1911', 1912); Koorders, Exkurs.-Fl. Java 2: 650 (1912). – Type: not located.

Begonia geniculata Jack, Malayan Misc. 1(7): 15 (1822); Klotzsch, Begoniac. 124 (1855); Candolle, Prodr. 15(1): 321 (1864). – *Petermannia geniculata* (Jack) Klotzsch, Monatsber. Kon. Preuss. Akad. Wiss. Berlin 1854: 124 (1854); Klotzsch, Abh. Kon. Akad. Wiss. Berlin 1854: 196 (1855); Klotzsch, Begoniac. 76 (1855). – Type: not located.

Begonia repanda Blume, Enum. Pl. Javae 1: 97 (1827); Klotzsch, Abh. Kon. Akad. Wiss. Berlin 1854: 192 (1855); Candolle, Prodr. 15(1): 321 (1864). – *Diploclinium repandum* (Blume) Klotzsch, Abh. Kon. Akad. Wiss. Berlin 1854: 192 (1855); Klotzsch, Begoniac. 72 (1855); Miquel, Fl. Ned. Ind. 1(1): 688 (1856). – Type: not located.

Begonia bombycina Blume, Enum. Pl. Javae 1: 97 (1827); Miquel, Pl. Jungh. 417 ('1855', 1857); Klotzsch, Abh. Kon. Akad. Wiss. Berlin 1854: 192 (1855); Klotzsch, Begoniac. 72 (1855); Candolle, Prodr. 15(1): 321 (1864). – *Diploclinium bombycinium* (Blume) Klotzsch, Abh. Kon. Akad. Wiss. Berlin 1854: 192 (1855); Klotzsch, Begoniac. 72 (1855); Miquel, Fl. Ned. Ind. 1(1): 687 (1856). – Type: not located.

Begonia angustifolia Blume, Enum. Pl. Javae 1: 97 (1827); Miquel, Fl. Ned. Ind. 1(1): 687 (1856); Candolle, Prodr. 15(1): 396 (1864); Koorders, Exkurs.-Fl. Java 2: 651 (1912). – *Diploclinium angustifolium* (Blume) Miq., Fl. Ned. Ind. 1(1): 687 (1856). – Type: not located.

■ **SUMATRA:** Sumatra, *J.E. Teijsmann s.n.* (K); Sumatra, 1877 – 1878, *unknown s.n.* (L); Sumatra, *J.E. Teijsmann s.n.* (K); Lampung Distr., Gunung Rate Telanggaran, 17 xi 1921, *Iboet* 131 (BO) [as *Begonia aff. isoptera*]; Lampongs, Tangamoes, 9 vii 1928, *C.N.A. de Voogd* 160 (BO, L) [as *Begonia aff. isoptera*]; Lampung Distr., Wai Lima Estate, 6 xii 1920 – 1930, *Anon.* 360 (BO) [as *Begonia aff. isoptera*]; Palembang, Ranaumeer, 7 ii 1929, *C.N.A. de Voogd* 450 (BO) [as *Begonia aff. isoptera*]; Padang, Ajer Mantjoer, viii 1878, *O. Beccari* PS610 (K, L); Riau Province, Bukit Tiga Puluh, 30 vii 2006, *D. Girmansyah, A. Poulsen, I. Hatta, F. Antoni* 793 (BO, E[2]); *H.S. Yates* 1138 (B, P) [as *Begonia isoptera ?*]; 1881 – 1882, *H.O. Forbes* 1250 (B); Bencoolen, 10 iii 1932, *C.N.A. de Voogd* 1325 (L); Lampung Distr., Penanggungan, 3 ix 1880, *H.O. Forbes* 1704 (K); Raisers Peak, 10 x 1880, *H.O. Forbes* 1850 (BM, K); Passumah plateau, *H.O. Forbes* 2379 (BM, L); Mt. Singalan, 28 v 1918, *H.A.B. Bunnemeijer* 2661 (L); G. Talang, 25 x 1918, *H.A.B. Bunnemeijer* 5009 (L); Enggano, 26 vi 1936, *W.J. Lutjeharms* 5115 (A, K, P); Pulau Sebesie, 25 iv 1921, *Drs. Leeuwen-Reijnvaan* 5264 (BO) [as *Begonia aff. isoptera*]; G. Talang, 29 x 1918, *H.A.B. Bunnemeijer* 5362 (L); Bencoolen, Lebong Tandai, vi 1922, *C.J. Brooks* 7609 (K); Lampongs, Mt. Tanggamus, 27 iv 1968, *M. Jacobs* 8096 (A, L) [as *Begonia cf. isoptera*].

JAVA: Java, *C.L. von Blume s.n.* (L; 5 sheets); Java, *W. Lobb s.n.* (K; 2 sheets); Java, *unknown s.n.* (P; 2 sheets); Java, *J.B.L.T. Leschenault de la Tour s.n.* (P); Java, *C.L. von Blume s.n.* (P); Java, *J.G. Boerlage s.n.* (L); Java, *unknown s.n.* (L; 5 sheets); Java, *P.W. Korthals s.n.* (L); Java, *F.W. Junghuhn s.n.* (L); Geger Bintang, *J.G. Boerlage s.n.* (L); *J.G. Boerlage s.n.* (L); Preanger, Tjibodas, *unknown s.n.* (L); Buitenzorg, G. Gede, 19 xii 1940, *S. Bloembergen s.n.* (L); Geger Bintang, 21 vi 1888, *J.G. Boerlage s.n.* (L); Java, *T. Horsfield s.n.* (K); Java, *C.L.v. Blume s.n.* (K); Preanger, Tjibodas, 2 viii 1918, *L.G.d. Berger* (BO); Batavia, Tjampea, 1918, *T. Valeton s.n.* (BO); 22 ii 1896, *E.H. Hallier s.n.* (BO); Batavia, Tjampea, 1914, *Burck & de Monchy s.n.* (BO); G. Windoe, 30 iv 1909, *Soegandiredjo s.n.* (L[2]); Preanger, Gunung Gede, 14 ii 1915, *H.N. Ridley s.n.* (K); iii 1920, *C.B. Kloss s.n.* (K); Mt. Salak, 17 ii 1915, *H.N. Ridley s.n.* (K); *T. Horsfield s.n.* (K); *T. Horsfield s.n.* (K); *C. Millet s.n.* (K); *J.B. Spanoghe s.n.* (K); Preanger, Tjibodas, *W.J.C. Kooper s.n.* (U); Preanger, Tjibodas, 26 xii 1975, *H. Ern s.n.* (B); *W.B. Hillebrand s.n.* (B); Preanger, Gunung Gede, *H.N. Ridley s.n.* (BM); Banjoemas, Gunung Slamat, 13 iii 2004, *D. Girmansyah* 31 (A); 1 xii 1842, *H. Zollinger* 65 (P); Gunung Malabar, vii 1916, *C.J.F. Denker* 69 (BO); G. Windoe, 24 iv 1909, *Soegandiredjo* 90 (BO); G. Windoe, 30 iv 1909, *Soegandiredjo* 92

(L[2]); *L.R. Lanjouw* 106 (BO) [as *Begonia aff. isoptera*]; Preanger, Tjibodas, *Sapiin* 178 (U); *L.R. Lanjouw* 194 (BO, K); xii 1879, *H.O. Forbes* 223a (A[2], K); Batavia, G. Salak, 1 vi 1894, *H. Raap* 232 (L); Mt. Pajung, Udjung Kulon F.R., 7 i 1964, *N. Wirawan* 238 (BO, L); *W. Lobb* 249 (BM); *Nagel* 271 (B[2]); 27 x 1861, *T. Anderson* 274 (K); Preanger, Tjibodas, 6 vii 1941, *S. Bloembergen* 281 (BO); Preanger, Tjibodas, 2 vi 1896, *H. Raap* 296 (L); *J.C. Ploem* 304 (B); Gunung Tjibodas, 2 viii 1918, *L.G.d. Berger* 347 (BO); Java, *unknown* 400 (B); Java, 1858, *A.F. Jagor* 447 (B[3]); Preanger, Gunung Gede, 17 vii 1941, *S. Bloembergen* 454 (BO); Preanger, Tjadas Malang, 23 viii 1922, *W.F. Winckel* 531B (BO); Java, *H.O. Forbes* 548 (BO); Java, 1880, *H.O. Forbes* 557 (L); Bantam, Bodjongmanik, 15 vi 1912, *S.H. Koorders* 628? (BO); Gunung Tjibodas, 18 iii 1928, *C.G.G.J.v. Steenis* 641 (BO); Preanger, Gunung Gede, 26 vi 1924, *C.N.A. de Voogd* 649 (L); Preanger, Tjibodas, 6 viii 1896, *H. Raap* 719 (L); Java, 1880, *H.O. Forbes* 756a (L); SE Java, *H.O. Forbes* 766 (BM *n.v.*); Java, *H. Zollinger* 784 (P); Java, 18 x 1842, *H. Zollinger* 785 (B, BM, K, P); SE Java, 1880, *H.O. Forbes* 786a (A, B[2], BM, FI); SE Java, 1880, *H.O. Forbes* 857 (A, B, BM, FI); Mt. Halimun N.P., 15 iv 2001, *W.S. Hoover, J.M. Hunter, H. Wiriadinata, D. Girmansyah* 938 (A); West Java, Lake Telagabodas, 21 iii 2001, *W.S. Hoover, J.M. Hunter, H. Wiriadinata, D. Girmansyah* 942 (A) [as *Begonia cf. isoptera*]; West Java, Mt. Cikurai, 22 iii 2001, *W.S. Hoover, J.M. Hunter, H. Wiriadinata, D. Girmansyah* 947 (A) [as *Begonia cf. isoptera*]; West Java, Mt. Bukittunggu, 28 iii 2001, *W.S. Hoover, J.M. Hunter, H. Wiriadinata, D. Girmansyah* 959 (A) [as *Begonia cf. isoptera*]; West Java, Mt. Abig, 7 iv 2001, *W.S. Hoover, J.M. Hunter, H. Wiriadinata, D. Girmansyah* 963 (A); West Java, Mt. Tilu, 8 iv 2001, *W.S. Hoover, J.M. Hunter, H. Wiriadinata, D. Girmansyah* 970 (A); West Java, Mt. Kancana, 8 iv 2001, *W.S. Hoover, J.M. Hunter, H. Wiriadinata, D. Girmansyah* 974 (A); West Java, Mt. Waringin, 9 iv 2001, *W.S. Hoover, H. Wiriadinata* 978 (A); West Java, Mt. Talaga Warna, 11 iv 2001, *W.S. Hoover, J.M. Hunter, H. Wiriadinata, D. Girmansyah* 987 (A) [as *Begonia cf. repanda*]; West Java, Mt. Talaga Warna, 11 iv 2001, *W.S. Hoover, J.M. Hunter, H. Wiriadinata, D. Girmansyah* 988 (A); West Java, Mt. Talaga Warna, 11 iv 2001, *W.S. Hoover, H. Wiriadinata* 989 (A); Priangan, G. Pangrango, 11 iv 2001, *W.S. Hoover, J.M. Hunter, H. Wiriadinata, D. Girmansyah* 990 (A); Buitenzorg, ix 1932, *W.J.C. Kooper* 1035 (U); 11 v 1843, *H. Zollinger* 1291 (P); Indragiri, G. Patoea, 4 vi 1905, *G.v.d. Brink* 1399 (BO); Batavia, Megamendoeng, *C.G.C. Reinwardt* 1721 (L[2]); 15 xi 1843 – 19 xi 1843, *H. Zollinger* 1785 (A, B, P[2]); Gunung Tjibodas, 25 ix 1922, *R.C. Bakhuizen van den Brink* 1796 (U); Tjidadap, 9 vi 1919, *R.C. Bakhuizen van den Brink* 1899 (L); Tjidadap, 11 vi 1919, *R.C. Bakhuizen van den Brink* 1900 (L); Tjibeureum, 23 viii 1928, *C.G.G.J.v. Steenis* 2013 (BO) [as *Begonia aff. isoptera*]; Geger Bintang, 26 vii 1928, *C.G.G.J.v. Steenis* 2101 (BO); Batavia, Megamendoeng, 5 i 1894, *V.F. Schiffner* 2259 (L); Batavia, Megamendoeng, 5 i 1894, *V.F. Schiffner* 2263 (L); Batavia, G. Liiang, 31 x 1928, *C.G.G.J.v. Steenis* 2434 (BO); Preanger, Gunung Gede, 15 ix 1911, *C.A.B. Backer* 3147 (BO); Preanger, Paroengkoeda, 6 i 1920, *R.C. Bakhuizen van den Brink* 3169 (BO, U); Mt. Pajung, Udjung Kulon F.R., 8 v 1992, *McDonald, J.J. Afriastini* 3318 (A) [as *Begonia cf. isoptera*]; Preanger, *O. Warburg* 3331 (B[2]); Preanger, Gunung Gede, 14 ix 1911, *C.A.B. Backer* 3339 (BO); G. Reganis, 17 vii 1912, *C.A.B. Backer* 4068 (BO); Wanajasa, 2 ix 1920, *G.v.d. Brink* 4672 (BO, L); Preanger, Tjibodas, 20 i 1906, *H.G.A. Engler* 4754 (B); Preanger, Tjibodas, 30 i 1921, *R.C. Bakhuizen van den Brink* 5152 (K, L); *C.G.C. Reinwardt* 5153 (L[2]); 9 ii 1906, *H.G.A. Engler* 5238 (B[2]); Papandajan, 26 xi 1912, *C.A.B. Backer* 5503 (B, BO); Preanger, G. Limo, 15 ix 1933, *C.G.G.J.v. Steenis* 5632 (BO); Batavia, Tjampea, 13 iv 1899, *Anon.* 6093 (BO); Batavia, G. Pantjar, 17 iv 1924, *R.C. Bakhuizen van den Brink* 6096 (BO); *R.H.C.C. Scheffer* 6099 (BO); Preanger, Gunung Gede, *R.H.C.C. Scheffer* 6108 (BO); Preanger, G. Limo, Poentjakpas, 29 ix 1935, *C.G.G.J.v. Steenis* 6827 (BO); Preanger, G. Limo, Poentjakpas, 27 iv 1941, *P. Buwalda* 8072 (BO); Mt. Salak, 19 xii 1974, *P.v. Royen* 10673 (L) [as *Begonia cf. isoptera*]; Mt. Halimun N.P., 20 v 2002, *H. Wiriadinata et al.* 10679 (A); *O. Warburg* 11299 (B); Mt. Cendana, 17 iii 2004, *H. Wiriadinata et al.* 11329 (A); Preanger, Tjibodas, 28 iii 1950, *S.J.v. Ooststroom* 13256 (L); Preanger, Tjibodas, 28 iii 1950, *S.J.v. Ooststroom* 13860 (L); Preanger,

Tjiandjoer, Takoka, 1 iii 1894, *S.H. Koorders* 14964B (BO); G. Tjiparaj, 27 vii 1914, *C.A.B. Backer* 14997 (B, BO); Preanger, Gunung Gede, 30 xii 1977, *A. Nitta* 15136 (K); Priangan, G. Pangrango, 29 v 1950, *C.G.G.J.v. Steenis* 17599 (BO, K, L) [as *Begonia aff. isoptera*]; *S.H. Koorders* 20803B (BO); Batavia, G. Sanggaboewana, 29 iii 1918, *C.A.B. Backer* 23687 (BO); Preanger, Pengalangan, 14 x 1918, *C.A.B. Backer* 26197 (BO); Preanger, Tjisondari, Tjigenteng, 22 i 1897, *S.H. Koorders* 26335B (BO); Preanger, Tjisondari, Tjigenteng, 23 i 1897, *S.H. Koorders* 26399B (BO); Preanger, Wanaradja, Pangentjongan-Telagabodas, 19 i 1897, *S.H. Koorders* 26719B (BO); i 1912, *S.H. Koorders* 31356B (BO); Preanger, Tjibodas, 19 x 1898, *S.H. Koorders* 31866 (BO); Preanger, Tjibodas, 1 xi 1898, *S.H. Koorders* 32002B (BO); Bantam, Goenoeng Karang, 29 v 1912, *S.H. Koorders* 40720B (BO) [as *Begonia aff. isoptera*]; Bantam, Lebakkidoel, Gunung Kantjana, 10 vi 1912, *S.H. Koorders* 40868B (BO); Bantam, Lebakkidoel, Gunung Kantjana, 13 vi 1912, *S.H. Koorders* 41297B (BO); Batavia, Tjampea, 17 xi 1926, *S.H. Koorders* 41648 (BO).

LESSER SUNDA ISLANDS: Lombok: Sembaloen, *I. Rensch* 299 (B); Rinjani Nat. Park, 23 iv 2001, *W.S. Hoover, J.M. Hunter, H. Wiriadinata, D. Girmansyah* 10009 (A). **Flores:** Rana Mese, 18 iv 1966, *P.E. Schmutz* 55 (L); Manau, 24 iv 1965, *A.J.G.H. Kostermans, N. Wirawan* 580 (L); Rana Mese, 1927, *I. Rensch* 1120 (B); Koeting, 11 vi 1930, *O. Jaag* 1556a (L); Waas, 28 viii 1962, *P.J.J. Loetens* 1863 (L); 5 xi 1932, *O. Posthumus* 3076 (L[2]); Rana Mese, 14 xi 1932, *O. Posthumus* 3282 (BO, K, L) [as *Begonia aff. isoptera*]; 14 xi 1932, *O. Posthumus* 3382 (A, L); Ruteng, Lusang Pass, 16 vi 1975, *J.F. Veldkamp* 7029 (L). **Sumba:** Sumba, 18 vii 1974, *J.A.J. Verheijen* 4040 (L *n.v.*).

MOLUCCAS: Obi Island: Anggai, Gunung Batu Putih, 13 xi 1974, *E.F. de Vogel* 4005 (L) [as *Begonia cf. isoptera*]. **Seram:** Seram, v 1911 – viii 1911, *E. Stresemann* 131 (L) [as *Begonia isoptera* ?]; Pileana, 28 x 1937, *P.J. Eyma* 1828 (A *n.v.*, L *n.v.*) [as *Begonia cf. isoptera*]; Nanoesa, Rembatoe, Honitetoe, 25 i 1938 – 26 i 1938, *P.J. Eyma* 2666 (L); Kp. Kiandarat, G. Kilia, 19 viii 1938, *P. Buwalda* 5661 (A, L); Kp. Kiandarat, 31 viii 1938, *P. Buwalda* 5883 (L).

- **Notes:** A name applied to any nondescript member of § *Petermannia* from Sumatra, Java, the Lesser Sunda Islands and the Moluccas. In Sumatra and Java, this species probably covers several taxa, and some of the species currently in synonymy could be reinstated. The specimens from the Moluccas certainly represent another taxon, and the species complex is in need of a thorough monographic treatment.
- **IUCN category: DD**. This species is likely to be least concern, but really needs to be further studied and lectotypified before we can be sure it is widespread.

Begonia isopteroidea King [§ Petermannia], J. Asiat. Soc. Bengal, Pt. 2, Nat. Hist. 71: 59 (1902); Ridley, Fl. Malay Penins. 1: 856 (1922); Irmscher, Mitt. Inst. Allg. Bot. Hamburg 8: 112 (1929); Kiew, Begonias Penins. Malaysia 116 (2005). – Type: Peninsular Malaysia, Perak, Gunung Berumban, *L. Wray Jr.* 1548 (holo K *n.v.*).

- **PENINSULAR MALAYSIA:** Perak, Gunung Berumban, *L. Wray Jr.* 1548 (holo K *n.v.*) – Type of *Begonia isopteroidea* King.
- **IUCN category: DD**. Insufficient specimens could be georeferenced with certainty.

Begonia jagorii Warb. in Perkins **[§ Petermannia]**, Fragm. Fl. Philipp. 54 (1904); Merrill, Philipp. J. Sci. 6: 382 ('1911', 1912); Merrill, Enum. Philipp. Fl. Pl. 3: 123 (1923). – Type: Philippines, Luzon, 1861, *A.F. Jagor* 889 (syn B); Philippines, Luzon, 1861, *A.F. Jagor* 890 (syn B).

- **PHILIPPINES: Luzon:** 1861, *A.F. Jagor s.n.* (B[3]); 1861, *A.F. Jagor* 889 (syn B) – Type of *Begonia jagorii* Warb.; 1861, *A.F. Jagor* 890 (syn B) – Type of *Begonia jagorii* Warb. **Mindoro:** Baco River, iv 1903, *E.D. Merrill* 1787 (B); 12 iii 1997, *P. Wilkie et al.* 29036 (E) [as *Begonia aff. jagorii*].
- **IUCN category: VU D2**. Known only from three collections.

Begonia jayaensis Kiew [§ Diploclinium], Begonias Penins. Malaysia 197 (2005). – Type: Peninsular Malaysia, Kelantan, Sungai Bring, *R. Kiew, S. Anthonysamy* RK2906 (holo SING *n.v.*; iso K *n.v.*, KEP *n.v.*, L *n.v.*).

- **PENINSULAR MALAYSIA:** Kelantan, Sungai Nenggiri, Gua Jaya, *R. Kiew, S. Anthonysamy* RK2890 (para SING *n.v.*) – Type of *Begonia jayaensis* Kiew; Kelantan, Sungai Nenggiri, Gua Chawan, *R. Kiew, S. Anthonysamy* 2892 (para SING *n.v.*) – Type of *Begonia jayaensis* Kiew; Kelantan, Sungai Bring, *R. Kiew, S. Anthonysamy* RK2906 (holo SING *n.v.*; iso K *n.v.*, KEP *n.v.*, L *n.v.*) – Type of *Begonia jayaensis* Kiew; Kelantan, Sungai Nenggiri, Gua Jaya, *R. Kiew, S. Anthonysamy* RK4912 (para KEP *n.v.*, L *n.v.*, SING *n.v.*) – Type of *Begonia jayaensis* Kiew.
- **IUCN category: EN B2ab(iii).** Known only from three sites on limestone, with an extremely local distribution (Kiew, 2005).

Begonia jiewhoei Kiew [§ Petermannia], Begonias Penins. Malaysia 123 (2005). – Type: Peninsular Malaysia, Kelantan, Gua Musang, 14 iii 2000, *R. Kiew* 4916 (holo SING *n.v.*; iso K *n.v.*, KEP *n.v.*).

- **PENINSULAR MALAYSIA:** Kelantan, Bertam, *UNESCO Limestone Expedition s.n.* (para SING *n.v.*) – Type of *Begonia jiewhoei* Kiew; Kelantan, Gua Musang, 14 iii 2000, *R. Kiew* 4916 (holo SING *n.v.*; iso K *n.v.*, KEP *n.v.*) – Type of *Begonia jiewhoei* Kiew.
- **IUCN category: EN B2ab(iii).** Known only from two localities near a road, and described as 'very rare' (Kiew, 2005).

Begonia josephii A. DC. [§ Diploclinium], Ann. Sci. Nat. Bot., IV 11: 126 (1859); Candolle, Prodr. 15(1): 313 (1864); Clarke, Fl. Brit. Ind. 2: 639 (1879); Hara, Fl. E. Himalaya 214 (1966); Hara, Fl. E. Himalaya 2: 84 (1971); Grierson, Fl. Bhutan 2: 240 (1991). – Type: not located.

Begonia scutata Wall. *nom. nud.*, Num. List 129: 3686 (1831); Golding, Phytologia 40: 17 (1978). – Type: not located.

- **BURMA:** Rit Jawing, 12 viii 1919, *F.K. Ward* 3508 (E).
- **IUCN category: DD.** Insufficient specimens could be georeferenced with certainty.

Begonia kachak K.G. Pearce [§ Petermannia], Gard. Bull. Singapore 55: 74 (2003). – Type: Borneo, Sarawak, Gua Niah, *J.A.R. Anderson, S. Tan, E. Wright* S26074 (holo SAR *n.v.*).

- **BORNEO: Sarawak:** Niah Caves, *Alphonso, Samsuri* A217 (para SING *n.v.*) – Type of *Begonia kachak* K.G. Pearce; Niah Caves, *Alphonso, Samsuri* A222 (para SING *n.v.*) – Type of *Begonia kachak* K.G. Pearce; Niah Caves, *Alphonso, Samsuri* A248 (para SING *n.v.*) – Type of *Begonia kachak* K.G. Pearce; Gunung Subis, *B.L. Burtt, P.J.B. Woods* B2010 (para SAR *n.v.*) – Type of *Begonia kachak* K.G. Pearce; *Mohidin* S21603 (para SAR *n.v.*) – Type of *Begonia kachak* K.G. Pearce; Gua Niah, *J.A.R. Anderson, S. Tan, E. Wright* S26074 (holo SAR *n.v.*) – Type of *Begonia kachak* K.G. Pearce; Gua Pangomah, *J.A.R. Anderson* S31691 (para SAR *n.v.*, SING *n.v.*) – Type of *Begonia kachak* K.G. Pearce; Gunung Brangin, *Yii Puan Ching* S40168 (para SAR *n.v.*) – Type of *Begonia kachak* K.G. Pearce; Great Cave, *K.G. Pearce, Remlanto* S78536 (para SAR *n.v.*) – Type of *Begonia kachak* K.G. Pearce; Great Cave, *Jemree Sabli* S89062 (para SAR *n.v.*) – Type of *Begonia kachak* K.G. Pearce; Sg. Subis, *K.G. Pearce* S89463 (para SAR *n.v.*) – Type of *Begonia kachak* K.G. Pearce.
- **IUCN category: LC.** Occurs within the Niah National Park, where it is described as 'locally abundant' (Pearce, 2003), and also known from two other limestone localities (G. Brangin and Gua Pangomah).

Begonia kaniensis Irmsch. [§ Diploclinium], Bot. Jahrb. Syst. 50: 373 (1913). – Type: New Guinea, Papua New Guinea, Kani Range, 7 x 1907, *F.R.R. Schlechter* 16645 (syn B); New Guinea, Papua New Guinea, Kani Range, *F.R.R. Schlechter* 20370 (syn B, P).

- **NEW GUINEA: Papua:** Mt. Jaya, *T.M.A. Utteridge et al.* 348 (L); Eipomek-Tal, 1 iii 1976, *P. Hiepko, W. Schultze-Motel* 1186 (B[2]); Gunung Tanampi, 12 viii 1997, *E. Widjaja* 6817 (L). **Papua New Guinea:** Morobe Distr., Yungziang, iv 1936, *J. & M.S. Clemens s.n.* (B); Doma, 29 x 1962, *P.J.B. Woods* 192 (E); Doma, 23 xi 1962, *P.J.B. Woods* 378 (E); Enga Prov., ii 1982, *T.M. Reeve* 590 (E); Madang Prov., Kaironk Village, 12 vi 1995, *G. Weiblen* 964 (L); Morobe Distr., Yungziang, 16 iv 1936, *Clemens* 2306 (B); Morobe Distr., Yungziang, 6 vii 1936, *Clemens* 3556 (B[2]); Morobe Distr., Sambanga, 26 xi 1937, *Clemens* 7775b (B); Morobe Distr., Samanzing, 15 x 1938, *Clemens* 8983 (B); Kaiser-Wilhelmsland, Schraderberg, 1 vi 1913, *C.L. Ledermann* 11900a (B); Chimbu Province, Crater Mt. Wildlife Management area, 22 vii 1998, *W. Takeuchi* 12458 (L); Lala River, 1935, *C.E. Carr* 14000 (B, BM); Kani Range, 7 x 1907, *F.R.R. Schlechter* 16645 (syn B) – Type of *Begonia kaniensis* Irmsch.; Eastern Highlands Distr., Warrapuri River, 2 ix 1963, *P.v. Royen* 18129 (E); Kani Range, *F.R.R. Schlechter* 20370 (syn B, P) – Type of *Begonia kaniensis* Irmsch.; Morobe Distr., Kasanombe, 29 viii 1973, *P. Katik, K. Taho* 36996 (E); Southern Highlands Distr., Mt. Giluwe, 28 xii 1973, *J. Croft et al.* 60882 (A, E).
- **Notes:** A lianescent species which shows interesting convergences with other climbing *Begonia*, such as symmetric leaves (§ *Gobenia* from South America and § *Cristasemen* from Africa) and seeds with air-sacs at the ends (§ *Cristasemen*) (see illustration on page 63).
- **IUCN category: LC**. One of the most widespread and frequently collected species in New Guinea.

Begonia kasutensis K.G. Pearce [§ Petermannia], Gard. Bull. Singapore 55: 82 (2003). – Type: Borneo, Sarawak, Gua Niah, *J.A.R. Anderson* S31940 (holo SAR *n.v.*; iso SING *n.v.*).

- **BORNEO: Sarawak:** Subis, *Ahmad* 3 (para SAR *n.v.*, SING *n.v.*) – Type of *Begonia kasutensis* K.G. Pearce; Niah Caves, *Ahmad* 65 (para SING *n.v.*) – Type of *Begonia kasutensis* K.G. Pearce; Gunung Subis, *S. Tan, E. Wright* S27269 (para SAR *n.v.*, SING *n.v.*) – Type of *Begonia kasutensis* K.G. Pearce; Gunung Subis, 30 iv 1972, *J.A.R. Anderson* 31940 (E); Gua Niah, *J.A.R. Anderson* S31940 (holo SAR *n.v.*; iso SING *n.v.*) – Type of *Begonia kasutensis* K.G. Pearce; Bukit Kasut, *K.G. Pearce, Remlanto* S78596 (para SAR *n.v.*) – Type of *Begonia kasutensis* K.G. Pearce; Gunung Subis, *Jemree Sabli* S89049 (para SAR *n.v.*) – Type of *Begonia kasutensis* K.G. Pearce.
- **IUCN category: LC**. Occurs within the Niah National Park, and so should be considered LC as long as there is no degradation or encroachment.

Begonia keeana Kiew [§ Petermannia], Gard. Bull. Singapore 53: 265 (2001). – Type: Borneo, Sabah, Lahad Datu District, Segama River, Tempadong, *R. Kiew* RK4766 (holo SAN *n.v.*; iso K *n.v.*, SING *n.v.*).

- **BORNEO: Sabah:** Kinabatangan, Tabin Wildlife Reserve, 17 x 2000, *A. Poulsen* 1668 (E, L); Lahad Datu District, Segama River, Tempadong, *R. Kiew* RK4766 (holo SAN *n.v.*; iso K *n.v.*, SING *n.v.*) – Type of *Begonia keeana* Kiew.
- **IUCN category: LC**. The main stronghold for this species is the Tabin Wildlife Reserve. Any logging or degradation of this area would immediately lead to a threat level of VU or EN.

Begonia keithii Kiew [§ Petermannia], Gard. Bull. Singapore 50: 189 (1998). – Type: Borneo, Sabah, Semporna, Batu Tengar Cave, *R. Kiew* RK4327 (holo SING *n.v.*; iso K *n.v.*, KEP *n.v.*, L *n.v.*, SAN *n.v.*, SAR *n.v.*).

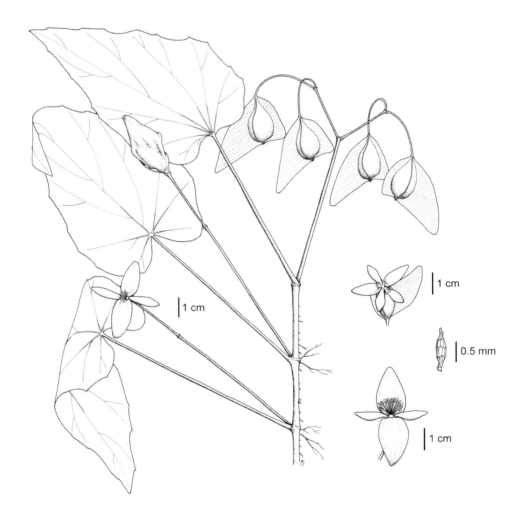

Begonia kaniensis

■ **BORNEO: Sabah:** Semporna, Batu Tengar Cave, *R. Kiew* RK4327 (holo SING; iso K *n.v.*, KEP *n.v.*, L *n.v.*, SAN *n.v.*, SAR *n.v.*) – Type of *Begonia keithii* Kiew.
■ **IUCN category: CR B2ab(iii)**. Known from only one locality, which has been proposed in the past as a commercial source of limestone. Until the status of this area within the Semporan Reserve is clarified, this species must be considered CR.

Begonia kelliana Irmsch. [§ Petermannia], Notizbl. Bot. Gart. Berlin–Dahlem 7: 102 (1917). – Type: New Guinea, Papua, 1910, *K. Gjellerup s.n.* (holo not located).
■ **NEW GUINEA: Papua:** 1910, *K. Gjellerup s.n.* (holo not located) – Type of *Begonia kelliana* Irmsch.; Koeria, i 1914, *R.F. Janowski* 584 (L) [as *Begonia cf. kelliana*]; Idenburg River, Bernhard Camp, iii 1939, *L.J. Brass* 13336 (A, BO); Idenburg River, Bernhard Camp, iv 1939, *L.J. Brass* 13748 (A, BM, L). **Papua New Guinea:** Aitape Subdistr., Wantipe Village, 26 vii 1961, *P.J. Darbyshire, R.D. Hoogland* 8260 (A, B, BM, E, L); Aitape Subdistr., Wantipe Village, 4 viii 1961, *P.J. Darbyshire, R.D. Hoogland* 8390 (L).
■ **IUCN category: LC**. Described as 'plentiful in rainforests of lower mountain slopes' (*Brass* 13748 (A)).

Begonia kerrii Craib [§ Diploclinium], Bull. Misc. Inform. Kew 1911: 57 (1911); Gagnepain, in Lecomte (ed.), Fl. Indo-Chine 2: 1118 (1921); Craib, Fl. Siam. 1: 775 (1931). – Type: Thailand, Ban Kan, 10 xii 1909, *A.F.G. Kerr* 508a (holo K; iso BM).
■ **THAILAND:** Ban Kan, 10 xii 1909, *A.F.G. Kerr* 508a (holo K; iso BM) – Type of *Begonia kerrii* Craib; Chiangmai, Doi Chiang Dao Animal Sanctuary, 8 x 1995, *J.F. Maxwell* 95-854 (A); Kang Tana N.P., 22 viii 2001, *R. Pooma, W.J.J.O. de Wilde, B.E.E. Duyfjes* 2362 (L) [as *Begonia cf. kerrii*]; Meh Ping Rapids, 22 xii 1913, *A.F.G. Kerr* 3043a (BM, K).
■ **IUCN category: DD**. Insufficient specimens could be georeferenced with certainty.

Begonia kerstingii Irmsch. [§ Petermannia], Bot. Jahrb. Syst. 50: 345 (1913). – Type: New Guinea, Papua New Guinea, Ortzen Mountains, Tajomanna, 15 v 1896, *O. Kersting* 2132 (syn B); New Guinea, Papua New Guinea, Ortzen Mountains, 17 v 1896, *C.A.G. Lauterbach* 2154 (syn B); New Guinea, Papua New Guinea, Bismarck Mts., 7 vii 1899, *H. Rodatz, T.F. Bevan* 243 (syn B[2]).
Begonia spilotophylla auct. non F. Muell.: Warburg, Fl. Deutsch. Schutzgeb. Südsee 547 (1901).
■ **NEW GUINEA: Papua New Guinea:** Bismarck Mts., 7 vii 1899, *H. Rodatz, T.F. Bevan* 243 (syn B[2]) – Type of *Begonia kerstingii* Irmsch.; Ortzen Mountains, Tajomanna, 15 v 1896, *O. Kersting* 2132 (syn B) – Type of *Begonia kerstingii* Irmsch.; Ortzen Mountains, 17 v 1896, *C.A.G. Lauterbach* 2154 (syn B) – Type of *Begonia kerstingii* Irmsch.; Hunsteinspitze, 9 viii 1912, *C.L. Ledermann* 8173 (B[2]).
■ **IUCN category: LC**. Occurs at a range of altitudes (200–1100 m) on the central mountains of New Guinea.

Begonia kinabaluensis Sands [§ Petermannia], Kew Mag. 7: 72 (1990); Sands, Pl. Mt. Kinabalu 158 (2001). – Type: Borneo, Sabah, Mt. Kinabalu, Liwagu River, 16 viii 1984, 6°5′N, 116°33′E, *M.J.S. Sands* 3890 (holo K).
■ **BORNEO: Sabah:** Mt. Kinabalu, Marai Parai, 18 ix 1958, *S.H. Collenette* A53 (BM); Mt. Kinabalu, 24 iii 1982, *I.W.J. Sinclair* 183 (E); Mt. Kinabalu, Ulu Langanani, 4 viii 1961, *W.L. Chew, E.J.H. Corner, A. Stainton* 1234 (K); Mt. Kinabalu, Tenompok F.R., 7 vi 1952, *Kadir* 1677 (K); Mt. Kinabalu, Liwagu River, 16 viii 1984, *M.J.S. Sands* 3890 (holo K) – Type of *Begonia kinabaluensis* Sands; Kiau, ii 1910, *L.S. Gibbs* 3948 (BM); Tenom, Crocker Range, 3 iv 1984, *J.H. Beaman et al.* 9166 (L); Mt. Kinabalu, Penosok Plateau, 21 vii 1963, *H.P. Fuchs et al.* 21105 (L); Mt. Kinabalu, Penosok Plateau,

8 ix 1963, *H.P. Fuchs et al.* 21679 (L); Ranau District, Liwagu Kundasang, 21 ii 1962, *W. Meijer* 27673 (K); Mt. Kinabalu, Tenompok F.R., 5 ii 1932, *J. & M.S. Clemens* 28248 (BM, K, L); Mt. Kinabalu, Penibukan, Pinokkok Falls, 1 iii 1932, *J. & M.S. Clemens* 31875 (BM); Mt. Kinabalu, Dehobang Falls, 11 ix 1933, *J. & M.S. Clemens* 40288 (BM, K); Mt. Kinabalu, Penibukan, Pinokkok Falls, *J. & M.S. Clemens* 50005A (BM).
- **IUCN category: LC**. Commonly collected on Mt. Kinabalu and also recorded from the Crocker Range.

Begonia kingdon-wardii Tebbitt [§ Sphenanthera], Kew Bull. 62: 143 (2007). – Type: Burma, Kachin Hills, 27°20′N, 97°30′E, *F.K. Ward* 7341 (holo K *n.v.*; iso K *n.v.*).
- **BURMA:** Kachin Hills, 10 xii 1937, *F.K. Ward* 13569 (para BM *n.v.*) – Type of *Begonia kingdon-wardii* Tebbitt; *ibid.*, *F.K. Ward* 7341 (holo K *n.v.*; iso K *n.v.*) – Type of *Begonia kingdon-wardii* Tebbitt.
- **IUCN category: DD** (Tebbitt, 2007).

Begonia kingiana Irmsch. [§ Ridleyella], Mitt. Inst. Allg. Bot. Hamburg 8: 106 (1929); Kiew, Begonias Penins. Malaysia 37 (2005). – Type: Peninsular Malaysia, Perak, Kuala Dipang, ix 1890, *H.N. Ridley* 9689 (lecto SING *n.v.*; isolecto B).
- **PENINSULAR MALAYSIA:** Perak, *C. Curtis s.n.* (K); Kelantan, Gua Musang, 6 viii 1962, *Unesco Limestone Expedition* 395 (L); Perak, Tambun, 10 iii 1971, *S.C. Chin* 845 (L); Selangor, Batu Caves, viii 1984, *R. Kiew* 1341 (L); Pahang, Merapoh, Gua, 13 viii 1971, *S.C. Chin* 1490 (L); Kelantan, Gua Musang, 16 viii 1971, *S.C. Chin* 1598 (L); Perak, *B. Scortechini* 1607 (B, BM, L); P. Langgun, 8 xii 1974, *M.M.J.v. Balgooy* 2230 (L) [as *Begonia cf. kingiana*]; Kelantan, Gunung Renayang, 8 v 1990, *R. Kiew, S. Anthonysamy* 2865 (L); ix 1885, *Kings Collector* 8245 (P); Perak, Kuala Dipang, ix 1890, *H.N. Ridley* 9689 (lecto SING *n.v.*; isolecto B) – Type of *Begonia kingiana* Irmsch.
- **IUCN category: LC**. Stated as being 'common and found on most limestone hills' (Kiew, 2005). Quarrying of any of these sites would immediately lead to a threat level of VU or EN for this species.

Begonia klemmei Merr. [§ Diploclinium], Philipp. J. Sci. 6: 402 ('1911', 1912); Merrill, Enum. Philipp. Fl. Pl. 3: 123 (1923). – Type: Philippines, Luzon, Lepanto District, Sagada, 8 xi 1906, *W. Klemme* BS5677 (not located).
- **PHILIPPINES: Luzon:** Benguet, Baguio, 24 ix 1904, *R.S. Williams* 1318 (para K) – Type of *Begonia klemmei* Merr.; Lepanto District, Sagada, 8 xi 1906, *W. Klemme* BS5677 (not located) – Type of *Begonia klemmei* Merr.
- **IUCN category: DD**. Insufficient specimens could be georeferenced with certainty.

Begonia klossii Ridl. [§ Platycentrum], J. Linn. Soc., Bot. 41: 290 (1913); Ridley, Fl. Malay Penins. 1: 861 (1922); Irmscher, Mitt. Inst. Allg. Bot. Hamburg 8: 133 (1929); Kiew, Begonias Penins. Malaysia 164 (2005). – Type: Peninsular Malaysia, Selangor, Menuang Gasing, ii 1912, *C.B. Kloss s.n.* (holo K *n.v.*).
- **PENINSULAR MALAYSIA:** Selangor, Menuang Gasing, ii 1912, *C.B. Kloss s.n.* (holo K *n.v.*) – Type of *Begonia klossii* Ridl.; Selangor, Batu Caves, *R. Kiew* 5200 (SING *n.v.*); Selangor, Sungei Pisang, *R. Kiew* 3251 (SING *n.v.*); Selangor, Ulu Yam-Batu Dam Road, *R. Kiew* 5285 (SING *n.v.*).
- **IUCN category: VU D2**. A narrow endemic in the Gombak Valley.

Begonia koksunii Kiew [§ Platycentrum], Begonias Penins. Malaysia 153 (2005). – Type: Peninsular Malaysia, Gerik, Sungai Mangga, *A. Piee, K.-S. Yap* RK5212 (holo SING *n.v.*).

- **PENINSULAR MALAYSIA:** Gerik, Sungai Mangga, viii 2002, *A. Piee, K.-S. Yap s.n.* (para SING *n.v.*) – Type of *Begonia koksunii* Kiew; Gerik, Sungai Mangga, *A. Piee, K.-S. Yap* RK5212 (holo SING *n.v.*) – Type of *Begonia koksunii* Kiew.
- **IUCN category: CR B2ab(iii).** Known from only one locality, which is threatened by logging (Kiew, 2005).

Begonia koordersii Warb. ex L.B. Sm. & Wassh. [§ Petermannia], Phytologia 52(7): 444 (1983). – Type: Sulawesi, Minahassa, 1 iii 1895, *S.H. Koorders* 16246B (lecto K; isolecto B, BO).
Begonia koordersii Warb. *nom. nud.*, Natuurw. Tijdschr. Ned.-Indië 63: 91 (1904); Koorders-Schumacher, Suppl. Fl. Celebes 3: 47 (1922); Smith & Wasshausen, Phytologia 52(7): 444 (1983).
- **SULAWESI:** *J.G.F. Riedel s.n.* (K, P); *A.B. Meyer s.n.* (B); Goeroepahi, 18 iii 1917, *Kauderns* 18 (L) [as *Begonia koordersii* ?]; Manado, *unknown* 41 (B); Gorantalo, Gunung Boliohutu, 22 iv 2002, *M. Mendum, H.J. Atkins, M. Newman, Hendrian, A. Sofyan* 141 (E); Minahassa, Wiau complex, 29 vi 1956, *L.L. Forman* 319 (K); Minahassa, Tomohon, 11 vii 1894, *K.F. & P.B. Sarasin* 400 (B); Sidaoento, 12 vii 1939, *S. Bloembergen* 4200 (L); Manado, *J.E. Teijsmann* 5338 (BO); Batui River, 15 x 1989, *M. Coode* 5945 (L); Bolaang-Mongondow Distr., Dumoga Bone National Park, 15 iii 1985, *E.F. de Vogel, J.J. Vermeulen* 6556 (L); Sungai Pehoeia, 25 iv 1975, *W. Meijer* 9603 (L); Minahassa, Bojong, *O. Warburg* 15189 (B); Ratatotok, 20 iii 1895, *S.H. Koorders* 16242B (BO *n.v.*); Minahassa, 1 iii 1895, *S.H. Koorders* 16246B (lecto K; isolecto B, BO) – Type of *Begonia koordersii* Warb. ex L.B. Sm. & Wassh.
- **Notes:** See notes under *B. rieckei* Warb.
- **IUCN category: LC.** This species is widespread in Sulawesi.

Begonia kui C.-I Peng [§ Coelocentrum], Botanical Studies 48: 127 (2007). – Type: Vietnam, Thai Nguyen Prov., 17 vii 2006, *C.I. Peng* 20847 (holo HAST *n.v.*; iso A *n.v.*, HN *n.v.*, IBK *n.v.*, MO *n.v.*, NY *n.v.*).
- **VIETNAM:** Thai Nguyen Prov., 17 vii 2006, *C.I. Peng* 20847 (holo HAST *n.v.*; iso A *n.v.*, HN *n.v.*, IBK *n.v.*, MO *n.v.*, NY *n.v.*) – Type of *Begonia kui* C.-I Peng.
- **IUCN category: DD.** Insufficient specimens could be georeferenced with certainty.

Begonia labordei H. Lév. [§ Diploclinium], Bull. Soc. Agric. Sarthe 39: 323 (1904). – Type: China, Kouy-Tcheou, 8 x 1897, *Martin, E. Bodinier* 1952 (syn E *n.v.*); China, Kouy-Tcheou, 11 ix 1899, *J. Laborde, E. Bodinier* 1952bis (syn P *n.v.*).
- **BURMA:** Sause Gorge – Sadon, ix 1912, *G. Forrest* 9125 (E); Myitkyina District, 15 ix 1938, *Naw Mu Pa* 17464 (K); Hpimaw Pass, viii 1924, *G. Forrest* 24835 (E).
   VIETNAM: Laoke, Chapa, Song ta Van, viii 1936, *P.A. Petelot* 7086 (B *n.v.*).
- **IUCN category: LC.** Widely collected in Yunnan as well as occurring in Burma and Vietnam, at altitudes ranging from 800 to 3300 m.

Begonia laccophora Sands [not placed to section], in Coode *et al.*, Checkl. Fl. Pl. Gymnosperms Brunei Darussalam App. 2: 434 (1997). – Type: Borneo, Brunei, Temburong, 25 iv 1992, *R.J. Johns* 7303 (holo K; iso BRUN *n.v.*).
- **BORNEO: Brunei:** Temburong, 25 iv 1992, *R.J. Johns* 7303 (holo K; iso BRUN *n.v.*) – Type of *Begonia laccophora* Sands.
- **IUCN category: VU D2.** Known from only one locality.

Begonia lacera Merr. [§ Petermannia], Philipp. J. Sci. 10: 49 (1915); Merrill, Enum. Philipp. Fl. Pl. 3: 123 (1923). – Type: Philippines, Mindanao, Basilan, Cumalarang River, *J. Reillo* 16111 (syn B).

- **PHILIPPINES: Mindanao:** Basilan, Cumalarang River, *J. Reillo* 16111 (syn B) – Type of *Begonia lacera* Merr.
- **IUCN category: DD.** Insufficient specimens could be georeferenced with certainty.

Begonia laevis Ridl. [§ Platycentrum], J. Fed. Malay States Mus. 8(4): 39 (1917). – Type: Sumatra, Korinchi, Kormeli, Sungei Kumbang, 18 iii 1914, *H.C. Robinson, C.B. Kloss s.n.* (syn BM, K).
- **SUMATRA:** Korinchi, Kormeli, Sungei Kumbang, 18 iii 1914, *H.C. Robinson, C.B. Kloss s.n.* (syn BM, K) – Type of *Begonia laevis* Ridl.; Korinchi, Kormeli, Sungei Kumbang, 1 iv 1914, *H.C. Robinson, C.B. Kloss s.n.* (BM); Gunong Batu Lopang, 8 vii 1972, *W.J.J.O. de Wilde* 13497 (BO).
- **Notes:** Originally placed in § *Petermannia* in error; this species patently belongs in § *Platycentrum* and is close to *B. teysmanniana* and *B. altissima*.
- **IUCN category: LC.** Occurs within the Kerinci-Sebalt National Park, at relatively high altitudes where forest cover is still largely intact.

Begonia lagunensis Elmer [§ Petermannia], Leafl. Philipp. Bot. 2: 735 (1910); Merrill, Philipp. J. Sci. 6: 406 ('1911', 1912); Merrill, Enum. Philipp. Fl. Pl. 3: 123 (1923). – Type: Philippines, Luzon, Tayabas, Lucban, *A.D.E. Elmer* 7467 (syn E, L); Philippines, Luzon, Tayabas, Lucban, v 1907, *A.D.E. Elmer* 9327 (syn BM, E, L).
- **PHILIPPINES: Luzon:** Zambales, ii 1906, *A. Loher* 6092 (B) [as *Begonia cf. lagunensis*]; Laguna, Makiling, ix 1910, *E.D. Merrill* 7136 (E); Tayabas, Lucban, *A.D.E. Elmer* 7467 (syn E, L) – Type of *Begonia lagunensis* Elmer; Tayabas, Lucban, v 1907, *A.D.E. Elmer* 9327 (syn BM, E, L) – Type of *Begonia lagunensis* Elmer; Tayabas, Infanta, viii 1909, *C.B. Robinson* 9440 (B); Laguna, Makiling, xi 1912, *V. Servinas* 16893 (BM[2], K, L, P); Laguna, Makiling, 6 xii 1912 – 9 xii 1912, *C.B. Robinson* 17063 (P); Laguna, Makiling, *A.D.E. Elmer* 18017 (BM); Laguna, Makiling, vi 1917 – vii 1917, *A.D.E. Elmer* 18429 (BM[2], K, L, P, U); Camarines, Mt. Isaro(g), xi 1913 – xii 1913, *M. Ramos* 22110 (BM, K, L, P); Albay Province, Mt. Malinao, 3 ii 1956, *G. Edano* 34506 (K); Sorsogon, Mt. Bulusan, v 1957, *G. Edano, H. Gutierrez* 37847 (K). **Mindoro:** Mt. Halcon, *C.E. Ridsdale* 1751 (K); Mt. Halcon, xi 1906, *E.D. Merrill* 6135 (B, P). **Negros:** Canlaon Volcano, iv 1910, *E.D. Merrill* 6981 (B). **Leyte:** *M. Ramos* 15216 (B, P) [as *Begonia lagunensis* ?]. **Samar:** iii 1914 – iv 1914, *M. Ramos* 17561 (BM, K).
- **Notes:** See notes under *B. aequata* A. Gray.
- **IUCN category: DD.** Taxonomic uncertainty.

Begonia lailana Kiew & Geri [§ Petermannia], Gard. Bull. Singapore 55: 117 (2003). – Type: Borneo, Sarawak, Bau, Gunung Kawa, *C. Geri* SBC3753 (holo SAR *n.v.*; iso K *n.v.*, L *n.v.*, SAN *n.v.*, SBC *n.v.*, SING *n.v.*).
- **BORNEO: Sarawak:** Bau, Gunung Poing, *Julia et al.* SBC353 (para SBC *n.v.*) – Type of *Begonia lailana* Kiew & Geri; Kuching – Bau, 4 iv 1984, *K.G. Pearce* 427 (K); Bau, Gunung Poing, *Malcolm et al.* SBC1555 (para SBC *n.v.*) – Type of *Begonia lailana* Kiew & Geri; Bau, Gunung Doya, *Julia et al.* SBC2055 (para SBC *n.v.*) – Type of *Begonia lailana* Kiew & Geri; Bau, Gunung Batu, *Meekiong et al.* SBC2136 (para SBC *n.v.*) – Type of *Begonia lailana* Kiew & Geri; Bau, Gunung Lanyang, *Julia et al.* SBC2901 (para SBC *n.v.*, SING *n.v.*) – Type of *Begonia lailana* Kiew & Geri; Bau, Gunung Kawa, *Meekiong et al.* SBC3114 (para SBC *n.v.*) – Type of *Begonia lailana* Kiew & Geri; Bau, Gunung Aup, *Malcolm et al.* SBC3303 (para SBC *n.v.*) – Type of *Begonia lailana* Kiew & Geri; Bau, Gunung Podam, *Malcolm et al.* SBC3364 (para SBC *n.v.*) – Type of *Begonia lailana* Kiew & Geri; Bau, Gunung Tabai, *Julia et al.* SBC3414 (para SBC *n.v.*) – Type of *Begonia lailana* Kiew & Geri; Bau, Gunung Doya, *Raymond, M. Mendum* SBC3451 (para SBC *n.v.*) – Type of *Begonia lailana* Kiew & Geri; Bau, Gunung Doya, *Raymond, M. Mendum* SBC3473 (para SBC *n.v.*) – Type of *Begonia lailana* Kiew & Geri; Bau, Gunung Kawa, *C. Geri et al.* SBC3592 (para SBC *n.v.*) – Type of *Begonia*

lailana Kiew & Geri; Bau, Gunung Kawa, *C. Geri* SBC3753 (holo SAR *n.v.*; iso K *n.v.*, L *n.v.*, SAN *n.v.*, SBC *n.v.*, SING *n.v.*) – Type of *Begonia lailana* Kiew & Geri; Bau, Gunung Poing, *C. Geri* SBC3755 (para SBC *n.v.*) – Type of *Begonia lailana* Kiew & Geri; Bau, Gunung Tabai, *C. Geri* SBC3757 (para SBC *n.v.*) – Type of *Begonia lailana* Kiew & Geri; Bau, Gunung Aup, *C. Geri et al.* SBC6756 (para SBC *n.v.*) – Type of *Begonia lailana* Kiew & Geri.

- **IUCN category: NT**. Although has a very restricted distribution in the Kuching limestone area, the protologue lists it as locally common at three localities. There is a record from an orchard, indicating some tolerance of secondary habitats. However, the area it is recorded from is not under formal protection.

Begonia lambii Kiew [§ Petermannia], Gard. Bull. Singapore 53: 267 (2001). – Type: Borneo, Sabah, Pensiangan District, Batu Tinahas, *R. Kiew, S. Anthonysamy* RK4405 (holo SAN *n.v.*; iso K *n.v.*, SING *n.v.*).

- **BORNEO: Sabah:** Pensiangan District, Batu Punggul, *A. Ibrahim* 135 (para SING *n.v.*) – Type of *Begonia lambii* Kiew; Pensiangan District, Sapulut, *R. Kiew, S. Anthonysamy* 4345 (para SING *n.v.*) – Type of *Begonia lambii* Kiew; Pensiangan District, Batu Tinahas, *R. Kiew, S. Anthonysamy* RK4405 (holo SAN *n.v.*; iso K *n.v.*, SING *n.v.*) – Type of *Begonia lambii* Kiew; Pensiangan District, Batu Tinahas, *R. Kiew, S. Anthonysamy* 4408 (para K *n.v.*, SAN *n.v.*, SING *n.v.*) – Type of *Begonia lambii* Kiew; Pensiangan District, Batu Punggul, *S. Jimpin* SAN135991 (para SAN *n.v.*) – Type of *Begonia lambii* Kiew; Lahad Datu District, Ulu Segama, 13 iii 1987, *G.C.G. Argent et al.* 1987148 (E[2]) [as *Begonia aff. lambii*].
- **Notes:** The collection by Argent *et al.* (1987148) probably represents a related but undescribed taxon.
- **IUCN category: EN B2ab(iii)**. Known from four localities in lowland forests, none of which is under formal protection.

Begonia lancifolia Merr. [§ Petermannia], Philipp. J. Sci. 10: 48 (1915); Merrill, Enum. Philipp. Fl. Pl. 3: 123 (1923). – Type: Philippines, Mindanao, Basilan, Cumalarang River, *J. Reillo* 16162 (syn B).

- **PHILIPPINES: Mindanao:** Basilan, Cumalarang River, *J. Reillo* 16162 (syn B) – Type of *Begonia lancifolia* Merr.
- **IUCN category: DD**. Insufficient specimens could be georeferenced with certainty.

Begonia lancilimba Merr. [§ Diploclinium], Philipp. J. Sci. 14: 424 (1919); Merrill, Enum. Philipp. Fl. Pl. 3: 123 (1923). – Type: Philippines, Panay, Antique Province, Culasi, vi 1914 – vii 1914, *R.C. McGregor* 32232 (holo P).

- **PHILIPPINES: Panay:** Antique Province, Culasi, vi 1914 – vii 1914, *R.C. McGregor* 32232 (holo P) – Type of *Begonia lancilimba* Merr.; Antique Province, v 1918 – viii 1918, *R.C. McGregor* 32286 (para BM, BO, K) – Type of *Begonia lancilimba* Merr.; Antique Province, v 1918 – viii 1918, *R.C. McGregor* 32570 (para K) – Type of *Begonia lancilimba* Merr.
- **IUCN category: DD**. Insufficient specimens could be georeferenced with certainty.

Begonia langbianensis Baker f. [§ Platycentrum], J. Nat. Hist. Soc. Siam. 4: 133 (1921). – Type: Vietnam, Annam, Lang Bian, iv 1918, *C.B. Kloss s.n.* (holo BM).

- **VIETNAM:** Annam, Lang Bian, iv 1918, *C.B. Kloss s.n.* (holo BM) – Type of *Begonia langbianensis* Baker f.
- **IUCN category: DD**. Insufficient specimens could be georeferenced with certainty.

Begonia latistipula Merr. [§ Petermannia], Philipp. J. Sci. 10: 51 (1915); Merrill, Enum. Philipp. Fl. Pl. 3: 123 (1923). – Type: Philippines, Leyte, Dagami, viii 1912, *M. Ramos* 15367 (syn B, BM).

- **PHILIPPINES: Leyte:** iii 1914, *C.A. Wenzel* 1024 (BM); Dagami, viii 1912, *M. Ramos* 15367 (syn B, BM) – Type of *Begonia latistipula* Merr.; Biliran, vi 1914, *R.C. McGregor* 18734 (K); Biliran, vi 1914, *R.C. McGregor* 18736 (P); Mt. Abucayan, ii 1923, *G. Edano* 41728 (B, K, P); Mt. Abucayan, ii 1923, *G. Edano* 41771 (P). **Samar:** iii 1914 – iv 1914, *M. Ramos* 17516 (BM); Catubig River, ii 1916 – iii 1916, *M. Ramos* 24484 (K, P).
- **IUCN category: EN B2ab(iii).** The Mindanao–Eastern Visayas rain forest eco-region to which this species is endemic has been severely degraded.

Begonia lauterbachii Warb. [§ Petermannia], Fl. Deutsch. Schutzgeb. Südsee 458 (1901); Irmscher, Bot. Jahrb. Syst. 78: 175 (1959). – Type: New Guinea, Papua New Guinea, Kaiser-Wilhelmsland, 7 vi 1896, *C.A.G. Lauterbach* 275 (syn not located); New Guinea, Papua New Guinea, Kaiser-Wilhelmsland, 7 vi 1896, *C.A.G. Lauterbach* 285 (syn not located); New Guinea, Papua New Guinea, Kaiser-Wilhelmsland, 3 v 1890, *C.A.G. Lauterbach* 64 (syn B); New Guinea, Papua New Guinea, Kaiser-Wilhelmsland, Nuru Basin, 7 vi 1896, *C.A.G. Lauterbach* 2275 (syn B, L).

- **NEW GUINEA: Papua New Guinea:** Kaiser-Wilhelmsland, 3 v 1890, *C.A.G. Lauterbach* 64 (syn B) – Type of *Begonia lauterbachii* Warb.; Kaiser-Wilhelmsland, Simbang, 7 i 1894, *L. Karnbach* 113 (B[2]); Kaiser-Wilhelmsland, 1889 – 1891, *K. Weinland* 168 (B[2], L); Kaiser-Wilhelmsland, 7 vi 1896, *C.A.G. Lauterbach* 275 (syn not located) – Type of *Begonia lauterbachii* Warb.; Kaiser-Wilhelmsland, 7 vi 1896, *C.A.G. Lauterbach* 285 (syn not located) – Type of *Begonia lauterbachii* Warb.; Kaiser-Wilhelmsland, Simbang, viii 1899, *E. Nyman* 797 (B); Kaiser-Wilhelmsland, Nuru Basin, 7 vi 1896, *C.A.G. Lauterbach* 2275 (syn B, L) – Type of *Begonia lauterbachii* Warb.; Kaiser-Wilhelmsland, 8 vi 1896, *C.A.G. Lauterbach* 2285 (B); Kaiser-Wilhelmsland, 24 vi 1907, *F.R.R. Schlechter* 16175 (B).
- **IUCN category: DD.** Insufficient specimens could be georeferenced with certainty.

Begonia layang-layang Kiew [§ Petermannia], Gard. Bull. Singapore 53: 272 (2001). – Type: Borneo, Sabah, Pensiangan District, Sapulut, *R. Kiew, S. Anthonysamy* RK4441 (holo SAN *n.v.*; iso BRUN *n.v.*, K *n.v.*, L *n.v.*, SAR *n.v.*, SING *n.v.*).

- **BORNEO: Sabah:** Pensiangan District, Sapulut, *R. Kiew, S. Anthonysamy* RK4441 (holo SAN *n.v.*; iso BRUN *n.v.*, K *n.v.*, L *n.v.*, SAR *n.v.*, SING *n.v.*) – Type of *Begonia layang-layang* Kiew.
- **IUCN category: EN B2ab(iii).** Known only from one locality, not yet under formal protection.

Begonia lazat Kiew & Reza Azmi [§ Petermannia], Gard. Bull. Singapore 50: 45 (1998). – Type: Borneo, Sabah, Kinabatangan, Kampung Buang Sayang, *Reza Azmi* 206 (holo SAN *n.v.*).

- **BORNEO: Sabah:** Kinabatangan, Kampung Buang Sayang, *Reza Azmi* 206 (holo SAN *n.v.*) – Type of *Begonia lazat* Kiew & Reza Azmi.
- **IUCN category: EN B2ab(iii).** Known only from one locality, not yet under formal protection.

Begonia ledermannii Irmsch. [§ Petermannia], Bot. Jahrb. Syst. 50: 344 (1913). – Type: New Guinea, Papua New Guinea, Northeast New Guinea, *C.L. Ledermann* 7093 (holo B; iso B[2]).

- **NEW GUINEA: Papua New Guinea:** Northeast New Guinea, *C.L. Ledermann* 7093 (holo B; iso B[2]) – Type of *Begonia ledermannii* Irmsch.
- **IUCN category: DD.** Insufficient specimens could be georeferenced with certainty.

Begonia lengguanii Kiew [§ Reichenheimia], Begonias Penins. Malaysia 211 (2005). – Type: Peninsular Malaysia, Pahang, Bukit Rengit, 1 vii 1988, *Saw Leng Guan* FRI36295 (holo KEP *n.v.*).

- **PENINSULAR MALAYSIA:** Pahang, Lancang Forest Reserve, ix 1986, *B.H. Kiew s.n.* (para SING *n.v.*) – Type of *Begonia lengguanii* Kiew; Pahang, Lancang, Bukit Tapah, *A. Zainuddin* C24 (para UKMB *n.v.*) – Type of *Begonia lengguanii* Kiew; Pahang, Bukit Rengit, *R. Kiew* 5201 (para SING *n.v.*) – Type of *Begonia lengguanii* Kiew; Pahang, Bukit Rengit, 1 vii 1988, *Saw Leng Guan* FRI36295 (holo KEP *n.v.*) – Type of *Begonia lengguanii* Kiew.
- **IUCN category: VU D2**. Known only from two areas, one of which is in the Lanjang Forest Reserve.

Begonia lepida Blume [§ Bracteibegonia], Enum. Pl. Javae 1:98 (1827); Candolle, Prodr. 15(1): 317 (1864); Koorders, Exkurs.-Fl. Java 2:645 (1912); Golding & Karegeannes, Phytologia 54(7):494 (1984). – *Diploclinium lepidum* (Blume) Miq., Fl. Ned. Ind. 1(1):686 (1856). – Type: Java, *J.B. Spanoghe s.n.* (syn L).

    Knesebeckia bracteata Hassk., Hort. Bogor. Descr. 316 (1858); Candolle, Prodr. 15(1): 317 (1864). – Type: not located.

- **SUMATRA:** Sumatra, *unknown s.n.* (K); Sumatra, *unknown s.n.* (K); Siberut Island, 8 ix 1924, *C.B. Kloss* 10582 (K).

    JAVA: Java, 1851, *P.F.W. Goering s.n.* (P); Java, 1836, *unknown s.n.* (P); Java, *unknown s.n.* (L); Java, *J.B. Spanoghe s.n.* (syn L) – Type of *Begonia lepida* Blume; Java, *W. Lobb s.n.* (K); Java, *J.B. Spanoghe s.n.* (K); Mt. Salak, 17 ii 1915, *H.N. Ridley s.n.* (K); *unknown s.n.* (B); *H. Kuhl, J.C.v. Hasselt s.n.* (B); *C.L.v. Blume s.n.* (B); *O. Warburg s.n.* (B); Mt. Salak, 1912, *Noerkas s.n.* (BO); Tjidadap, 1918, *W.F. Winckel s.n.* (U); Tjidadap, 1918, *W.F. Winckel* 14B (BO); Tjidadap, *W.F. Winckel* 84B (BO); Preanger, Gunung Gede, iii 1929, *L.v.d. Pijl* 113 (BO); Preanger, Gunung Gede, viii 1922 – ix 1922, *F.A. Kramer* 124 (BO); *L.R. Lanjouw* 197 (BO); *H.O. Forbes* 223a (BM); vi 1917, *H.O. Forbes* 548 (BO); 29 v 1848, *H. Zollinger* 1270 (A, P); Preanger, Tjadas Malang, 5 ix 1916, *G.v.d. Brink* 1380 (BO); *H. Zollinger* 1495 (B[2], BM[2], K); Batavia, Megamendoeng, 1 ix 1919, *H.J. Lam* 1729 (BO); Batavia, Megamendoeng, 23 xii 1928, *C.G.G.J.v. Steenis* 2235 (BO); Tjibeber, Tjikaroem, 25 vi 1916, *Doctors v. Leeuwen-Reynvaan* 2388 (BO); Preanger, Tjadas Malang, 19 vi 1923, *R.C. Bakhuizen van den Brink* 2775 (U); Gunong Salak, *C.L.v. Blume* 6070 (BO); G. Ganiisan, 9 xii 1923, *R.C. Bakhuizen van den Brink* 6191 (BO); Batavia, G. Tjipoetik, *R.C. Bakhuizen van den Brink* 7177 (BO); Pasir Telaga Estate, 4 i 1954, *A.H.G. Alston* 12903 (A); Preanger, Tjiandjoer, Takoka, 28 ii 1894, *S.H. Koorders* 14978 (BO); Mt. Menapa, Nanggoeng, 18 xii 1940, *C.G.G.J.v. Steenis* 17406 (BO); *S.H. Koorders* 41697 (BO).

- **Notes:** See notes under *B. bracteata* Jack.
- **IUCN category: DD**. Taxonomic uncertainty.

Begonia lepidella Ridl. [§ Petermannia], J. Fed. Malay States Mus. 8(4): 40 (1917). – Type: Sumatra, Korinchi, Sandaran Agong, 29 v 1914, *H.C. Robinson, C.B. Kloss s.n.* (syn BM); Sumatra, Korinchi, Siolak Dras, 19 iii 1914, *H.C. Robinson, C.B. Kloss s.n.* (syn BM); Sumatra, Korinchi, Sungei Kumbang, 4 iv 1914, *H.C. Robinson, C.B. Kloss s.n.* (syn BM, K) – Type of *Begonia lepidella* Ridl.

- **SUMATRA:** Korinchi, Sandaran Agong, 29 v 1914, *H.C. Robinson, C.B. Kloss s.n.* (syn BM) – Type of *Begonia lepidella* Ridl.; Korinchi, Siolak Dras, 19 iii 1914, *H.C. Robinson, C.B. Kloss s.n.* (syn BM) – Type of *Begonia lepidella* Ridl.; Korinchi, Siolak Dras, 16 iii 1914, *H.C. Robinson, C.B. Kloss s.n.* (BM); Korinchi, Siolak Dras, 16 iii 1914, *H.C. Robinson, C.B. Kloss s.n.* (K); Korinchi, Sungei Kumbang, 4 iv 1914, *H.C. Robinson, C.B. Kloss s.n.* (syn BM, K) – Type of *Begonia lepidella* Ridl.; Atjeh, Gunung Leuser Nature Reserve, 16 v 1972, *de Wilde, de Wilde-Duyfies* 12010 (K); Atjeh, Gunung Leuser Nature Reserve, 16 v 1972, *de Wilde, de Wilde-Duyfies* 12129 (K); Atjeh, Gunung Leuser Nature Reserve, 16 v 1972, *de Wilde, de Wilde-Duyfies* 12578 (K).

- **IUCN category: LC**. There is still forest cover on Mt. Kerinci at the altitudes stated in the protologue (c. 1400–3000 m).

Begonia leptantha C.B. Rob. [§ Petermannia], Philipp. J. Sci. 6: 211 (1911); Merrill, Philipp. J. Sci. 6: 381 ('1911', 1912); Merrill, Enum. Philipp. Fl. Pl. 3: 124 (1923). – Type: Philippines, Luzon, Quezon, Polillo Island, viii 1909, *C.B. Robinson* 6857 (syn B, K).
- **PHILIPPINES: Luzon:** Quezon, Polillo Island, viii 1909, *C.B. Robinson* 6857 (syn B, K) – Type of *Begonia leptantha* C.B. Rob.; Quezon, Polillo Island, x 1909 – xi 1909, *R.C. McGregor* 10322 (para B, K) – Type of *Begonia leptantha* C.B. Rob.; Tayabas, Mt. Binuang, *M. Ramos, G. Edano* 28729 (BO, P); Tayabas, Mt. Binuang, v 1917, *M. Ramos, G. Edano* 28770 (BM, L, P); Tayabas, Mt. Tulaog, v 1917, *M. Ramos, G. Edano* 29151 (P).
- **IUCN category: EN B2ab(iii)**. The Luzon rain forest eco-region to which this species is endemic has been reduced to less than 25% of its original area.

Begonia leucochlora Sands [§ Petermannia], in Coode *et al.*, Checkl. Fl. Pl. Gymnosperms Brunei Darussalam App. 2: 432 (1997). – Type: Borneo, Brunei, Temburong, 23 iii 1991, 4°33′N, 115°9′E, *M.J.S. Sands* 5566 (holo K; iso BRUN *n.v.*).
- **BORNEO: Brunei:** Temburong, 23 iii 1991, *M.J.S. Sands* 5566 (holo K; iso BRUN *n.v.*) – Type of *Begonia leucochlora* Sands.
- **IUCN category: VU D2**. Although endemic to well-forested Brunei, this species is known only from one collection in lowland forest at the edge of the Ulu Temburong National Park.

Begonia leucosticta Warb. in Perkins **[§ Petermannia]**, Fragm. Fl. Philipp. 55 (1904); Merrill, Philipp. J. Sci. 6: 385 ('1911', 1912); Merrill, Enum. Philipp. Fl. Pl. 3: 124 (1923). – Type: Philippines, Luzon, Isabela Province, *O. Warburg* 12004 (holo B).
- **PHILIPPINES: Luzon:** Isabela Province, *O. Warburg* 12004 (holo B) – Type of *Begonia leucosticta* Warb.; Apayao, Mt. Sulu, v 1917, *E. Fenix* 28396 (P). **Mindanao:** Bucas Island, 4 x 1906, *E.D. Merrill* 5274 (B).
- **IUCN category: DD**. Insufficient specimens could be georeferenced with certainty, and whether the collection from Bucas is identified correctly needs further investigation.

Begonia leucotricha Sands [§ Petermannia], in Coode *et al.*, Checkl. Fl. Pl. Gymnosperms Brunei Darussalam App. 2: 434 (1997). – Type: Borneo, Brunei, Labi Subdistr., Mendaram Valley, 18 iii 1991, 4°20′N, 114°27′E, *M.J.S. Sands* 5452 with R.J. Johns (holo K; iso BRUN *n.v.*).
- **BORNEO: Brunei:** Labi Subdistr., Mendaram Valley, 18 iii 1991, *M.J.S. Sands* 5452 with R.J. Johns (holo K; iso BRUN *n.v.*) – Type of *Begonia leucotricha* Sands.
- **IUCN category: VU D2**. Although endemic to well-forested Brunei, this species is known only from one collection in lowland forest.

Begonia littleri Merr. [§ Petermannia], Philipp. J. Sci. 6: 379 ('1911', 1912); Merrill, Enum. Philipp. Fl. Pl. 3: 124 (1923). – Type: Philippines, Mindanao, Basilan, iv 1903, *de Vore, Hoover* 94 (syn not located).
- **PHILIPPINES: Mindanao:** Basilan, iv 1903, *de Vore, Hoover* 94 (syn not located) – Type of *Begonia littleri* Merr.; Basilan, vi 1910, *C.B. Robinson* 11512 (para BM, K) – Type of *Begonia littleri* Merr.; Basilan, viii 1912, *J. Reillo* 15483 (B, BM, K); Basilan, viii 1912 – ix 1912, *J. Reillo* 16161 (BM); Zamboanga, Mt. Tubuan, x 1919, *M. Ramos, G. Edano* 36648 (B, L, P).

- **Notes:** Assigned to § *Platycentrum* in Doorenbos *et al.* (1998) presumably in error; I have assigned it here to § *Petermannia* as it is obviously related to other Philippine species in that section with short petioles.
- **IUCN category: CR B2ab(iii).** Less than 2% of forest cover remains on Basilan.

Begonia lobbii A. DC. [§ Reichenheimia], Prodr. 15(1): 390 (1864). – Type: not located.
   Mitscherlichia lobbii Hassk., Hort. Bogor. Descr. 331 (1858); Candolle, Prodr. 15(1): 390 (1864). – Type: not located.
- **JAVA:** *Fide* de Candolle (1864). No specimens seen.
- **IUCN category: DD.** Insufficient specimens could be georeferenced with certainty.

Begonia loheri Merr. [§ Petermannia], Philipp. J. Sci. 6: 382 ('1911', 1912); Merrill, Enum. Philipp. Fl. Pl. 3: 124 (1923). – Type: Philippines, Luzon, Rizal, Angilog, 15 iii 1906, *A. Loher* 6090 (syn B, K); Philippines, Luzon, Rizal, Angilog, 16 iii 1906, *A. Loher* 6098 (syn K).
- **PHILIPPINES: Luzon:** Rizal, Angilog, 15 iii 1906, *A. Loher* 6090 (syn B, K) – Type of *Begonia loheri* Merr.; Rizal, Angilog, 16 iii 1906, *A. Loher* 6098 (syn K) – Type of *Begonia loheri* Merr.; Rizal Province, Mt. Irig, ii 1923, *M. Ramos* 41999 (BM); Rizal, Mt. Angilog, ii 1923, *G. Lopez* 42043 (B, BO, L, P).
- **IUCN category: DD.** Insufficient specimens could be georeferenced with certainty.

Begonia longibractea Merr. [§ Petermannia], Philipp. J. Sci. 17: 293 (1920); Merrill, Enum. Philipp. Fl. Pl. 3: 124 (1923). – Type: Philippines, Mindanao, Siargo Island, *M. Ramos, J. Pascasio* BS34870 (syn not located).
- **PHILIPPINES: Mindanao:** Siargo Island, *M. Ramos, J. Pascasio* BS34870 (syn not located) – Type of *Begonia longibractea* Merr.
- **IUCN category: VU D2.** Known only from one collection, possibly within the Siargao Wildlife Sanctuary forests, though this needs confirming.

Begonia longicarpa K.-Y Guan & D.K. Tian [§ Leprosae], Acta Bot. Yunnan. 22(2): 130 (2000). – Type: China, Yunnan, *T. Daike* 9729 (holo KUN).
- **VIETNAM:** Tonkin, Pho Lu, 3 ii 1936, *E. Poilane* 25151 (P).
- **IUCN category: DD.** Insufficient specimens could be georeferenced with certainty.

Begonia longicaulis Ridl. [§ Platycentrum], J. Straits Branch Roy. Asiat. Soc. 75: 35 (1917); Ridley, Fl. Malay Penins. 1: 120 (1922); Irmscher, Mitt. Inst. Allg. Bot. Hamburg 8: 120 (1929); Kiew, Begonias Penins. Malaysia 188 (2005). – Type: Peninsular Malaysia, Pahang, Gunong Tahan, vii 1911, *H.N. Ridley s.n.* (lecto SING *n.v.*; isolecto K *n.v.*).
- **PENINSULAR MALAYSIA:** Pahang, Gunong Tahan, vii 1911, *H.N. Ridley s.n.* (lecto SING *n.v.*; isolecto K *n.v.*) – Type of *Begonia longicaulis* Ridl.; Pahang, Gunong Tahan, 11 ix 1937, *E.J.H. Corner s.n.* (L).
- **IUCN category: VU D2.** Although endemic to a locality within the Taman Negra National Park, it is known only from one gully. Any disturbance to this site would lead immediately to a threat category of CR.

Begonia longifolia Blume [§ Sphenanthera], Catalogus 102 (1823); Blume, Enum. Pl. Javae 1: 97 (1827); Candolle, Prodr. 15(1): 398 (1864); Koorders, Exkurs.-Fl. Java 2: 650 (1912); Tebbitt, Brittonia 55(1): 25 (2003); Tebbitt, Begonias 168 (2005); Kiew, Begonias Penins. Malaysia 107 (2005). – *Diploclinium longifolium* (Blume) Miq., Fl. Ned. Ind. 1(1): 687 (1856). – Type: Java, Salak, *C.L.v. Blume* 740 (holo B).

Casparya trisulcata A. DC., Ann. Sci. Nat. Bot., IV 11: 119 (1859); Candolle, Prodr. 15(1): 277 (1864). – *Begonia trisulcata* (A. DC.) Warb., Nat. Pflanzenfam. 3(6A): 142 (1894). – Type: Java, Mt. Jojing, 1 v 1845, *H. Zollinger* 2850 (holo G-DC; iso B, BM, P[2]).

Begonia inflata C.B. Clarke in Hook. f., Fl. Brit. Ind. 2: 636 (1879); Clarke, J. Linn. Soc., Bot. 18: 115 (1881); Craib, Fl. Siam. 1: 774 (1931); Grierson, Fl. Bhutan 2: 242 (1991); Tebbitt, Brittonia 55(1): 25 (2003). – Type: India, Darjeeling, *C.B. Clarke* 12312A (syn K *n.v.*); India, Darjeeling, *C.B. Clarke* 12312C (syn K *n.v.*); Bhutan, *W. Griffith* 2587 (syn B, GH *n.v.*, K, P).

Begonia tricornis Ridl., J. Straits Branch Roy. Asiat. Soc. 75: 35 (1917); Tebbitt, Brittonia 55(1): 25 (2003). – Type: Peninsular Malaysia, Pahang, Telom, xi 1900, *H.N. Ridley* 14123 (holo SING *n.v.*; iso K).

Begonia crassirostris Irmsch., Mitt. Inst. Allg. Bot. Hamburg 10: 513 (1939); Tebbitt, Brittonia 55(1): 25 (2003). – Type: China, Kwangtun, Lo-fau-Shan, 15 viii 1883, *Ford* 1 (syn K *n.v.*); China, Yunnan, Szemao, *Henry* 12251 (syn B, K *n.v.*); China, Yunnan, Shi Ping, *Henry* 13600 (syn K *n.v.*); China, Kwangtun, Su-liu-kwan, 25 xii 1928, *T. Ying* 1744 (syn CAL *n.v.*, E *n.v.*); China, Hainan, Lam Ko District, Lin Fa Shan, 2 viii 1927, *Tsang Wai Tak* 278 (syn CAL *n.v.*, E *n.v.*); China, Hainan, Sha Po Shan, 27 v 1928, *Tsang Wai Tak* 536 (syn CAL *n.v.*); China, Kwangsi, 31 viii 1928, *Ching* 7281 (syn W *n.v.*); China, Hainan, Five Finger Mt., 28 iv 1922, *F.A. McClure* 9325 (syn K *n.v.*, P, PNH *n.v.*).

Begonia roxburghii auct. non (Miq.) A. DC.: Ridley, J. Fed. Malay States Mus. 4: 20 (1909); Ridley, Fl. Malay Penins. 1: 854 (1922).

■ **BURMA:** Kachin Hills, *S.M. Toppin* 4339 (K *n.v.*); Theronliang Tidding Valley, *F.K. Ward* 7936 (K *n.v.*).

THAILAND: Pan Paung River Valley, 13 vi – 16 vi 1946, *A.J.G.H. Kostermans* 838 (A, K, L); Kao Nawng, 10 iv 1927, *A.F.G. Kerr* 13258 (ABD, K).

VIETNAM: Laoke, Chapa, *Anon. s.n.* (B *n.v.*); South Vietnam, 1 iii 1959, *P. Tixier s.n.* (P); Dalat, Tazan, xi 1953, *M. Schmid s.n.* (P); Dalat, xi 1953, *M. Schmid s.n.* (P); Dalat, Daa Tria, xi 1953, *M. Schmid s.n.* (P); Phu-Cho, *Anon.* 65 (P *n.v.*) [as *Begonia aff. longifolia*]; Binh Tri Thien, A Luoi, 7 ix 1980, *A. Phuakam* 122 (HN *n.v.*); Do Huay Phue, 9 ix 1980, *Aluoi* 164 (HN *n.v.*); Cuc Phuong National Park, 21 vii 1999, *N.M. Cuong* 286 (A); Hoa Bin Province, 2 iv 1982, *N.H. Nguyen* 949 (HNU *n.v.*); Annam, Lang Bian, Ninh Thuan, *P.A. Eberhardt* 1764 (P *n.v.*); Annam, Thua-thien Province, Song Valley, *P.A. Eberhardt* 3107 (P *n.v.*); Tonkin, Mt. Bavi, ix 1885, *B. Balansa* 3757 (P); Vinh Yen Province, Tam Dao, *P.A. Eberhardt* 4900 (K *n.v.*); Vinh Yen Province, Tam Dao, *P.A. Eberhardt* 4990 (K); Tonkin, 7 iv 1936, *E. Poilane* 25559 (P); Annam, 14 ix 1938, *E. Poilane* 27810 (P); Annam, *E. Poilane* 31978 (P); Cuc Phuong National Park, 18 vii 1999, *T.B. Croat, V.D. Nguyen* 78043 (A).

PENINSULAR MALAYSIA: Selangor, Ginting Bedai, 24 ix 1914, *H.N. Ridley s.n.* (K *n.v.*, SING *n.v.*); Selangor, Ginting Bedai, *H.N. Ridley s.n.* (K); Selangor, Ginting Sempah, iii 1917, *H.N. Ridley, H.C. Robinson, C.B. Kloss s.n.* (K); Selangor, Ginting Bedai, iii 1917, *C.B. Kloss s.n.* (K); Selangor, Ginting Bedai, 24 ix 1914, *C.B. Kloss s.n.* (K); Selangor, Menuang Gasing, *C.B. Kloss s.n.* (K *n.v.*); Selangor, Ginting Bedai, iii 1917, *C.B. Kloss s.n.* (K *n.v.*); Selangor, Mt. Menuang Gasing, ii 1912, *C.B. Kloss s.n.* (BM); Pahang, Telom, xi 1900, *H.N. Ridley s.n.* (14123?) (holo SING *n.v.*; iso K) – Type of *Begonia tricornis* Ridl.; Pahang, Cameron Highlands, Tanah Rata, 27 vii 2002, *S. Neale, G. Bramley* 11 (E[2]); Gombak Road, 12 vii 1984, *S. Anthonysamy* 432 (L); Pahang, Cameron Highlands, Kampung Raja, 28 vii 1991, *B.H. Kiew* 3244 (L); Selangor, Ginting Sempah, 6 viii 1922, *I.H. Burkill* 9989 (B, K[2], K *n.v.*[2], SING *n.v.*); Pahang, Telom, xi 1900, *H.N. Ridley* 14123 (holo SING *n.v.*; iso K) – Type of *Begonia tricornis* Ridl.; Pahang, Pulau Tioman, 8 v 1927, *Md. Nur* SFN18883 (K *n.v.*); Pahang, Cameron Highlands, 30 iv 1937, *Md. Nur* SFN32961 (BM, K, L).

SUMATRA: Dolok Sibual Buali, 15 i 2000, *S.J. Davies, S.K. Rambe* 2000-44 (A); Berastagi Woods, 10 vi 1928, *C. Hanel, Rahmat Si Boeea* 580 (A); West Sumatra Prov., Gunung Merapi, 19 vii 2006, *D. Girmansyah, A. Poulsen, I. Hatta, R. Neivita* 759 (E); Baboeli – Paekas, 9 i 1932, *Bangham* 776 (A);

Sibayak Volcano, 15 ii 1932, *Bangham* 1018 (A); Asahan, Hoeta Bagasan, 7 ix 1934 – 4 ii 1935, *Rahmat Si Boeea* 1082 (A); Lintang, Helling, *H.A.B. Bunnemeijer* 3551 (L *n.v.*); Lintang, Helling, *H.A.B. Bunnemeijer* 3746 (L *n.v.*); Pajakumbuh, Mt. Sago, 30 xii 1907, *E. Meijer Drees* 7446 (L) [as *Begonia longifolia* ?]; Asahan, Aek Si Tamboerak, 28 x 1936, *Rhamat Si Boeea* 10653 (A).

JAVA: Java, 1836, *unknown s.n.* (P); Buitenzorg, 22 ii 1896, *E.H. Hallier s.n.* (BO); Java, *A. Zippelius s.n.* (B); Java, *unknown s.n.* (B); Java, 1858, *Nagel* 272 (B *n.v.*[2]); Salak, *C.L.v. Blume* 740 (holo B) – Type of *Begonia longifolia* Blume; Besoeki Panjoer Idjen, *S.H. Koorders* 1912 (L *n.v.*); Batavia, G. Salak, 27 i 1894, *V.F. Schiffner* 2268 (A *n.v.*, BO *n.v.*, L *n.v.*); Batavia, G. Liiang, 29 x 1928, *C.G.G.J.v. Steenis* 2382 (BO); Mt. Jojing, 1 v 1845, *H. Zollinger* 2850 (holo G-DC; iso B, BM, P[2]) – Type of *Casparya trisulcata* A. DC.; Preanger, Tjadas Malang, *R.C. Bakhuizen van den Brink* 2899 (BO *n.v.*, L *n.v.*); Poerwakarta, Tjihandjawar, *R.C. Bakhuizen van den Brink* 4343 (BO); Pasin Walang, 1913, *Backun* 8724 (B *n.v.*); West Java, Mt. Bodas, 17 iv 2001, *W.S. Hoover, J.M. Hunter, H. Wiriadinata, D. Girmansyah* 10003 (A); Preanger, Gunung Gede, 3 iii 1959, *J. Sinclair* 10080 (E); Idjen Mts., Banjoepait waterfall, 29 iv 1940, *C.G.G.J.v. Steenis* 12046 (BO); Batavia, Megamendoeng, Gunong Kendeng, 30 v 1940 – 31 v 1940, *C.G.G.J.v. Steenis* 12200 (BO); Gunung Telaja, 20 x 1940, *C.G.G.J.v. Steenis* 12311 (BO); Batavia, G. Salak, 20 ix 1896, *S.H. Koorders* 24153B (BO); Telomojo, 15 vi 1897, *S.H. Koorders* 27663B (BO); 15 vi 1897, *S.H. Koorders* 28516B (BO); *S.H. Koorders* 44300B (BO).

LESSER SUNDA ISLANDS: Bali: Mt. Batukau, *A.J.G.H. Kostermans* KK&SS92 (L *n.v.*); Batukau National Reserve, iii 1992, *J.J. Afriastini* 152 (L); Batu Kau, 22 iii 1964, *Dilmy* 991 (L *n.v.*); Batockaoe, *C.N.A. de Voogd* 2142 (L *n.v.*); Mt. Lesung, *McDonald, Ismail* 4850 (A, E, K, L); Bedugul, vi 1976, *E. Meijer Drees* 10456 (L); Lake Bratan, 20 vi 1976, *W. Meijer* 10550 (L); Lake Bratan, 20 vi 1976, *W. Meijer* 10551 (L *n.v.*). **Lombok:** Rinjani Nat. Park, 23 i 2001, *W.S. Hoover, J.M. Hunter, H. Wiriadinata, D. Girmansyah* 10007 (A); Rinjani Nat. Park, 24 iv 2001, *W.S. Hoover, J.M. Hunter, H. Wiriadinata, D. Girmansyah* 10010 (A). **Timor:** *H.O. Forbes* 3863 (BM).

SULAWESI: Bunta sub-distr., Sungai Hek, Cabang Tiga, 27 ii 2004, *Hendrian, M. Newman, S. Scott, M. Nazre Saleh, D. Supriadi* 939 (E, L).

MOLUCCAS: Seram: Sikeu Walala – Wae Tapakasitam, 20 xii 1996, *M. Kato et al.* 1120 (A).

- **Notes:** One of the most widespread species of *Begonia*. Closely allied to *B. acetosella*, *B. aptera*, *B. hayatae*, *B. renifolia*, *B. sarcocarpa* and *B. turbinata*, and is possibly a widespread progenitor of these more narrowly distributed taxa (Tebbitt, 2003b). The paraphyly of *B. longifolia* has been confirmed by molecular studies (Tebbitt *et al.*, 2006). The record from the Moluccas (Seram, *Kato et al.* 1120) could possibly be referred to *B. aptera*.
- **IUCN category: LC.** Widespread and ecologically tolerant.

Begonia longinoda Merr. [§ Diploclinium], Philipp. J. Sci. 6: 397 ('1911', 1912); Merrill, Enum. Philipp. Fl. Pl. 3: 124 (1923). – Type: Philippines, Luzon, Tayabas, Tagcauayan, iii 1911, *M. Ramos* BS13372 (holo PNH).

- **PHILIPPINES: Luzon:** Tayabas, Tagcauayan, iii 1911, *M. Ramos* BS13372 (holo PNH) – Type of *Begonia longinoda* Merr.
- **IUCN category: DD.** Insufficient specimens could be georeferenced with certainty.

Begonia longipedunculata Golding & Kareg. [§ Platycentrum], Phytologia 54(7): 496 (1984). – *Begonia longipetiolata* Baker f., J. Bot. 62(suppl.): 44 (1924); Golding & Karegeannes, Phytologia 54(7): 496 (1984). – Type: Sumatra, Palembang, Mt. Dempo, 1878 – 1883, *H.O. Forbes* 2423a (holo BM).

- **SUMATRA:** Palembang, Mt. Dempo, 1878 – 1883, *H.O. Forbes* 2423a (holo BM) – Type of *Begonia longipetiolata* Baker f.
- **IUCN category: VU D2.** Located within a Protection Forest on Mt. Dempo.

Begonia longiscapa Warb. in Perkins **[§ Diploclinium]**, Fragm. Fl. Philipp. 52 (1904); Merrill, Philipp. J. Sci. 6: 393 ('1911', 1912); Merrill, Enum. Philipp. Fl. Pl. 3: 124 (1923). – Type: Philippines, Leyte, 1861, *A.F. Jagor* 1009 (holo B; iso B).

■ **PHILIPPINES: Luzon:** Sorsogon, Mt. Juban, 14 vi 1956, *G. Edano* 37142 (BM). **Bohol:** viii 1923 – x 1923, *M. Ramos* 42670 (B, BM[2], K); viii 1923 – x 1923, *M. Ramos* 43373 (B, BM[2], K, P). **Leyte:** 16 ix 1913, *C.A. Wenzel* 452 (BM); 1861, *A.F. Jagor* 1009 (holo B; iso B) – Type of *Begonia longiscapa* Warb.; Cabalian, xii 1922, *M. Ramos* 41497 (K). **Samar:** iii 1914 – iv 1914, *M. Ramos* 17502 (K); Laquilacon, vi 1924, *R.C. McGregor* 43798 (K). **Mindanao:** Surigao, 27 vi 1927, *C.A. Wenzel* 2938 (K); Davao Province, Mt. Apo, viii 1908, *A.D.E. Elmer* 11787 (BM, E, K); Surigao, iv 1919, *M. Ramos, J. Pascasio* 34414 (K, P).

■ **IUCN category: NT.** Known from a range of localities, one of which is in or near a national park (Mt. Apo). However, forest cover has been severely depleted in the Mindanao–Eastern Visayas eco-region.

Begonia longiseta Irmsch. [§ Petermannia], Webbia 9: 499 (1954). – Type: Borneo, Sarawak, Colline del Bellaga, ix 1867, *O. Beccari* PB3800 (holo FI).

■ **BORNEO: Sarawak:** Poi Range, 16 viii 1962, *B.L. Burtt, P.J.B. Woods* 2868 (E[2]) [as *Begonia* cf. *longiseta*]; Colline del Bellaga, ix 1867, *O. Beccari* PB3800 (holo FI) – Type of *Begonia longiseta* Irmsch.

■ **IUCN category: DD.** Insufficient specimens could be georeferenced with certainty.

Begonia longistipula Merr. [§ Petermannia], Philipp. J. Sci. 6: 379 ('1911', 1912); Merrill, Enum. Philipp. Fl. Pl. 3: 124 (1923). – Type: Philippines, Mindanao, Surigao, ii 1906 – iv 1906, *F.H. Bolster* 248 (syn not located).

■ **PHILIPPINES: Samar:** Catubig River, ii 1916 – iii 1916, *M. Ramos* 24183 (BM, K). **Mindanao:** Surigao, ii 1906 – iv 1906, *F.H. Bolster* 248 (syn not located) – Type of *Begonia longistipula* Merr.; Surigao, *M. Ramos, J. Pascasio* 24783 (BM, BO); Surigao, *M. Ramos, J. Pascasio* 34783 (B, L, P).

■ **IUCN category: EN B2ab(iii).** Said to be common by the collector (Bolster) in 1906 at altitudes of 125 m. However, it is known from only two localities and the forest cover at this altitude is severely depleted in the Mindanao–Eastern Visayas eco-region.

Begonia longovillosa A. DC. [§ Diploclinium], Ann. Sci. Nat. Bot., IV 11: 130 (1859); Candolle, Prodr. 15(1): 324 (1864); Fernández-Villar, Noviss. App. 98 (1880); Merrill, Philipp. J. Sci. 6: 393 ('1911', 1912); Merrill, Enum. Philipp. Fl. Pl. 3: 124 (1923). – Type: Philippines, Luzon, Manilla, *B.K.A.A.F.v. Hügel* 4170 (holo not located).

■ **PHILIPPINES: Luzon:** Manilla, *B.K.A.A.F.v. Hügel* 4170 (holo not located) – Type of *Begonia longovillosa* A. DC.

■ **IUCN category: DD.** Insufficient specimens could be georeferenced with certainty.

Begonia lowiana King [§ Diploclinium], J. Asiat. Soc. Bengal, Pt. 2, Nat. Hist. 71: 67 (1902); Ridley, Fl. Malay Penins. 1: 864 (1922); Irmscher, Mitt. Inst. Allg. Bot. Hamburg 8: 137 (1929); Kiew, Begonias Penins. Malaysia 192 (2005). – Type: Peninsular Malaysia, Pahang, Cameron Highlands, Gunung Berumban, *L. Wray Jr.* 1567 (lecto K *n.v.*); Peninsular Malaysia, Perak, Gunong Batu Puteh, *L. Wray Jr.* 316 (syn not located).

■ **PENINSULAR MALAYSIA:** Pahang, Cameron Highlands, Gunong Brinchang, 26 vii 2002, *S. Neale* 5 (E[2]); Perak, *unknown* 13 (BM); Perak, Gunong Batu Puteh, *L. Wray Jr.* 316 (syn not located) – Type of *Begonia lowiana* King; Pahang, Cameron Highlands, Gunung Berumban, 20 iii 1992, *Klackenberg, Lundin* 701 (L); Pahang, Cameron Highlands, Gunung Berumban, *L. Wray*

Jr. 1567 (lecto K *n.v.*) – Type of *Begonia lowiana* King; Pahang, Gunung Brinchang, 9 iv 1987, *Leng Guan* 34360 (L).
■ **IUCN category: LC**. Recorded as 'common in bamboo forests on Gunung Brinchang' (Kiew, 2005).

Begonia lunatistyla Irmsch. [§ Petermannia], Webbia 9: 503 (1953). – Type: Borneo, Sarawak, Colline del Bellaga, ix 1867, *O. Beccari* PB3784 (holo FI).
■ **BORNEO: Sarawak:** Colline del Bellaga, ix 1867, *O. Beccari* PB3784 (holo FI) – Type of *Begonia lunatistyla* Irmsch.
■ **IUCN category: DD**. Insufficient specimens could be georeferenced with certainty.

Begonia luzonensis Warb. in Perkins **[§ Diploclinium]**, Fragm. Fl. Philipp. 52 (1904); Merrill, Philipp. J. Sci. 6: 403 ('1911', 1912); Merrill, Enum. Philipp. Fl. Pl. 3: 124 (1923). – Type: Philippines, Luzon, Rizal Province, Montalban, *O. Warburg* 13087 (holo B).
Begonia bakeri Elmer *nom. nud.*, Leafl. Philipp. Bot. 10: 3706 (1939); Smith & Wasshausen, Phytologia 52: 441 (1983). – Type: Philippines, Luzon, Sorsogon, Mt. Bulusan, x 1915, 12°46′14″N, 124°3′7″E, *A.D.E. Elmer* 14597 (syn B, BM[2], K, L, P, U).
■ **PHILIPPINES: Luzon:** Cagayan Province, Abulug River, i 1912, *C.M. Weber* 1531 (E, K, P); Bulacan Province, Angat, iii 1886, *W. Lobb* 2924 (K); Bataan, Mt. Mariveles, 1 i 1904, *E.D. Merrill* 3734 (B[2], BM); Bataan, Mt. Mariveles, xi 1904, *A.D.E. Elmer* 6849 (K); Sorsogon, Irosin – Lake Bulusan, 17 vi 1958, *J. Sinclair* 9596 (E); Rizal Province, Montalban, *O. Warburg* 13087 (holo B) – Type of *Begonia luzonensis* Warb.; Sorsogon, Mt. Bulusan, x 1915, *A.D.E. Elmer* 14597 (B, BM[2], K, L, P, U) – Type of *Begonia bakeri* Elmer *nom. nud.*; Laguna, iii 1916 – iv 1916, *C. Mabesa* 25387 (K, P); Rizal Province, Mt. Irig, ii 1923, *M. Ramos* 41187 (P); Rizal Province, Mt. Irig, ii 1923, *M. Ramos* 41887 (B, BM, K, L).
■ **Notes:** The synonym *B. bakeri* Elmer is not a valid name as no Latin diagnosis was included in the description, necessary from 1935.
■ **IUCN category: NT**. Although *B. luzonensis* has an area of occupancy of c. 20,000 km^2, the Luzon rain forest eco-region is highly fragmented and modified. The species can colonise roadside banks however (*J. Sinclair* 9596 (E)).

Begonia macgregorii Merr. [§ Petermannia], Philipp. J. Sci. 7: 310 (1912); Merrill, Enum. Philipp. Fl. Pl. 3: 124 (1923). – Type: Philippines, Luzon, Nueva Vizacaya, Dupax, iii 1912 – iv 1912, *R.C. McGregor* 11334 (syn B, BM, BO, L, P).
■ **PHILIPPINES: Luzon:** Nueva Vizacaya, Dupax, iii 1912 – iv 1912, *R.C. McGregor* 11334 (syn B, BM, BO, L, P) – Type of *Begonia macgregorii* Merr.; Nueva Vizacaya, Mt. Alzapan, *M. Ramos, G. Edano* 45598 (B, BM[2], BO, P); Nueva Vizacaya, Mt. Alzapan, v 1925 – vi 1925, *M. Ramos, G. Edano* 45698 (K); Mountain Province, Banaue, Bayinan, 9 xii 1962, *H. Conklin, Buwaya* 78637 (K). **Samar:** Concord, Bagacay, iv 1948 – v 1948, *M.D. Sulit* 6305 (L).
■ **IUCN category: DD**. Insufficient specimens could be georeferenced with certainty.

Begonia macintyreana M. Hughes [§ Petermannia], Edinburgh J. Bot. 63: 194 (2006). – Type: Sulawesi, Gorantalo, Nr Tulabolo, 4 iv 2002, 0°29′38″N, 123°16′22″E, *M. Mendum, H.J. Atkins, M. Newman, Hendrian, A. Sofyan* 2 (holo E; iso A, E, L).
■ **SULAWESI:** Gorantalo, Nr Tulabolo, 4 iv 2002, *M. Mendum, H.J. Atkins, M. Newman, Hendrian, A. Sofyan* 2 (holo E; iso A, E, L) – Type of *Begonia macintyreana* M. Hughes.
■ **IUCN category: VU D2** (Hughes, 2006).

Begonia macrotoma Irmsch. [§ Platycentrum], Notes Roy. Bot. Gard. Edinburgh 21: 41 (1951). – Type: China, Yunnan, 29 ix 1938, *T.T. Yu* 7778 (holo KUN *n.v.*).
- **VIETNAM:** Kontum Prov., Ngoc Linh Mountain, 9 iii 1995, *L. Averyanov et al.* VH586 (P).
- **IUCN category: DD.** Insufficient specimens could be georeferenced with certainty.

Begonia madaiensis Kiew [§ Petermannia], Gard. Bull. Singapore 53: 273 (2001). – Type: Borneo, Sabah, Lahad Datu District, Gunung Madai, *P. Kiew* RK5057 (holo SAN *n.v.*; iso K *n.v.*, SAR *n.v.*, SING *n.v.*).
- **BORNEO: Sabah:** Lahad Datu District, Gunung Madai, *R. Kiew* RK5057 (holo SAN *n.v.*; iso K *n.v.*, SAR *n.v.*, SING *n.v.*) – Type of *Begonia madaiensis* Kiew.
- **IUCN category: VU D2.** Known only from the type locality.

Begonia malachosticta Sands [§ Petermannia], Kew Mag. 7: 61 (1990); Kiew, Gard. Bull. Singapore 53: 276 (2001). – Type: Borneo, Sabah, Kinabatangan, Bukit Dulung Lambu, Gomantong Cave, *M.J.S. Sands* 3933 (holo K).
- **BORNEO: Sabah:** Kinabatangan, Bukit Dulung Lambu, Gomantong Cave – Bukit Dulung Lambu, 10 ix 1976, *M. Tamura, M. Hotta* 596 (L); Kinabatangan, Bukit Dulung Lambu, Gomantong Cave, *M.J.S. Sands* 3933 (holo K) – Type of *Begonia malachosticta* Sands.
- **IUCN category: LC.** Known only from the type locality, but appears to be restricted to inaccessible rock faces in the Gomantong Reserve.

Begonia malindangensis Merr. [§ Petermannia], Philipp. J. Sci. 6: 391 ('1911', 1912); Merrill, Enum. Philipp. Fl. Pl. 3: 124 (1923). – Type: Philippines, Mindanao, Misamis Occidental, Mt. Malindang, v 1906, 8°13′N, 123°38′E, *E.A. Mearns, W.I. Hutchinson* 4563 (syn B[2], BO, P).
- **PHILIPPINES: Mindanao:** Davao Province, Todaya, iv 1904, *E.B. Copeland* 1284 (para B) – Type of *Begonia malindangensis* Merr.; Misamis Occidental, Mt. Malindang, v 1906, *E.A. Mearns, W.I. Hutchinson* 4563 (syn B[2], BO, P) – Type of *Begonia malindangensis* Merr.; Davao Province, Mt. Apo, ix 1909, *A.D.E. Elmer* 11808 (BM, E, K, L); Bukidnon subprovince, Mt. Lipa, vi 1920 – vii 1920, *M. Ramos, G. Edano* 38492 (K, L); Bukidnon subprovince, Mt. Camates, vi 1920 – vii 1920, *M. Ramos, G. Edano* 38594 (B, K, L, P).
- **Notes:** Allied to *B. merrittii* Merr.
- **IUCN category: LC.** There are currently 25,000 ha of forest in the Malindang Reserve. Any further disturbance to this forest will immediately lead to an increase in threat level.

Begonia malmquistiana Irmsch. [§ Petermannia], Bot. Jahrb. Syst. 50: 337 (1913). – Type: New Guinea, Papua New Guinea, Northeast New Guinea, *C.L. Ledermann* 8362 (syn B[2]); New Guinea, Papua New Guinea, Northeast New Guinea, *C.L. Ledermann* 9328 (syn B[4]); New Guinea, Papua New Guinea, Kaiser-Wilhelmsland, Etappenberg, 18 x 1912, *C.L. Ledermann* 9363 (syn B[4]); New Guinea, Papua New Guinea, Northeast New Guinea, *C.L. Ledermann* 9454a (syn B).
- **NEW GUINEA: Papua New Guinea:** Northeast New Guinea, *C.L. Ledermann* 8362 (syn B[2]) – Type of *Begonia malmquistiana* Irmsch.; Northeast New Guinea, *C.L. Ledermann* 9328 (syn B[4]) – Type of *Begonia malmquistiana* Irmsch.; Kaiser-Wilhelmsland, Etappenberg, 18 x 1912, *C.L. Ledermann* 9363 (syn B[4]) – Type of *Begonia malmquistiana* Irmsch.; Northeast New Guinea, *C.L. Ledermann* 9454a (syn B) – Type of *Begonia malmquistiana* Irmsch.
- **Notes:** Irmscher described two formas (f. *angustifolia* and f. *latifolia*), but did not assign any specimens to these names.
- **IUCN category: DD.** Insufficient specimens could be georeferenced with certainty.

Begonia mamutensis Sands [§ Petermannia], Pl. Mt. Kinabalu 158 (2001). – Type: Borneo, Sabah, Mt. Kinabalu, Mamut copper mine, 3 v 1984, 6°1′59″N, 116°39′59″E, *M.J.S. Sands* 3956 (holo K).
- **BORNEO: Sabah:** Ranau District, Langanan Falls, *M.J.S. Sands* 3924 (para K *n.v.*) – Type of *Begonia mamutensis* Sands; Mt. Kinabalu, Mamut copper mine, 3 v 1984, *M.J.S. Sands* 3956 (holo K) – Type of *Begonia mamutensis* Sands; Mt. Kinabalu, Lohan/Mamut Copper Mine, *J.H. Beaman* 10598 (para K *n.v.*) – Type of *Begonia mamutensis* Sands.
- **IUCN category: VU D2**. Known only from the type locality on Mt. Kinabalu, on the edge of the reserve area.

Begonia manillensis A. DC. [§ Diploclinium], Ann. Sci. Nat. Bot., IV 11: 129 (1859); Candolle, Prodr. 15(1): 323 (1864); Fernández-Villar, Noviss. App. 98 (1880); Merrill, Philipp. J. Sci. 5: 365 (1910); Merrill, Philipp. J. Sci. 6: 404 ('1911', 1912); Merrill, Enum. Philipp. Fl. Pl. 3: 125 (1923). – Type: Philippines, Luzon, Manilla, *G.S. Perrottet* (holo P).
- **PHILIPPINES: Luzon:** Manilla, *G.S. Perrottet* (holo P) – Type of *Begonia manillensis* A. DC.; *W. Lobb s.n.* (K) [as *Begonia manillensis* ?]; Rizal Province, iii 1905, *Aherns collector* 2699 (P); Bataan, Mt. Mariveles, 1 i 1904, *E.D. Merrill* 3734 (K); Rizal Province, Montalban, 26 x 1890, *A. Loher* 4310 (K); 13 xi 1890, *A. Loher* 4311 (K) [as *Begonia cf. manillensis*]; Benguet, Baguio, iii 1907, *A.D.E. Elmer* 8427 (BM); Benguet, Baguio, iii 1907, *A.D.E. Elmer* 9326 (BM, K); Rizal Province, Mt. Irig, ii 1923, *M. Ramos* 41920 (B[3], BM, K).
- **IUCN category: VU D2**. Has a low area of occupancy in the Luzon rain forest eco-region, which is highly fragmented and modified.

Begonia martabanica A. DC. [§ Parvibegonia], Ann. Sci. Nat. Bot., IV 11: 136 (1859); Candolle, Prodr. 15(1): 354 (1864); Clarke, Fl. Brit. Ind. 651 (1879); Gagnepain, in Lecomte (ed.), Fl. Indo-Chine 2: 1118 (1921); Craib, Fl. Siam. 1: 776 (1931). – Type: Burma, Martaban, *W. Lobb* 393 (iso BM, E, K).

Begonia martabanica var. martabanica
- **BURMA:** Martaban, *W. Lobb* 393 (iso BM, E, K) – Type of *Begonia martabanica* A. DC.; Tenasserim Division, Tavoy District, Paungdaw Power Station, viii 1962, *J. Keenan et al.* 750 (A, E, K) [as *Begonia cf. martabanica*]; Tenasserim Division, Tavoy District, Paungdaw Power Station, viii 1961, *J. Keenan et al.* 772 (A, E, K) [as *Begonia cf. martabanica*]; Tenasserim Division, Tavoy District, Paungdaw Power Station, viii 1961, *J. Keenan* 1323 (E); Tenasserim Division, Tavoy District, Paungdaw Power Station, viii 1961, *J. Keenan et al.* 1378 (E, K) [as *Begonia cf. martabanica*]; Tenasserim Division, Tavoy District, Paungdaw Power Station, x 1964, *J. Keenan* 1974 (E); Maymyo Plateau, Ani Sakan, 15 x 1911, *J.H. Lace* 5503 (E).
  THAILAND: Rachaburi, Kanburi, Hin Dat, 3 vii 1926, *N. Put* 78 (BM, K); Rachaburi, Kanburi, Hin Dat, 14 vii 1926, *N. Put* 135 (ABD, BM, K); Khao Yai National Park, 14 viii 2000, *J.F. Maxwell* 00-392 (A); Lampang Prov., Doi Kuhn Dahn National Park, 29 vii 1994, *J.F. Maxwell* 94-822 (A, L); Kwae Noi River Basin, 23 vii 1946, *A.J.G.H. Kostermans* 1304 (A) [as *Begonia cf. martabanica*]; Ayuthia, Saraburi, Muak Lek, 2 x 1963, *T. Smitinand, H. Sleumer* 1309 (BO) [as *Begonia aff. martabanica*]; Chiangmai, 25 ix 1910, *A.F.G. Kerr* 1412 (ABD, K); Chiangmai, Doi Sootep, 25 ix 1910, *A.F.G. Kerr* 1412a (ABD, BM, K); Loei Province, Phu Luang, 6 vii 1968, *K. Bunchuai* 1714 (L); Rachaburi, Kanburi, Sai Yok, 2 viii 1928, *N. Put* 1837 (ABD, BM); Chiangmai, Doi Sootep, 26 viii 1911, *A.F.G. Kerr* 1969 (BM, K, P); Rachaburi, Kanburi, Sai Yok, 1 viii 1928, *A. Marcan* 2393 (ABD, K); Rachaburi, Kanburi, Sai Yok, 2 viii 1928, *N. Put* 2402 (BM); Rachaburi, Kanburi, Sai Yok, 2 viii 1928, *A. Marcan* 2893 (BM); Tung Salaeng Luang Nat. Park, 21 x 1984, *Murata et al.* 38264 (L) [as *Begonia cf. martabanica*].
  LAOS: *unknown* 135 (E).

Begonia martabanica var. pseudoclivalis Irmsch., Mitt. Inst. Allg. Bot. Hamburg 8: 145 (1929); Kiew, Begonias Penins Malaysia 68 (2005). – Type: Peninsular Malaysia, Perak, Temengoh, vii 1909, *H.N. Ridley* 14606 (holo SING *n.v.*; iso BM, K).
- **PENINSULAR MALAYSIA:** Perak, Temengoh, vii 1909, *H.N. Ridley* 14606 (holo SING *n.v.*; iso BM, K) – Type of *Begonia martabanica* var. *pseudoclivalis* Irmsch.
- **Notes:** Distinct in § *Parvibegonia* in having a combination of caulescent habit and symmetric cordiform leaves.
- **IUCN category: LC.** Widespread in Thailand, and also recorded from localities in Burma and Laos.

Begonia masarangensis Irmsch. [§ Petermannia], Bot. Jahrb. Syst. 50: 368 (1913). – Type: Sulawesi, Masarang, 23 iv 1894, *K.F. & P.B. Sarasin* 269 (holo B; iso K).
- **SULAWESI:** Masarang, 23 iv 1894, *K.F. & P.B. Sarasin* 269 (holo B; iso K) – Type of *Begonia masarangensis* Irmsch.; Menado, G. Ngilalaki, *S. Bloembergen* 4014 (BO); Tomohon, Gunung Masarang, 10 i 1895, *S.H. Koorders* 16240B (BO).
- **Notes:** Possibly synonymous with *B. hispidissima* Zipp. ex Koord., but see notes under that species.
- **IUCN category: DD.** Taxonomic uncertainty.

Begonia maxwelliana King [§ Platycentrum], J. Asiat. Soc. Bengal, Pt. 2, Nat. Hist. 71: 66 (1902); Ridley, Fl. Malay Penins. 1: 863 (1922); Irmscher, Mitt. Inst. Allg. Bot. Hamburg 8: 118 (1929); Craib, Fl. Siam. 1: 776 (1931); Henderson, Malayan Wild Flowers – Dicotyledons 168 (1959); Kiew, Begonias Penins. Malaysia 174 (2005). – Type: Peninsular Malaysia, Perak, Maxwells Hill, *L. Wray Jr.* 119 (lecto K *n.v.*); Peninsular Malaysia, Perak, Maxwells Hill, *B. Scortechini* 1607 (syn not located); Peninsular Malaysia, Perak, Maxwells Hill, *B. Scortechini* 1798 (syn BM, L); Peninsular Malaysia, Perak, Maxwells Hill, *Kings Collector* 2038 (syn P); Peninsular Malaysia, Perak, Maxwells Hill. vi 1888, *L. Wray Jr.* 2199 (syn B, FI); Sumatra, Nupal Siljur, *H.O. Forbes* 3119a (syn BM, K, L[2]).
- **PENINSULAR MALAYSIA:** Perak, Thaiping, *unknown s.n.* (BM); Perak, Maxwells Hill, *L. Wray Jr.* 119 (lecto K *n.v.*) – Type of *Begonia maxwelliana* King; Pahang, Cameron Highlands, 19 iii 1992, *Klackenberg, Lundin* 611 (L); Perak, Gunong Bubu, 17 viii 1966, *D. Hou* 643 (L); Pahang, Cameron Highlands, Gunung Berumban, 20 iii 1992, *Klackenberg, Lundin* 702 (L); Perak, Maxwells Hill, Perak, *B. Scortechini* 1607 (syn not located) – Type of *Begonia maxwelliana* King; Perak, Maxwells Hill, *B. Scortechini* 1798 (syn BM, L) – Type of *Begonia maxwelliana* King; Perak, Maxwells Hill, *Kings Collector* 2038 (syn P) – Type of *Begonia maxwelliana* King; Perak, Maxwells Hill, vi 1888, *L. Wray Jr.* 2199 (syn B, FI) – Type of *Begonia maxwelliana* King; Negeri Sembilan, 21 xi 1977, *A.G. Piggot* 2231 (L); Perak, Gunung Bubu, 26 iv 1987, *R. Kiew* 2573 (L); Perak, Bukit Kinta F.R., 28 iv 1987, *R. Kiew* 2594 (L); Kedah, G. Inan F.R., 8 ii 1968, *T.C. Whitmore* 4671 (L); Perak, Maxwells Hill, 16 ix 1949, *J. Sinclair* 6179 (E).
  SUMATRA: Nupal Siljur, *H.O. Forbes* 3119a (syn BM, K, L[2]) – Type of *Begonia maxwelliana* King.
- **Notes:** Reported from Thailand (Pattani, *Kerr* 7879 (ABD, BM, K)), but the collection has distinctly hairy petioles and therefore is likely to represent a new taxon. The syntypes remaining from Sumatra probably also represent a different taxon, but further investigation is needed.
- **IUCN category: LC.** Described as 'common in Perak' (Kiew, 2005), and occurs at altitudes of 300–1300 m.

Begonia mearnsii Merr. [§ Petermannia], Philipp. J. Sci. 6: 383 ('1911', 1912); Merrill, Enum. Philipp. Fl. Pl. 3: 125 (1923). – Type: Philippines, Mindanao, Mt. Malindang, v 1906, *Davison, W.I. Hutchinson* 4749 (syn not located).

- **PHILIPPINES: Mindanao:** Mt. Malindang, v 1906, *Davison, W.I. Hutchinson* 4749 (syn not located) – Type of *Begonia mearnsii* Merr.; Zamboanga, xi 1911 – xii 1911, *E.D. Merrill* 8196 (B, BM[2]); Lanao, iii 1916, *R.J. Alvarez* 25257 (K).
- **IUCN category: VU D2.** Known from three localities, one of which is in the Mt. Malindang Reserve.

Begonia media Merr. & L.M. Perry [§ Petermannia], J. Arnold Arbor. 24: 54 (1943); Gilli, Ann. Naturhist. Mus. Wien 83: 422 (1980). – Type: New Guinea, Papua New Guinea, Central Division, Mafulu, xi 1933, *L.J. Brass* 5435 (holo NY; iso B).
- **NEW GUINEA: Papua New Guinea:** Morobe Distr., Wau Subprov., Bulldog Road, x 1977, *W.S. Hoover* 54 (L); Mt. Kolorong Conservation Area, 6 vi 1992, *R. Hoft* 2265 (L); Central Division, Mafulu, xi 1933, *L.J. Brass* 5435 (holo NY; iso B) – Type of *Begonia media* Merr. & L.M. Perry; Morobe Distr., Wau-Salamaua Road, 9 i 1956, *J.S. Womersley, A. Millar* 8717 (BM) [as *Begonia aff. media*]; Southern Highlands Distr., Mt. Ialibu, 23 vi 1969, *M. Coode, P. Wardle, P. Katik* 40343 (E) [as *Begonia aff. media*]; Morobe Distr., Wau Subprov., Kaindi, 19 xi 1983, *K. Kerenga, C.N. Dao* 56621 (L); Goroka Subprov., Gahavisuka Provincial Park, *K. Kerenga, N. Cruttewell* 56695 (L) [as *Begonia cf. media*].
- **IUCN category: LC.** Occurs across a range of 880 km between the Central Range and the Owen Stanley Range.

Begonia megacarpa Merr. [§ Petermannia], Philipp. J. Sci. 9: 378 (1914); Merrill, Enum. Philipp. Fl. Pl. 3: 125 (1923). – Type: Philippines, Leyte, Dagami, ix 1913, *C.A. Wenzel* 457 (syn BM).
- **PHILIPPINES: Luzon:** Rizal, Balacbac, v 1916, *A. Loher* 12997 (BM); Rizal, Balacbac, v 1916, *A. Loher* 13006 (P); Rizal Province, Mt. Irig, ii 1923, *M. Ramos* 41889 (B, P); Tayabas, Casiguran, v 1925 – vi 1925, *M. Ramos, G. Edano* 45444 (B, BM, P). **Leyte:** Dagami, ix 1913, *C.A. Wenzel* 457 (syn BM) – Type of *Begonia megacarpa* Merr.; Cabalian, v 1922, *G. Lopez* 40805 (B, BM). **Mindanao:** Butuan, Bunauan, ix 1913, *E.H. Taylor s.n.* (para not located) – Type of *Begonia megacarpa* Merr.
- **IUCN category: VU B2ab(iii).** Despite its apparent large range (Luzon to Mindanao), this species occurs in low altitude forest (c. 60 m), which is severely fragmented.

Begonia megalantha Merr. [§ Petermannia], Philipp. J. Sci. 10: 47 (1915); Merrill, Enum. Philipp. Fl. Pl. 3: 125 (1923). – Type: Philippines, Luzon, Ifugao, Mt. Polis, ii 1913, *R.C. McGregor* BS19860 (syn not located).
- **PHILIPPINES: Luzon:** Ifugao, Mt. Polis, ii 1913, *R.C. McGregor* BS19857 (para not located) – Type of *Begonia megalantha* Merr.; Ifugao, Mt. Polis, ii 1913, *R.C. McGregor* 19858 (para BM, L, P) – Type of *Begonia megalantha* Merr.; Ifugao, Mt. Polis, ii 1913, *R.C. McGregor* BS19859 (para not located) – Type of *Begonia megalantha* Merr.; Ifugao, Mt. Polis, ii 1913, *R.C. McGregor* BS19860 (syn not located) – Type of *Begonia megalantha* Merr.; Kalinga Sub Prov., Lubuagan, Mt. Masingit, ii 1920, *M. Ramos, G. Edano* 37502 (L).
- **IUCN category: CR B2ab(iii).** Known only from the type locality on Mt. Polis, which has little remaining forest cover and receives no legal protection (Collar *et al.*, 1999).

Begonia megaptera A. DC. [§ Platycentrum], Ann. Sci. Nat. Bot., IV 11: 134 (1859); Candolle, Prodr. 15(1): 348 (1864); Clarke, Fl. Brit. Ind. 2: 646 (1879); Hara, Fl. E. Himalaya 214 (1966). – Type: not located.
- **BURMA:** Southern Shan States, *R.W. MacGregor* 1068 (E).
- **IUCN category: DD.** Insufficient specimens could be georeferenced with certainty.

Begonia melikopia Kiew [§ Petermannia], Gard. Bull. Singapore 53:277 (2001). – Type: Borneo, Sabah, Gua Melikop, *R. Kiew* 5011 (holo SAN *n.v.*; iso BRUN *n.v.*, K *n.v.*, L *n.v.*, SAR *n.v.*, SING *n.v.*).
- **BORNEO: Sabah:** Gua Melikop, *R. Kiew* 5011 (holo SAN *n.v.*; iso BRUN *n.v.*, K *n.v.*, L *n.v.*, SAR *n.v.*, SING *n.v.*) – Type of *Begonia melikopia* Kiew.
- **IUCN category: DD**. Insufficient specimens could be georeferenced with certainty.

Begonia mendumae M. Hughes [§ Petermannia], Edinburgh J. Bot. 63: 196 (2006). – Type: Sulawesi, Gorantalo, Gunung Ali, 28 iv 2002, *M. Mendum, H.J. Atkins, M. Newman, Hendrian, A. Sofyan* 240 (holo E).
- **SULAWESI:** Gorantalo, Gunung Ali, 28 iv 2002, *M. Mendum, H.J. Atkins, M. Newman, Hendrian, A. Sofyan* 240 (holo E) – Type of *Begonia mendumae* M. Hughes.
- **IUCN category: VU D2** (Hughes, 2006).

Begonia merrittii Merr. [§ Petermannia], Philipp. J. Sci. 5: 365 (1910); Merrill, Philipp. J. Sci. 6: 386 ('1911', 1912); Merrill, Enum. Philipp. Fl. Pl. 3: 125 (1923). – Type: Philippines, Luzon, Benguet, Mt. Santo Tomas, v 1904, *A.D.E. Elmer* 6254 (iso B[2], P); Philippines, Luzon, Benguet, Mt. Santo Tomas, v 1904, *A.D.E. Elmer* 6254 (syn BO, K).
- **PHILIPPINES: Luzon:** Bontoc (sub-province), v 1911, *M. Vanoverbergh* 513 (L); Benguet, Mt. Tonglon, *R.S. Williams* 1211 (para not located) – Type of *Begonia merrittii* Merr.; iv 1887, *J.E. Vidal* 1439 (Fl); Bontoc (sub-province), x 1912, *M. Vanoverbergh* 1524 (P); Benguet, Mt. Santo Tomas, 29 xi 1904, *R.S. Williams* 1532 (K); Mt. Pulog, 9 iii 1961, *M.L. Dr. Steiner* 1974 (L); Benguet, *A. Loher* 4318 (K); Benguet, Pauai, *E.A. Mearns* FB4385 (para not located) – Type of *Begonia merrittii* Merr.; Benguet, x 1905 – xi 1905, *E.D. Merrill* 4781 (para B, K) – Type of *Begonia merrittii* Merr.; Benguet, Mt. Tonglon, x 1905 – xi 1905, *E.D. Merrill* 4823 (para B, K) – Type of *Begonia merrittii* Merr.; Benguet, Mt. Tonglon, *H.M. Curran* 4996 (para BM) – Type of *Begonia merrittii* Merr.; Benguet, Mt. Tonglon, *M. Ramos* BS5455 (para B) – Type of *Begonia merrittii* Merr.; Mt. Data range, 12 v 1961, *P.v. Royen, H. Sleumer* 5611 (L); Benguet, xii 1908, *M. Ramos* 5839 (L, P); Benguet, Mt. Santo Tomas, v 1904, *A.D.E. Elmer* 6254 (iso B[2], P; syn BO, K) – Type of *Begonia merrittii* Merr.; Mt. Pulog, *E.D. Merrill* 6502 (para not located) – Type of *Begonia merrittii* Merr.; Mt. Pulog, 21 i 1968, *M. Jacobs* 7028 (K); Mt. Tabayoc, 15 ii 1968, *M. Jacobs* 7460 (K, L[2]); Mountain Province, Mt. Pauai, i 1948, *M.D. Sulit* 7494 (L); Mt. Nangaoto, ii 1948, *M.D. Sulit* 7777 (K); Benguet, Pauai, vi 1906, *R.C. McGregor* 8496 (para B, K) – Type of *Begonia merrittii* Merr.; Benguet, Baguio, iii 1907, *A.D.E. Elmer* 8521 (K); Benguet, Baguio, iii 1907, *A.D.E. Elmer* 8593 (BM, BO, E, K, L, U); Mt. Pulog, Kabayan, 7 x 1992, *E.B. Barbon et al.* 8807 (K); Mt. Pulog, Kabayan, 23 ii 1992, *G.C.G. Argent et al.* 9516 (L); Benguet, v 1914, *E.D. Merrill* 9663 (BM, L, P); Benguet, Mt. Santo Tomas, Mt. Tonglon, *H.N. Whitford* 11107 (para B[2]) – Type of *Begonia merrittii* Merr.; Benguet, Mt. Santo Tomas, Mt. Tonglon, iv 1908, *H.N. Whitford* 11107 (para B[2]) – Type of *Begonia merrittii* Merr.; Ifugao, Semegar, 26 iii 1994, *Barbon et al.* 12782 (L); Ifugao, Mt. Himi-o, 2 iv 1994, *Barbon et al.* 12912 (L); Benguet, Baguio, iii 1913, *A.D.E. Elmer* 14277 (B, BM, BO, E, K, L, P, U); Lepanto District, *F.W. Darling* FB14491 (para not located) – Type of *Begonia merrittii* Merr.; Mountain Province, Mt. Amuyao, 2 viii 1994, *Reynoso et al.* 14531 (L); Benguet, Mt. Lusod, *M.L. Merritt, H.M. Curran* FB15736 (para not located) – Type of *Begonia merrittii* Merr.; Mt. Pulog, 26 ix 1909, *H.M. Curran, M.L. Merritt, T.C. Zachokke* 16176 (para B, K) – Type of *Begonia merrittii* Merr.; Mt. Pulog, i 1909, *H.M. Curran, M.L. Merritt, T.C. Zachokke* 16176 (para B, K) – Type of *Begonia merrittii* Merr.; Ifugao, Mt. Polis, ii 1913, *R.C. McGregor* 19861 (BM, P); Benguet, Pauai, iv 1919 – vi 1919, *J.K. Santos* 31822 (L); Bontoc (sub-province), Mt. Puxis, iii 1920, *M. Ramos, G. Edano* 37759 (BO, L); Mt. Pulog, ii 1925 – iii 1925, *M. Ramos, G. Edano* 44929 (B).
- **Notes:** Merrill notes that this species is absolutely restricted to wet mossy forests above 2250 m, the altitudinal limit of *Pinus insularis* Endl. The dainty nature of the plant on a herbarium sheet

belies its true stature as a shrub sometimes up to 10 feet in height (see *Williams* 1532 (K) for an example of the woody basal stem).
- **IUCN category: LC**. Occurs in the Pulag National Park and the upper and lower Agno River Basin Resource Reserves. Forest at high altitude in these areas, to which this species is limited, is in reasonable condition.

Begonia microptera Hook. [not placed to section], Bot. Mag. 83: pl. 4974 (1857). – Type: not located.
- **Notes:** This species is known only from the protologue and accompanying illustration. It is stated as being native to Borneo, without precise locality.
- **IUCN category: DD**. Insufficient specimens could be georeferenced with certainty.

Begonia mindanaensis Warb. in Perkins **[§ Petermannia]**, Fragm. Fl. Philipp. 55 (1904); Merrill, Philipp. J. Sci. 6: 376 ('1911', 1912); Merrill, Enum. Philipp. Fl. Pl. 3: 125 (1923). – Type: Philippines, Mindanao, Davao Province, Mt. Dagatpan, *O. Warburg* 14633 (holo B).
- **PHILIPPINES: Leyte:** 29 iii 1913, *C.A. Wenzel* 53 (BM); Palo, i 1906, *A.D.E. Elmer* 7253 (E); Ormoc, Lake Danao, iii 1950, *G. Edano* 14235 (K); Mt. Abucayan, ii 1923, *G. Edano* 41730 (B, P). **Mindanao:** Butuan, Bunauan, 1913, *E.H. Taylor s.n.* (K); Butuan, Agusan River, v 1913, *unknown s.n.* (K); Davao Province, Todaya, iv 1904, *E.B. Copeland* 1246 (B); Surigao, 2 v 1927, *C.A. Wenzel* 2943 (B, K); Butuan, Agusan River, x 1910, *E.D. Merrill* 7310 (B, BM[2], L, P); Surigao, 16 v 1927, *C.A. Wenzel* 7691? (B); Zamboanga, xi 1911 – xii 1911, *E.D. Merrill* 8167 (B, K, P); Agusan, Mt. Urdaneta, vii 1912, *A.D.E. Elmer* 13367 (B, BM, E, K, L, P, U); Davao Province, Mt. Dagatpan, *O. Warburg* 14633 (holo B) – Type of *Begonia mindanaensis* Warb.; Surigao, iv 1919, *M. Ramos, J. Pascasio* 34454 (K, P); Surigao, iv 1919, *M. Ramos, J. Pascasio* 34457 (L); Surigao, iv 1919, *M. Ramos, J. Pascasio* 34508 (K); Zamboanga, Malangas, x 1919 – xi 1919, *M. Ramos, G. Edano* 37229 (K); Zamboanga, Malangas, x 1919 – xi 1919, *M. Ramos, G. Edano* 37246 (L).
- **IUCN category: DD**. Occurs in the lowland (<1000 m) Mindanao–Eastern Visayas rain forest eco-region, which is heavily deforested. Based solely on geographic range, this species does not meet any of the criteria for threat, but further investigation into population size and decline is needed.

Begonia mindorensis Merr. [§ Diploclinium], Philipp. J. Sci. 6: 396 ('1911', 1912); Merrill, Philipp. J. Sci. 6: 376 ('1911', 1912); Merrill, Enum. Philipp. Fl. Pl. 3: 125 (1923). – Type: Philippines, Mindoro, Mindoro Oriental, Lake Naujan, *M.L. Merritt* 6867 (holo not located).
    Begonia sordidissima Elmer, Leafl. Philipp. Bot. 7: 2557 (1915). – Type: Philippines, Mindanao, Agusan, Mt. Urdaneta, viii 1912, *A.D.E. Elmer* 13575 (syn B, BM, E, FI, L, P, U).
- **PHILIPPINES: Palawan:** Iwahig, iv 1906, *F.W. Foxworthy* 779 (para B, BO, P) – Type of *Begonia mindorensis* Merr.; Puerto Princesa, Mt. Pulgar, Mt. Pulgar – Puerto Princesa, iii 1911, *A.D.E. Elmer* 12857 (para B, BM, BO, E, FI, K, P, U) – Type of *Begonia mindorensis* Merr.; Coron Island, ix 1922, *M. Ramos* 41162 (B, K, P); Culion Island, x 1922, *M. Ramos* 41304 (B, K, P) [as *Begonia aff. mindorensis*]. **Luzon:** Tayabas, Kabibihan, iii 1911, *F.W. Foxworthy, M. Ramos* 13131 (para BM) – Type of *Begonia mindorensis* Merr.; Tayabas, Tagcauayan, iii 1911, *M. Ramos* 13361 (para B, K) – Type of *Begonia mindorensis* Merr.; Sorsogon, Mt. Bulusan, *A.D.E. Elmer* 15755 (B, BM[2], FI, L, P, U). **Mindoro:** Baco River, iv 1905, *R.C. McGregor* 284 (para not located) – Type of *Begonia mindorensis* Merr.; Mindoro Oriental, Lake Naujan, *M.L. Merritt* FB6777 (para not located) – Type of *Begonia mindorensis* Merr.; Mindoro Oriental, Lake Naujan, *M.L. Merritt* 6867 (holo not located) – Type of *Begonia mindorensis* Merr.; Mt. Calavite, iv 1921, *M. Ramos* 39446 (B, P); Puerto Galera, xi 1925 – xii 1925, *M. Ramos* 46340 (P). **Mindanao:** Surigao, *C.A. Wenzel*

3204 (K); Agusan, Mt. Urdaneta, viii 1912, *A.D.E. Elmer* 13575 (syn B, BM, E, FI, L, P, U) – Type of *Begonia sordidissima* Elmer.

■ **IUCN category:LC**.The distribution of this species covers three eco-regions (Mindoro rain forest, Mindanao–Eastern Visayas rain forest and Luzon rain forest) and overlaps two protected areas (Najuan Lake National Park and Mt. Calavite Wildlife Sanctuary).

Begonia minjemensis Irmsch. [§ Diploclinium], Bot. Jahrb. Syst. 50: 375 (1913). – Type: New Guinea, Papua New Guinea, Minjem, 27 iii 1908, *F.R.R. Schlechter* 17511 (holo B).

■ **NEW GUINEA: Papua New Guinea:** Minjem, 27 iii 1908, *F.R.R. Schlechter* 17511 (holo B) – Type of *Begonia minjemensis* Irmsch.

■ **IUCN category: DD**. Insufficient specimens could be georeferenced with certainty.

Begonia minutiflora Sands [§ Petermannia], Pl. Mt. Kinabalu 160 (2001). – Type: Borneo, Sabah, Mt. Kinabalu, Sosopodon, 2 ii 1962, *W. Meijer* SAN29025 (holo K).

■ **BORNEO: Sabah:** Mt. Kinabalu, Tenompok F.R., 28 i 1994, *K.M. Wong* 2593 (K); Mt. Kinabalu, Liwagu River Trail, *M.J.S. Sands, Gunsalam* 3902 (para K *n.v.*) – Type of *Begonia minutiflora* Sands; Mt. Kinabalu, Sosopodon, *Kokawa, Hotta* 5141 (para L *n.v.*) – Type of *Begonia minutiflora* Sands; Mt. Kinabalu, Liwagu River Trail, *Kokawa* 6217 (para L *n.v.*) – Type of *Begonia minutiflora* Sands; Mt. Kinabalu, Lugas Hill, *J.H. Beaman* 10533 (para K *n.v.*) – Type of *Begonia minutiflora* Sands; Mt. Kinabalu, Sosopodon, 2 ii 1962, *W. Meijer* SAN29025 (holo K) – Type of *Begonia minutiflora* Sands; Mt. Kinabalu, 30 xi 1933, *Clemens* 50591 (B).

■ **IUCN category: VU D2**. A narrow endemic, not recorded from the core of the Kinabalu National Park.

Begonia modestiflora Kurz [§ Diploclinium], Flora 54: 296 (1871); Clarke, Fl. Brit. Ind. 640 (1879); Tebbitt, Begonias 141 (2005). – Type: Burma, Arracan, Baronga Island, x 1869, *W.S. Kurz s.n.* (lecto K, here designated; isolecto K).

Begonia yunnanensis H. Lév., **syn. nov.**, Repert. Spec. Nov. Regni Veg. 7: 20 (1909); Craib, Fl. Siam. 1: 780 (1931). – Type: China, Yunnan, *Henry* 12403 (lecto E, here designated; isolecto B, BM, HBG, K, L).

Begonia sootepensis var. *thorelii* Gagnep., **syn. nov.**, in Lecomte, Fl. Indo-Chine 2: 1104 (1921). – *Begonia yunnanensis* var. *thorelii* (Gagnep.) Golding & Kareg., Phytologia 54(7):499 (1984). – Type: Laos, Bassac, 1866 – 1868, *Thorel s.n.* (lecto P, here designated; isolecto P); Ile de Khon, 1866 – 1868, *Thorel* 2666 (syn K, P[2]); Ile de Khon, 1866 – 1868, *Thorel* 2239 (syn BM, P).

Begonia lushaiensis C.E.C. Fisch., **syn. nov.**, Bull. Misc. Inform. Kew 1928: 273 (1928). – Type: India, Assam, Lushai Hills, vii 1926, *A.D. Parry* 39 (holo K).

Begonia modestiflora var. modestiflora

■ **BURMA:** Arracan, Baronga Island, x 1869, *W.S. Kurz s.n.* (lecto K; isolecto K) – Type of *Begonia modestiflora* Kurz; Upper Cheidwin Distr., Kindat, 24 viii 1908, *J.H. Lace s.n.* (E); Moulmein, 1862, *C.S.P. Parish* 11? (K); Katha Distr., Pile Forest Reserve, 21 vii 1915, *C. Gilbert Rogers* 993 (E); N. Shan State, Gokteik Gorge, 2 viii 1908, *J.H. Lace* 4159 (A, E, K); N. Shan State, 7 x 1911, *J.H. Lace* 5448 (E, K); West Central Burma, Mt. Victoria, 6 iv 1956, *F.K. Ward* 22680 (E).

THAILAND: Chiengmai, Doi Chiang Dao Animal Sanctuary, 12 ix 1995, *Maxwell* 95-675 (L); Chiengmai, Doi Sahng Liang, 21 ix 1997, *J.F. Maxwell* 97-1021 (L); Chiengmai, Doi Sootep, 1912, *A.F.G. Kerr s.n.* (ABD); Chiengmai, Doi Sootep, 26 viii 1911, *A.F.G. Kerr* 1970 (BM, K); Doi Pa Mawn, 20 ix 1929, *H.B.G. Garrett* 457 (K); Doi Sootep National Park, Bahn Gee Ruins, 16 ix 1993, *P. Palee* 156 (A, L); Doi Sootep National Park, Chang Kian Valley, 23 viii 1988, *J.F. Maxwell* 88-1033 (L); Doi

Sootep National Park, Kohntatahn Falls, 4 xi 1987, *J.F. Maxwell* 87-1363 (L); Doi Sootep National Park, Mae Sa Falls, 28 ix 1989, *J.F. Maxwell* 89-1151 (A, L); Lampang Prov., Chae Son Nat. Park, 22 viii 1995, *Maxwell* 95-531 (L); Lampang Prov., Doi Kuhn Dahn National Park, 13 x 1993, *J.F. Maxwell* 93-1284 (A, L); Lampang Prov., Doi Kuhn Dahn National Park, 30 vii 1994, *J.F. Maxwell* 94-839 (A, L); Lampang Prov., Doi Kuhn Dahn National Park, Mah Meun Station trail, 24 ix 1993, *J.F. Maxwell* 93-1090 (A, L); Payap, Doi Chiengdao, *N. Put* 304 (ABD, BM); Payap, Doi Chiengdao, 14 x 1826, *N. Put* 331 (ABD, BM, K); *ibid.*, 15 viii 1935, *H.B.G. Garrett* 976 (K); *ibid.*, 13 ix 1937, *M. Tagawa et al.* 9889 (L); Prae Prov., Mae Yom Nat. Park, 11 x 1991, *J.F. Maxwell* 91-879 (A, L); Tak Prov., Mu Kee Haw Village, 19 viii 1994, *J.F. Maxwell* 94-898 (A, L).

■ **LAOS:** Laos, iv 1913, *Joseph s.n.* (L) [as *Begonia* cf. *modestiflora*]; Bassac, Khon Island, 9 x 1997, *Maxwell* 97-1081 (L); Bassac, Khon Island, 1866 – 1868, *C. Thorel* 2226 (syn K, P[2]) – Type of *Begonia sootepensis* var. *thorelii* Gagnep.; Me-Kong, 1866 – 1888, *C. Thorel* 2239 (BM, P).

Begonia modestiflora var. sootepensis (Craib) Z. Badcock ex M. Hughes, comb. nov. – *Begonia sootepensis* Craib, Bull. Misc. Inform. Kew 1911: 57 (1911); Craib, Aberdeen Univ. Stud. 47: 96 (1912); Gagnepain, in Lecomte (ed.), Fl. Indo-Chine 2: 1103 (1921). – *Begonia yunnanensis* var. *sootepensis* (Craib) Craib, Aberdeen Univ. Stud. 47: 96 (1912). – Type: Chiengmai, Doi Sootep, *A.F.G. Kerr* 785 (lecto K, here designated; isolecto B, BM, E, P).

■ **THAILAND:** Chiengmai, *A.F.G. Kerr* 1420 (BM); Chiengmai, Doi Sootep, 18 ix 1958, *T. Sorensen et al.* 5096 (L); Chiengmai, Doi Sootep, *A.F.G. Kerr* 785 (lecto K; isolecto B, BM, P) – Type of *Begonia sootepensis* Craib; Doi Pa Mawn, 20 ix 1929, *H.B.G. Garrett* 457 (ABD, B, L[2]); Doi Sootep National Park, Ban Mae Sah Mai, 17 viii 1988, *J.F. Maxwell* 88-1010 (L).

■ **Notes:** The synonymy follows that of Badcock (1998; unpublished PhD thesis).

■ **IUCN category: LC.** Widespread.

Begonia mollis A. DC. [§ Reichenheimia], Prodr. 15(1): 391 (1864); Koorders, Exkurs.-Fl. Java 2: 648 (1912). – Type: Java, 1854, *H. Zollinger* 484 (syn B, K, P); Java, *J.B. Spanoghe s.n.* (syn K).

   Begonia repens Blume, Enum. Pl. Javae 1: 95 (1827); Candolle, Prodr. 15(1): 391 (1864); Merrill, Philipp. J. Sci. 3: 84 (1908); Merrill, Philipp. J. Sci. 6: 406 ('1911', 1912); Koorders, Exkurs.-Fl. Java 2: 649 (1912). – *Diploclinium repens* (Blume) Miq., Fl. Ned. Ind. 1(1): 686 (1856); Miquel, Fl. Ned. Ind., Suppl. 1: 333 (1861); Candolle, Prodr. 15(1): 391 (1864). – Type: not located.

   Scheidweilera repens Hassk., Hort. Bogor. Descr. 325 (1858); Candolle, Prodr. 15(1): 391 (1864). – Type: not located.

   Mitscherlichia repens Miq., Fl. Ned. Ind., Suppl. 1: 333 (1861); Candolle, Prodr. 15(1): 391 (1864). – Type: not located.

■ **SUMATRA:** Sumatra, *P.W. Korthals s.n.* (L); Sumatra, *unknown s.n.* (K); Sumatra, *W. Marsden s.n.* (K); Mt. Singalan, 1878, *O. Beccari* PS127 (K, L).

   JAVA: Java, *J.B. Spanoghe s.n.* (syn K) – Type of *Begonia mollis* A. DC.; Java, *unknown s.n.* (B); Java, iv 1879, *H.O. Forbes* 40 (BM); Java, 1858, *Nagel* 273 (B[2]); Java, 1854, *H. Zollinger* 484 (syn B, K, P) – Type of *Begonia mollis* A. DC.; *H. Zollinger* 749 (BM, K).

   MOLUCCAS: Ambon: Ambon, vii 1913 – xi 1913, *C.B. Robinson* 66 (P).

■ **Notes:** The record from Ambon is annotated as 'representing *Empetrum acetosum rubrum* Rumph.', and is included here as this name is currently considered a synonym of *B. mollis* A. DC. However, the two are almost certainly different taxa.

■ **IUCN category: DD.** The largely historic collections could not be georeferenced with certainty.

Begonia monantha Warb. in K. Schum. & Lauterb. **[§ Petermannia]**, Fl. Deutsch. Schutzgeb. Südsee 322 (1905); Irmscher, Bot. Jahrb. Syst. 50: 573 (1914). – Type: New Guinea, Papua

New Guinea, Kaiser-Wilhelmsland, Torricellia-Geb., iv 1902, *F.R.R. Schlechter* 14575 (holo B; iso B, BM, P).

- **NEW GUINEA: Papua New Guinea:** Kaiser-Wilhelmsland, Torricellia-Geb., iv 1902, *F.R.R. Schlechter* 14575 (holo B; iso B, BM, P) – Type of *Begonia monantha* Warb.; Kaiser-Wilhelmsland, 4 ix 1907, *F.R.R. Schlechter* 20027 (B).
- **IUCN category: DD.** Insufficient specimens could be georeferenced with certainty.

Begonia montis-bismarckii Warb. in K. Schum. & Lauterb. **[§ Petermannia]**, Fl. Deutsch. Schutzgeb. Südsee 322 (1905); Merrill & Perry, J. Arnold Arbor. 24: 49 (1943). – Type: New Guinea, Papua New Guinea, Kaiser-Wilhelmsland, Bismarck-Geb., i 1902, 5°30′S, 144°45′E, *F.R.R. Schlechter* 13988 (syn B); New Guinea, Papua New Guinea, Kaiser-Wilhelmsland, Bismarck-Geb., i 1902, *F.R.R. Schlechter* 14020 (syn B).

- **NEW GUINEA: Papua:** Idenburg River, iii 1939, *L.J.Brass* 13397 (L) [as *Begonia montis-bismarckii*?]. **Papua New Guinea:** Kaiser-Wilhelmsland, Bismarck-Geb., i 1902, *F.R.R. Schlechter* 13988 (syn B) – Type of *Begonia montis-bismarckii* Warb.; Kaiser-Wilhelmsland, Bismarck-Geb., i 1902, *F.R.R. Schlechter* 14020 (syn B) – Type of *Begonia montis-bismarckii* Warb.; Kaiser-Wilhelmsland, 1 xi 1908, *F.R.R. Schlechter* 18608 (B).
- **IUCN category: LC.** The species grows at elevations of 850–1600 m. 'Common in undergrowth of flood plain rain forest' (*Brass* 13397 (L)).

Begonia mooreana (Irmsch.) L.L. Forrest & Hollingsw. [§ Symbegonia], Plant Syst. Evol. 241: 208 (2003). – *Symbegonia mooreana* Irmsch., Bot. Jahrb. Syst. 50: 381 (1913); Forrest & Hollingsworth, Plant Syst. Evol. 241: 208 (2003). – Type: New Guinea, Papua New Guinea, Bololo, 10 ix 1907, *F.R.R. Schlechter* 16541 (syn B, P); Papua New Guinea, Ibo-Gebirges, 28 v 1908, *F.R.R. Schlechter* 17800 (syn B, P).

- **NEW GUINEA: Papua:** camp 3-8, i 1913, *C.B. Kloss s.n.* (K). **Papua New Guinea:** Mt. Hagen vicinity, xii 1977, *W.S. Hoover* 75 (E); Kaiser-Wilhelmsland, Torricellia-Geb., Mt. Somoro, 31 viii 1961, *P.J. Darbyshire* 310 (K); Simbu Prov., Gembogl, 10 x 1980, *J. Sterly* 80-320 (E) [as *cf. Symbegonia mooreana*]; Waro airstrip, 13 x 1973, *M. Jacobs* 9196 (L) [as *Begonia mooreana*?]; Eastern Highlands Distr., Kumul Mission, 12 i 1962, *J.S. Womersley* 14134 (E) [as *cf. Symbegonia mooreana*]; Bololo, 10 ix 1907, *F.R.R. Schlechter* 16541 (syn B, P) – Type of *Symbegonia mooreana* Irmsch.; Ibo-Gebirges, 28 v 1908, *F.R.R. Schlechter* 17800 (syn B, P) – Type of *Symbegonia mooreana* Irmsch.
- **IUCN category: LC.** Collected across 250 km in the Central Range of Papua New Guinea.

Begonia morelii Irmsch. ex Kareg. [§ Reichenheimia], Begonian 42: 295 (1975); Irmscher, Begonian 29: 47 (1962); Karegeannes, Begonian 43: 26 (1976). – Type: Unknown collector and number, 'in Herbario Irmscheris'.

- **Notes:** Described from a plant that appeared spontaneously in cultivation in the Station Centrale de Physiologie Végétale at Versailles. Its natural distribution is unknown, although if its sectional placement is correct it should occur somewhere in the region covered by this checklist.
- **IUCN category: DD.** Insufficient specimens could be georeferenced with certainty.

Begonia moszkowskii Irmsch. [§ Petermannia], Bot. Jahrb. Syst. 50: 341 (1913). – Type: New Guinea, Papua, Van Rees Mts., Naumoni, x 1910, *M. Moszkowski* 358 (holo B).

- **NEW GUINEA: Papua:** Van Rees Mts., Naumoni, x 1910, *M. Moszkowski* 358 (holo B) – Type of *Begonia moszkowskii* Irmsch.
- **IUCN category: DD.** Insufficient specimens could be georeferenced with certainty.

Begonia multangula Blume [§ Sphenanthera], Enum. Pl. Javae 1: 96 (1827); Candolle, Prodr. 15(1): 275 (1864); Koorders, Exkurs.-Fl. Java 2: 646 (1912); Doorenbos, Begonian 47: 213 (1980); Tebbitt, Begonias 205 (2005). – *Platycentrum multangulum* (Blume) Miq., Fl. Ned. Ind. 1(1): 695 (1856). – *Sphenanthera multangula* (Blume) Klotzsch, Bot. Zeitung 15: 181 (1857). – *Casparya multangula* (Blume) A. DC., Prodr. 15(1): 275 (1864). – Type: not located.

Platycentrum multangulum var. *glabrata* (Miq.) Miq., Fl. Ned. Ind. 1(1): 695 (1856); Klotzsch, Bot. Zeitung 15: 182 (1857); Candolle, Prodr. 15(1): 275 (1864). – *Begonia multangula* var. *glabrata* Miq., Pl. Jungh. 4: 418 ('1855', 1857); Candolle, Prodr. 15(1): 275 (1864). – *Casparya multangula* var. *glabrata* (Miq.) A. DC., Prodr. 15(1): 276 (1864). – Type: Java, Gunung Merapi, *F.W. Junghuhn s.n.* (holo not located).

Casparya robusta var. *glabriuscula* A. DC., **syn. nov.**, Prodr. 15(1): 275 (1864). – Type: Java, 1 v 1845, *H. Zollinger* 2844 (lecto B, here designated; isolecto BM, P).

Begonia discolor auct. non R. Br.: Blume, Enum. Pl. Javae 1: 96 (1827).

- **JAVA:** Bandoeng, G. Tangkoeban Prahoe, *J.G. Boerlage* (L); Java, *C.G.C. Reinwardt* (L); Buitenzorg, 20 vii 1888, *M. Fleischer* (B); Java, *C.G.C. Reinwardt s.n.* (L); Java, *unknown s.n.* (L); Java, *P.W. Korthals s.n.* (L); Java, *unknown s.n.* (L); *unknown s.n.* (L); *W.S. Kurz s.n.* (BO); Buitenzorg, 20 vii 1898, *M. Fleischer s.n.* (B); Gunung Merapi, *F.W. Junghuhn s.n.* (holo not located) – Type of *Platycentrum multangulum* var. *glabrata* Miq.; Pasoeroean Res., Tosari, *H.N. Ridley s.n.* (BM, K) [as *Begonia* cf. *multangula*]; Gunung Mandalagiri, 31 iii 1920, *H.J. Lam* 271 (L); *J.C. Ploem* 387 (B); SE Java, 1880, *H.O. Forbes* 857a (A, B, BM); Mt Windu, 7 iv 2001, *W.S. Hoover et al.* 960 (A); West Java, Mt. Malabar, 9 i 2001, *W.S. Hoover, J.M. Hunter, H. Wiriadinata, D. Girmansyah* 982 (A); Priangan, G. Pangrango, 11 iv 2001, *W.S. Hoover et al.* 992 (A); Priangan, Bukit Toenggoel, 3 viii 1924, *C.A. Wisse* 1123 (L); 1 xi 1922, *unknown* 1288 (L) [as *Begonia multangula* ?]; Preanger, Tjibodas, *Sapiin* 1935 (U); 1 v 1845, *H. Zollinger* 2844 (lecto B; isolecto BM, P) – Type of *Casparya robusta* var. *glabriuscula* A. DC.; Preanger, ix 1886, *O. Warburg* 3330 (B); Pasoeroean Res., 12 vi 1935, *C.G.G.J.v. Steenis* 7331 (A); West Java, Mt. Cikurai, 15 iii 2002, *H. Wiriadinata, D. Girmansyah* 10536 (A); Lamongan, G. Taroeb, 12 vii 1938, *C.G.G.J.v. Steenis* 10774 (L); Preanger, *O. Warburg* 11298 (B[2]); G. Tjarame (Tjeremai), 21 xii 1940, *C.G.G.J.v. Steenis* 12829 (L); Pekalongan, Jororedjo, 17 ix 1911, *C.A.B. Backer* 16240 (BO); Pasoeroean Res., Soembertankil, 18 vi 1896, *S.H. Koorders* 23064B (L); Preanger, Tjibodas, *S.H. Koorders* 32001B (L); Pasoeroean Res., Tosari, 31 x 1892, *S.H. Koorders* 37006B (B); West Java, Mt. Cikurai, 12 iii 2002, *H. Wiriadinata, D. Girmansyah* 105033 (A n.v.).
  LESSER SUNDA ISLANDS: Bali: G. Abang, *C.G.G.J.v. Steenis* 8059 (A, K). **Lombok:** Rinjani Nat. Park, 23 iv 2001, *W.S. Hoover, J.M. Hunter, H. Wiriadinata, D. Girmansyah* 10008 (A) [as *Begonia* cf. *multangula*]. **Sumbawa:** Mt. Batulanteh, 22 iv 1961, *A.J.G.H. Kostermans* 18390 (A, P). **Flores:** Rana Mese, *I. Rensch* 1159 (B).
- **Notes:** Here I have followed the taxonomy of Tebbitt (1997) and synonymised *Casparya robusta* var. *glabriuscula* A. DC. with *B. multangula* Blume. This species is very close to and indeed nomenclaturally tangled with *B. robusta* Blume. At their most 'typical' form, *B. robusta* may be distinguished from *B. multangula* thus (Tebbitt, 2005):
  - *B. robusta*: Leaf blades above with long red glandular hairs, margin entire or with short rounded lobes; peduncle of infructescence usually at least four times longer than the pedicels; fruit usually with one longer wing and two shorter wings or ribs.
  - *B. multangula*: Leaf blades glabrous above or with short to long white or red glandular hairs, margin with several short angular lobes; peduncle of infructescence usually less than three times as long as the pedicels; fruit with three equal ribs or short thickened wings.
  However, many intermediate (possibly hybrid) specimens exist.
- **IUCN category: LC.** Widespread.

Begonia multidentata Warb. in K. Schum. & Lauterb. **[§ Petermannia]**, Fl. Deutsch. Schutzgeb. Südsee 322 (1905). – Type: New Guinea, Papua New Guinea, Kaiser-Wilhelmsland, Torricellia-Geb., iv 1902, *F.R.R. Schlechter* 14458 (holo B; iso BM).

- **NEW GUINEA: Papua New Guinea:** Kaiser-Wilhelmsland, Torricellia-Geb., iv 1902, *F.R.R. Schlechter* 14458 (holo B; iso BM) – Type of *Begonia multidentata* Warb.
- **IUCN category: DD**. Insufficient specimens could be georeferenced with certainty.

Begonia muricata Blume [§ Reichenheimia], Catalogus 103 (1823); Candolle, Prodr. 15(1): 323 (1864); Koorders, Exkurs.-Fl. Java 2: 648 (1912). – Type: not located.

Begonia saxatilis Blume, Enum. Pl. Javae 1: 95 (1827); Candolle, Prodr. 15(1): 351 (1864); Koorders, Exkurs.-Fl. Java 2: 647 (1912); Backer & Brink, G. v.d., Fl. Java 1: 309 (1964); Smith & Wasshausen, Phytologia 54(7): 470 (1984). – *Diploclinium saxatile* (Blume) Miq., Fl. Ned. Ind. 1(1): 686 (1856); Candolle, Prodr. 15(1): 351 (1864). – Type: Java, *unknown s.n.* (syn L, P).

Begonia rubra Blume, Enum. Pl. Javae 1: 96 (1827); Miquel, Fl. Ned. Ind. 1(1): 689 (1856); Hasskarl, Hort. Bogor. Descr. 349 (1858); Candolle, Prodr. 15(1): 275 (1864); Hasskarl, Neu. Schuss. Rumph. 146 (1866). – *Diploclinium rubrum* (Blume) Miq., Fl. Ned. Ind. 1(1): 689 (1856); Candolle, Prodr. 15(1): 275 (1864); Hasskarl, Neu. Schuss. Rumph. 146 (1866); Merrill, Interp. Herb. Amboin. 379 (1917); Smith & Wasshausen, Phytologia 54(7): 471 (1984). – Type: not located.

Diploclinium tuberosum Miq., Fl. Ned. Ind. 1(1): 685 (1856); Candolle, Prodr. 15(1): 323 (1864); Merrill, Interp. Herb. Amboin. 379 (1917); Smith & Wasshausen, Phytologia 52: 446 (1983); Smith & Wasshausen, Phytologia 54(7): 471 (1984). – Type: not located.

Sphenanthera robusta var. *rubra* Hassk., Hort. Bogor. Descr. 349 (1858); Candolle, Prodr. 15(1): 275 (1864); Smith & Wasshausen, Phytologia 54(7): 472 (1984). – Type: not located.

Casparya robusta var. *rubra* A. DC., Prodr. 15(1): 275 (1864); Smith & Wasshausen, Phytologia 54(7): 471 (1984). – Type: not located.

Begonia rumphii Vuijck ex Koord., Exkurs.-Fl. Java 2: 649 (1912). – Type: not located.

Begonia forbesii Vuijck ex Koord., Exkurs.-Fl. Java 2: 649 (1912). – Type: not located.

- **SUMATRA:** *H.O. Forbes* 2255 (B, BM, K); Pajakumbuh, Mt. Sago, 4 iv 1983, *S. Danimihardja* 2303 (L); Panti Nature Reserve, 9 ix 1988, *H. Nagamasu* 3233 (L); Mt. Tandikat, 23 vii 1955, *W. Meijer* 3911 (BM[2]); Sipora Island, 9 x 1924, *C.B. Kloss* 14652 (K); Atjeh, Gunung Leuser Nature Reserve, 6 vii 1979, *de Wilde, de Wilde-Duyfies* 18473 (L).
  JAVA: *W. Lobb s.n.* (K); Mt. Megamendoeng, 10 xii 1940, *C.N.A. de Voogd s.n.* (BO, K); Tjisaroewa, 12 ix 1932, *J. & M.S. Clemens s.n.* (K); *C.L.v. Blume s.n.* (P); Mt. Salak, 17 ii 1915, *H.N. Ridley s.n.* (K); *unknown s.n.* (syn L, P) – Type of *Begonia saxatilis* Blume; 13 vii 1858, *A.F. Jagor* 335 (B); Gunung Goentoer, 22 i 1916, *R.C. Bakhuizen van den Brink* 381 (BO); Salak, G. Gadjah, 13 ii 1921, *R.C. Bakhuizen van den Brink* 579 (U); West Java, Mt. Cikurai, 22 iii 2001, *W.S. Hoover, J.M. Hunter, H. Wiriadinata, D. Girmansyah* 948 (A); West Java, Mt. Talaga Warna, 11 iv 2001, *W.S. Hoover, J.M. Hunter, H. Wiriadinata, D. Girmansyah* 986 (not located); Priangan, G. Pangrango, 11 iv 2001, *W.S. Hoover, J.M. Hunter, H. Wiriadinata, D. Girmansyah* 993 (A); Mt. Halimun N.P., 15 iv 2001, *W.S. Hoover, J.M. Hunter, H. Wiriadinata, D. Girmansyah* 996 (A); Preanger, Pateungteung, 10 v 1917, *Doctors v. Leeuwen-Reynvaan* 2604 (B, BO); Mt. Salak, Waroengloa, 2 viii 1929, *C.G.G.J.v. Steenis* 3003 (BO); Bogor, 24 viii 1845, *H. Zollinger* 3006 (A, B, BM, P); Salak, G. Gadjah, 8 viii 1920, *R.C. Bakhuizen van den Brink* 4000 (BO); Batavia, G. Semboeng, 10 ix 1932, *C.G.G.J.v. Steenis* 5162 (BO); Batavia, G. Salak, Pasir Pogor, 26 ix 1913, *C.A.B. Backer* 9377 (BO); West Java, Mt. Bodas, 17 iv 2001, *W.S. Hoover, J.M. Hunter, H. Wiriadinata, D. Girmansyah* 10004 (A); Tjisarua, v 1937, *C.G.G.J.v. Steenis* 10301 (L); West Java, Mt. Cikurai, 12 iii 2002, *H. Wiriadinata, D. Girmansyah* 10502 (A); West Java, Mt. Cikurai, 14 iii 2002, *H. Wiriadinata, D. Girmansyah* 10514 (A); West Java, Mt. Cikurai, 15 iii 2002, *H. Wiriadinata, D. Girmansyah* 10530 (A); Mt. Halimun N.P., 20 v 2002, *H. Wiriadinata et*

al. 10668 (A); Banjoemas, Gunung Slamat, 18 iii 2004, *H. Wiriadinata et al.* 11345 (A); Buitenzorg, G. Halimoen, 4 v 1941 – 5 v 1941, *C.G.G.J.v. Steenis* 12430 (BO); Garoet, Kamodjan, 21 iv 1930, *Doctors v. Leeuwen-Reynvaan* 13291 (BO); Geger Bintang, 1 v 1950, *S.J.v. Ooststroom* 13836 (L); Geger Bintang, 1 v 1950, *S.J.v. Ooststroom* 13854 (L); Preanger, Pengalangan, 1919, *C.A.B. Backer* 26225 (B, BO); Batavia, G. Salak, 1 xii 1918, *C.A.B. Backer* 26341 (BO).
LESSER SUNDA ISLANDS: Sumbawa: xi 1879, *P.F.A. Colfs* 299 (L); *A.J.G.H. Kostermans* 18400 (A). **Sumba:** Tarimbang, *J.E. Teijsmann* 8818 (BO).
- **Notes:** Specimens in L and P annotated by Blume as *B. saxatilis* (here listed as syntypes) appear to differ from *B. muricata* and hence this species could be reinstated after further investigation.
- **IUCN category: LC.** Widespread.

Begonia murina Craib [§ Diploclinium], Gard. Chron., III 83: 66 (1928); Craib, Fl. Siam. 1: 776 (1931). – Type: Thailand, Rachaburi, Kanburi, *A.F.G. Kerr* 113 (holo ABD; iso ABD).
- **THAILAND:** Rachaburi, Kanburi, *A.F.G. Kerr* 113 (holo ABD; iso ABD) – Type of *Begonia murina* Craib.
- **IUCN category: DD.** Insufficient specimens could be georeferenced with certainty.

Begonia murudensis Merr. [§ Petermannia], Sarawak Mus. J. 3: 530 (1928). – Type: Borneo, Sarawak, Mt. Murud, x 1922, *E.G. Mjoberg* 119 (syn BM, K).
- **BORNEO: Sarawak:** Mt. Murud, x 1922, *E.G. Mjoberg* 119 (syn BM, K) – Type of *Begonia murudensis* Merr.
- **IUCN category: DD.** Insufficient specimens could be georeferenced with certainty.

Begonia mystacina L.B. Sm. & Wassh. [§ Petermannia], Phytologia 54(7): 469 (1984). – *Begonia richardsoniana* Merr. & L.M. Perry, J. Arnold Arbor. 24: 48 (1943); Smith & Wasshausen, Phytologia 54(7): 469 (1984). – Type: New Guinea, Papua, Idenburg River, iii 1939, *L.J. Brass* 13423 (holo A *n.v.*; iso L).
- **NEW GUINEA: Papua:** Vogelkopf, Tohkiri Range, 7 xi 1961 – 8 xi 1961, *P.v. Royen, H. Sleumer* 7265 (L); Mamberamo, Albatros Camp, v 1926, *Doctors v. Leeuwen-Reynvaan* 9037 (L) [as *Begonia cf. richardsoniana*]; Van Rees Mts., v 1926, *Doctors v. Leeuwen-Reynvaan* 9174 (L); Wandammen Peninsula, Wondiwoi Mts., 24 ii 1962, *F.A.W. Schram* 10633 (L); Idenburg River, iii 1939, *L.J. Brass* 13423 (holo A *n.v.*; iso L) – Type of *Begonia richardsoniana* Merr. & L.M. Perry.
- **IUCN category: DD.** Insufficient specimens could be georeferenced with certainty.

Begonia naumoniensis Irmsch. [§ Petermannia], Bot. Jahrb. Syst. 50: 362 (1913). – Type: New Guinea, Papua, Van Rees Mts., Naumoni, x 1910, *M. Moszkowski* 300 (syn B); New Guinea, Papua, Van Rees Mts., Taua, vii 1911 – viii 1911, *M. Moszkowski* 325 (syn B).
- **NEW GUINEA: Papua:** Van Rees Mts., Naumoni, x 1910, *M. Moszkowski* 300 (syn B) – Type of *Begonia naumoniensis* Irmsch.; Van Rees Mts., Taua, vii 1911 – viii 1911, *M. Moszkowski* 325 (syn B) – Type of *Begonia naumoniensis* Irmsch.; Mamberamo, Koeroedoe Island, x 1913, *R.F. Janowski* 479 (L); Jappen-Biak, 20 ix 1397, *L.J.V. Dijk* 847 (L) [as *Begonia cf. naumoniensis*]; Mamberamo, Oktenriver, xi 1926, *Doctors v. Leeuwen-Reynvaan* 11379 (L). **Papua New Guinea:** Aitape Subdistr., Kaiye Village, 18 vii 1961, *P.J. Darbyshire, R.D. Hoogland* 8173 (BM, E, L); Sepik District, Puari Creek, 17 iii 1984, *C.D. Sayers* 18975 (L); Sepik District, Vanimo Subdistr., Krisa Village, 2 ii 1969, *H. Streimann, A. Kairo* 39318 (L).
- **Notes:** The apex of the fruit of this species is slightly retuse, giving the capsules a characteristic heart-shape when pressed.

- **IUCN category: LC**. Collected from a range covering 500 km. Also found in secondary forest (*H. Streimann, A. Kairo* 39318 (L)).

Begonia negrosensis Elmer [§ Petermannia], Leafl. Philipp. Bot. 2: 736 (1910); Merrill, Philipp. J. Sci. 6: 385 ('1911', 1912); Merrill, Enum. Philipp. Fl. Pl. 3: 125 (1923). – Type: Philippines, Negros, Negros Occidental, Cuernos Mts., vi 1908, *A.D.E. Elmer* 9903 (syn BM, BO, E, FI, K, L).

- **PHILIPPINES: Luzon:** Pampanga, Mt. Pinatubo, v 1927, *A.D.E. Elmer* 21492 (K) [as *Begonia negrosensis* ?]; Mountain Province, Banaue, Bayinan, 9 xii 1962, *H. Conklin, Buwaya* 70693 (L). **Panay:** Capiz, Libacao, v 1919 – vi 1919, *A. Martelino, G. Edano* 35291 (B, BM, P); Capiz, Libacao, v 1919 – vi 1919, *A. Martelino, G. Edano* 35299 (B, BM, K, P); Capiz, Libacao, v 1919 – vi 1919, *A. Martelino, G. Edano* 35425 (BM). **Negros:** Negros Occidental, 8 xi 1999, *P. Wilkie et al.* 76 (E[2]); Negros Oriental, Lake Danao, viii 1948, *G. Edano* 6670 (L); Canlaon Volcano, iv 1910, *E.D. Merrill* 7024 (B, BM, K, L, P); Negros Occidental, Cuernos Mts., vi 1908, *A.D.E. Elmer* 9903 (syn BM, BO, E, FI, K, L) – Type of *Begonia negrosensis* Elmer; Negros Occidental, Kinabkaban River, 25 iii 1954, *G. Edano* 21855 (L). **Samar:** Mt. Malingon, iv 1948 – v 1948, *M.D. Sulit* 9181 (L).
- **IUCN category: LC**. Described as 'very common' (*P. Wilkie et al.* 76 (E)). Also occurs in Mt. Kanlaon National Park.

Begonia neopurpurea L.B. Sm. & Wassh. nom. nud. [not placed to section], Phytologia 52: 445 (1983). – *Begonia purpurea* Elmer *nom. nud.*, Leafl. Philipp. Bot. 10: 3707 (1939); Smith & Wasshausen, Phytologia 52: 445 (1983). – Type: Philippines, Luzon, Sorsogon, Mt. Bulusan, vii 1916, *A.D.E. Elmer* 16565 (syn A).

- **PHILIPPINES: Luzon:** Sorsogon, Mt. Bulusan, vii 1916, *A.D.E. Elmer* 16565 (syn A) – Type of *Begonia purpurea* Elmer *nom. nud.*
- **Notes:** Tentatively assigned to § *Diploclinium* (Doorenbos *et al.*, 1998). Strictly speaking, this is not a valid name as no Latin diagnosis was included in the description, necessary from 1935. However, it is included here for completeness.
- **IUCN category: VU D2**. Known only from the type, from a rather imprecise locality which may well be inside the Mt. Bulusan National Park. However, the altitude of the type collection is c. 200 m, and the proportion of forest at this altitude inside the reserve is small.

Begonia nepalensis (A. DC.) Warb. in Engl. & Prantl **[§ Monopteron]**, Nat. Pflanzenfam. 3(6A): 142 (1894). – *Mezierea nepalensis* A. DC., Ann. Sci. Nat. Bot., IV 11: 144 (1859) – Type: not located. *Begonia gigantea* Wall. *nom. nud.*, Num. List 129: 3677 (1831).

- **BURMA:** *Fide* Kress *et al.* (2003).
- **IUCN category: DD**. Insufficient specimens could be georeferenced with certainty.

Begonia niahensis K.G. Pearce [§ Petermannia], Gard. Bull. Singapore 55: 77 (2003). – Type: Borneo, Sarawak, Niah Caves, 4 vi 1962, 3°49′N, 113°47′E, *B.L. Burtt, P.J.B. Woods* 2009 (holo SAR *n.v.*; iso E[2]).

- **BORNEO: Sarawak:** Niah Caves, 4 vi 1962, *B.L. Burtt, P.J.B. Woods* 2009 (holo SAR *n.v.*; iso E[2]) – Type of *Begonia niahensis* K.G. Pearce; Niah National Park, *G.D. Haviland, C. Hose* 3224 (not located); Gua Niah, *J.A.R. Anderson, S. Tan, E. Wright* S26075 (para A, SAR *n.v.*, SING *n.v.*) – Type of *Begonia niahensis* K.G. Pearce; Gunung Brangin, *Yii Puan Ching* S40166 (para SAR *n.v.*) – Type of *Begonia niahensis* K.G. Pearce; Gua Niah, *K.G. Pearce, Remlanto* S78537 (para SAR *n.v.*) – Type of *Begonia niahensis* K.G. Pearce; Sg. Subis, *K.G. Pearce* S89460 (para SAR *n.v.*) – Type of *Begonia niahensis* K.G. Pearce.
- **IUCN category: VU D2**. Only one of the localities is within a protected area (Niah National Park).

Begonia nigritarum (Kamel) Steud. [§ Diploclinium], Nom. Bot. 1: 104 (1821); Dryander, Trans. Linn. Soc. 1: 171 (1791); Candolle, Prodr. 15(1): 401 (1864); Fernández-Villar, Noviss. App. 99 (1880); Merrill, Philipp. J. Sci. 6: 393 ('1911', 1912); Merrill, Fl. Man. 338 (1912); Merrill, Sp. Blancoanae 277 (1918); Merrill, Enum. Philipp. Fl. Pl. 3: 126 (1923). – *Acetosa nigritarum* Kamel, in J. Ray, Herb. Philipp. 3(app.): 14 (1704). – Type: not located.
Begonia capensis Blanco, Fl. Filip. 724 (1837); Náves, Fl. Filip. 3: 127 (1879); Merrill, Philipp. J. Sci. 6: 393 ('1911', 1912). – Type: not located.
Begonia rhombicarpa A. DC., Ann. Sci. Nat. Bot., IV 11: 129 (1859); Candolle, Prodr. 15(1): 323 (1864); Merrill, Philipp. J. Sci. 6: 394 ('1911', 1912). – Type: Philippines, Luzon, 1841, *H. Cuming* 510 (holo G-DC; iso B, E, FI, K, L[2], P).
Begonia rhombicarpa var. *lobbii* A. DC., Prodr. 15(1): 323 (1864). – Type: Philippines, Luzon, *W. Lobb s.n.* (holo K).
Begonia merrillii Warb. in Perkins, Fragm. Fl. Philipp. 53 (1904); Merrill, Philipp. J. Sci. 6: 394 ('1911', 1912). – Type: not located.

- **PHILIPPINES: Palawan:** viii 1913, *L. Escritor* 21551 (K); viii 1913, *L. Escritor* 21577 (P); Busuanga Island, ix 1922, *M. Ramos* 41222 (B); Busuanga Island, ix 1922, *G. Lopez* 41374 (B, K, L, P). **Luzon:** *W. Lobb s.n.* (holo K) – Type of *Begonia rhombicarpa* var. *lobbii* A. DC.; Manilla, *unknown s.n.* (P); Cordon Nueva Viscaya, *E.D. Merrill* 143 (B *n.v.*[2], BO *n.v.*); Bataan, Mt. Mariveles, v 1904, *H.N. Whitford* 204 (B); Bataan, Mt. Mariveles, vii 1904, *H.N. Whitford* 492 (B); Bataan, Mt. Mariveles, vii 1904, *H.N. Whitford* 500 (B[2]); 1841, *H. Cuming* 510 (holo G-DC; iso B, E, FI, K, L[2], P) – Type of *Begonia rhombicarpa* A. DC.; Bataan, Mt. Mariveles, v 1904, *T.E. Borden* 753 (B); Cavite Province, Mendez Nunez, *L. Mangubat* 1286 (B, BO); Bataan, Mt. Mariveles, Lamao River, v 1905, *H.N. Whitford* 1297 (B, K, P); Bataan, Lamar River, ix 1905, *H.N. Whitford* 1345 (B, K); Rizal, Tanay, v 1903, *E.D. Merrill* 2343 (B, K); Bataan, Mt. Mariveles, xii 1903, *E.D. Merrill* 3124 (B[2], BM, K); Manilla, 1836, *C. Gaudichaud-Beaupré* 3301 (P); Rizal Province, ix 1905, *Aherns collector* 3308 (B); Bataan, Mt. Mariveles, viii 1904, *E.D. Merrill* 3862 (B, BM); Rizal Province, Montalban, 3 viii 1890, *A. Loher* 4313 (K); Bataan, Limay, 11 viii 1908 – 14 viii 1908, *C.B. Robinson* 6183 (B, P); Bataan, Mt. Mariveles, xi 1904, *A.D.E. Elmer* 6680 (E); Laguna, Los Banos, *C.B. Robinson* 9912 (BM); Rizal, Antipolo, vi 1910, *M. Ramos* 12105 (B, BM[2], K, P); Laguna, Makiling, xi 1912, *V. Servinas* 16854 (BM, K, P); Laguna, Makiling, vi 1917 – vii 1917, *A.D.E. Elmer* 17474 (BM[2], K, P, U); Pangasinam Province, Umingan, iv 1914 – vi 1914, *F. Otanes* 17947 (K); Bulacan Province, Angat, ix 1913, *M. Ramos* 21676 (BM, K, P); Pampanga, Mt. Pinatubo, v 1927, *A.D.E. Elmer* 21998 (BM, K, P); Laguna, Makiling, xi 1912, *C. Mabesa* 26174 (BM); Tayabas, iii 1917, *G. Edano* 26915 (K, P). **Mindoro:** Bulalacao, viii 1906 – ix 1906, *J. Bermejos* 1527 (B, BO, P); Puerto Galera, x 1903, *E.D. Merrill* 3324 (B, BM, K). **Panay:** Iloilo Province, 27 xii 1912 – 31 xii 1912, *C.B. Robinson* 18022 (K). **Bohol:** *M. Ramos* 43302 (BM). **Leyte:** v 1913, *C.A. Wenzel* 42 (E *n.v.*); 17 viii 1915, *C.A. Wenzel* 1568 (BM); Palo, *A.D.E. Elmer* 7115 (BO, E). **Mindanao:** 1838 – 1840, *E.J.F.I. Guillou s.n.* (P) [as *Begonia rhombicarpa* ?].
- **Notes:** One of the most widespread species in the Philippines, and is highly variable as a result.
- **IUCN category: LC.** Widespread.

Begonia nivea Parish ex Kurz [§ Reichenheimia], J. Asiat. Soc. Bengal 42(2): 81 (1873). – Type: not located.
- **BURMA:** Moulmein, 1862, *C.S.P. Parish s.n.* (K); Tenasserim Division, Sin Yat Koa Falls, 22 x 1998, *J.F. Maxwell* 98-1316 (L).
- **IUCN category: DD.** Insufficient specimens could be georeferenced with certainty.

Begonia notata Craib [§ Diploclinium], Gard. Chron., III 83: 66 (1928); Craib, Fl. Siam. 1: 777 (1931). – Type: Thailand, Surat, Ta Ngaw, *A.F.G. Kerr* 147 (holo ABD; iso ABD).

- **THAILAND:** Surat, Ta Ngaw, *A.F.G. Kerr* 147 (holo ABD; iso ABD) – Type of *Begonia notata* Craib.
- **IUCN category: DD.** Insufficient specimens could be georeferenced with certainty.

Begonia novoguineensis Merr. & L.M. Perry [§ Petermannia], J. Arnold Arbor. 24: 57 (1943). – Type: New Guinea, Papua, Hollandia, vi 1938, *L.J. Brass* 8841 (holo A; iso L).

- **NEW GUINEA: Papua:** 1903, *Atasrip* 235 (L); Cycloop Mts., Ifar – Ormoe, 4 xi 1954, *C. Versteegh* 939 (L); Cycloop Mts., 31 xii 1954, *H.S. McKee* 1843 (L); Cycloop Mts., 31 xii 1954, *H.S. McKee* 1846 (L); Cycloop Mts., Ifar – Ormoe, 13 x 1954, *P.v. Royen* 3778 (L); Sentani, 9 viii 1957, *C. Kalkman* 3783 (L); Cycloop Mts., Ifar – Ormoe, 18 xii 1954, *P.v. Royen* 5133 (L); Cycloop Mts., Faika River – Gawesar River, 10 vi 1961, *P.v. Royen, H. Sleumer* 5678 (L); Cycloop Mts., Ifar – Ormoe, 13 vi 1961, *P.v. Royen, H. Sleumer* 5817 (L); Cycloop Mts., Baimungun Creek – Klifon River, 8 viii 1961, *P.v. Royen, H. Sleumer* 6533 (L); Hollandia, vi 1938, *L.J. Brass* 8841 (holo A; iso L) – Type of *Begonia novoguineensis* Merr. & L.M. Perry. **Papua New Guinea:** Kar Kar Island, 25 i 1968, *C.E. Ridsdale* 36747 (L).
- **IUCN category: LC.** Described as 'relatively common up to 1000 m in altitude' (*P.v. Royen* 3778 (L)).

Begonia nurii Irmsch. [§ Reichenheimia], Mitt. Inst. Allg. Bot. Hamburg 8: 95 (1929); Kiew, Begonias Penins. Malaysia 233 (2005). – Type: Peninsular Malaysia, Kelantan, Sungai Ketch, 9 ii 1924, *Md. Nur, F.W. Foxworthy* SFN12026 (holo SING *n.v.*; iso K).

- **PENINSULAR MALAYSIA:** Pahang, Bukit Serdam limestone, 20 vi 1971, *S.C. Chin* 1121 (L); Pahang, Gua Tipus, Chegar Perah?, 14 x 1927, *M.R. Henderson* SFN19382 (BO); Kelantan, Gua Ikan, 14 v 1990, *R. Kiew, S. Anthonysamy* 2938 (L); Kelantan, Bertam, 5 viii 1962, *Unesco Limestone Expedition* 365 (L); Kelantan, Sungai Ketch, 9 ii 1924, *Md. Nur, F.W. Foxworthy* SFN12026 (holo SING *n.v.*; iso K) – Type of *Begonia nurii* Irmsch.
- **IUCN category: VU D2.** Restricted to limestone in the Gua Musang area (Kiew, 2005).

Begonia oblongata Merr. [§ Petermannia], Philipp. J. Sci. 7: 310 (1912); Merrill, Enum. Philipp. Fl. Pl. 3: 126 (1923). – Type: Philippines, Mindanao, Zamboanga, 6 xii 1911, *E.D. Merrill* 8166 (syn BM).

- **PHILIPPINES: Mindanao:** Zamboanga, 6 xii 1911, *E.D. Merrill* 8166 (syn BM) – Type of *Begonia oblongata* Merr.; Zamboanga, 6 xii 1911, *E.D. Merrill* 8175 (para B, BM) – Type of *Begonia oblongata* Merr.; Zamboanga, Malangas, x 1919 – xi 1919, *M. Ramos, G. Edano* 36797 (B, P); Zamboanga, Malangas, *M. Ramos, G. Edano* 37420 (P); Zamboanga, Malangas, x 1919 – xi 1919, *M. Ramos, G. Edano* 37424 (B).
- **IUCN category: EN B2ab(iii).** Known only from three historical collections in the heavily degraded Mindano–Eastern Visayas rain forest eco-region.

Begonia oblongifolia Stapf [§ Petermannia], Trans. Linn. Soc. London, Bot., II 4: 165 (1894); Ridley, J. Malayan Branch Roy. Asiat. Soc. 46: 251 (1906); Sands, in Beaman *et al.*, Pl. Mt. Kinabalu 161 (2001). – Type: Borneo, Sabah, Mt. Kinabalu, Dahombang River, *G.D. Haviland* 1308 (holo K).

- **BORNEO: Sabah:** Dahobang Ridge, 4 i 1933, *J. & M.S. Clemens* Suppl. (BM); Mt. Kinabalu, Dahombang River, *G.D. Haviland* 1308 (holo K) – Type of *Begonia oblongifolia* Stapf; Mt. Kinabalu, Dallas, 15 ix 1931, *J. & M.S. Clemens* 26397 (BM); Mt. Kinabalu, Dallas, 8 xii 1931, *J. & M.S. Clemens* 27463 (BM); Mt. Kinabalu, Penibukan, ix 1931 – xii 1931, *J. & M.S. Clemens* 30531 (BM, BO, K).
- **IUCN category: LC.** Occurs within the Mt. Kinabalu National Park.

Begonia obovoidea Craib [§ Sphenanthera], Bull. Misc. Inform. Kew 1930: 413 (1930); Craib, Fl. Siam. 1: 777 (1931). – Type: Thailand, Puket, Ranawng, Nam Chut, 18 i 1927, 10°6'N, 99°6'E, *A.F.G. Kerr* 12902 (holo ABD; iso K).

■ **BURMA:** Tenasserim Division, Tavoy District, Paungdaw Power Station, 6 ix 1961, *J. Keenan* 1425 (A, E).
 THAILAND: Lansagah Distr., Khao Luang National Park, 3 ii 1990, *W.S. Hoover* 742 (A); Puket, Ranawng, Nam Chut, 18 i 1927, *A.F.G. Kerr* 12902 (holo ABD; iso K) – Type of *Begonia obovoidea* Craib; Kao Nawng, *A.F.G. Kerr* 13367 (not located).
■ **IUCN category: LC.** Occurs in Tai Rom Yen and Khao Luang National Parks.

Begonia obtusifolia Merr. [§ Diploclinium], Philipp. J. Sci. 14: 425 (1919); Merrill, Enum. Philipp. Fl. Pl. 3: 126 (1923). – Type: Philippines, Panay, Capiz, Mt. Macosolon, 19 iv 1918, *M. Ramos, G. Edano* 30803 (iso P).

■ **PHILIPPINES: Panay:** Capiz, Mt. Macosolon, 19 iv 1918, *M. Ramos, G. Edano* 30803 (iso P) – Type of *Begonia obtusifolia* Merr.
■ **IUCN category: DD.** Insufficient specimens could be georeferenced with certainty.

Begonia oligandra Merr. & L.M. Perry [not placed to section], J. Arnold Arbor. 24: 44 (1943). – Type: New Guinea, Papua, Idenburg River, i 1939, *L.J. Brass* 12344 (holo A).

■ **NEW GUINEA: Papua:** Tenmasigin, Orion Mts., 22 v 1959, *C. Kalkman* 4106 (A); Bele River Valley, x 1938, *L.J. Brass* 10829 (para A, BO) – Type of *Begonia oligandra* Merr. & L.M. Perry; Idenburg River, i 1939, *L.J. Brass* 12344 (holo A) – Type of *Begonia oligandra* Merr. & L.M. Perry. **Papua New Guinea:** Telefomin Subdist., Star Mts., 8 iv 1975, *J. Croft, Y. Leiean* 65767 (A, E).
■ **Notes:** Tentatively assigned to § *Diploclinium* (Doorenbos *et al.*, 1998).
■ **IUCN category: LC.** Occurs within three protected or proposed protected areas (Mamberambo Pegunungan Foja Wildlife Sanctuary, Jaywijaya Game Reserve and Mt. Capella designated reserve). It is distributed in the relatively less disturbed higher altitude forests (1600–2200 m).

Begonia oligantha Merr. [§ Petermannia], Philipp. J. Sci. 10: 50 (1915); Merrill, Enum. Philipp. Fl. Pl. 3: 126 (1923). – Type: Philippines, Mindanao, Bukidnon subprovince, Sumilao, 4 viii 1912, *E. Fenix* 15733 (syn B).

■ **PHILIPPINES: Mindanao:** Bukidnon subprovince, Sumilao, 4 viii 1912, *E. Fenix* 15733 (syn B) – Type of *Begonia oligantha* Merr.; Bukidnon subprovince, Mt. Candoon, vi 1920 – vii 1920, *M. Ramos, G. Edano* 38808 (B, L, P).
■ **IUCN category: DD.** Insufficient specimens could be georeferenced with certainty.

Begonia orbiculata Jack [not placed to section], Malayan Misc. 2(7): 9 (1822). – *Diploclinium orbiculatum* (Jack) Miq., Fl. Ned. Ind. 1(1): 688 (1856); Candolle, Prodr. 15(1): 398 (1864). – Type: not located.

■ **SUMATRA:** *T. Horsfield s.n.* (BM).
■ **IUCN category: DD.** Insufficient specimens could be georeferenced with certainty.

Begonia oreodoxa Chun & F. Chun ex C.Y. Wu & T.C. Ku [§ Platycentrum], Acta Phytotax. Sin. 33(3): 274 (1995). – Type: China, Yunnan, 1 iv 1940, *X. Wang et al.* 100036 (holo IBSC).

■ **VIETNAM:** *Fide* Shui (2006).
■ **IUCN category: DD.** Insufficient specimens could be georeferenced with certainty.

Begonia oreophila Kiew [§ Platycentrum], Gard. Bull. Singapore 57: 131 (2005). – *Begonia nubicola* Kiew, Begonias Penins. Malaysia 275 (2005). – Type: Peninsular Malaysia, Trengganu, Bukit Labohan, 31 viii 2000, *R. Kiew* 5091 (holo SING *n.v.*; iso K *n.v.*, KEP *n.v.*).

■ **PENINSULAR MALAYSIA:** Trengganu, Bukit Labohan, *Davison* BL1 (para KEP *n.v.*, L) – Type of *Begonia nubicola* Kiew; Trengganu, Bukit Labohan, 31 viii 2000, *R. Kiew* 5091 (holo SING *n.v.*; iso K *n.v.*, KEP *n.v.*) – Type of *Begonia nubicola* Kiew; Pahang, Bukit Pelindung, *R. Kiew* 5291 (para SING *n.v.*) – Type of *Begonia nubicola* Kiew.

■ **IUCN category: VU D2.** Known only from two low altitude sites. Any disturbance of these areas will immediately lead to a threat level of EN.

Begonia otophora Merr. & L.M. Perry [§ Petermannia], J. Arnold Arbor. 24: 46 (1943). – Type: New Guinea, Papua, Idenburg River, iii 1938, *L.J. Brass* 13218 (holo A).

■ **NEW GUINEA: Papua:** Idenburg River, iii 1938, *L.J. Brass* 13218 (holo A) – Type of *Begonia otophora* Merr. & L.M. Perry.

■ **IUCN category: DD.** Insufficient specimens could be georeferenced with certainty.

Begonia oxysperma A. DC. [§ Baryandra], Ann. Sci. Nat. Bot., IV 11: 122 (1859); Candolle, Prodr. 15(1): 287 (1864); Fernández-Villar, Noviss. App. 98 (1888); Merrill, Philipp. J. Sci. 6: 399 ('1911', 1912); Merrill, Enum. Philipp. Fl. Pl. 3: 126 (1923); Tebbitt, Begonias 189 (2005). – Type: Philippines, Luzon, *W. Lobb* 465 (iso BM, E[2], FI, K).

■ **PHILIPPINES: Luzon:** *W. Lobb* 465 (iso BM, E[2], FI, K) – Type of *Begonia oxysperma* A. DC.; Tayabas, Mt. Banahao, 22 vi 1904, *W. Klemme* 886 (B); Benguet, Baguio, 21 vi 1904, *R.S. Williams* 986 (K); Ifugao, Mt. Polis, 20 vii 1994, *L. Co* 3767 (L); Benguet, *A. Loher* 4319 (K); Benguet, Tonglon, *A. Loher* 4320 (K); Benguet, Baguio, iii 1904, *A.D.E. Elmer* 6013 (B[3], K, P); Tayabas, Lucban, v 1907, *A.D.E. Elmer* 7533 (BM, E); Benguet, v 1911, *E.D. Merrill* 7688 (B, BM, E, K, L, P); Sierra Madre Mts., 19 iii 1968, *M. Jacobs* 7854 (K); Tayabas, Lucban, v 1907, *A.D.E. Elmer* 8036a (BM); Laguna, Makiling, vi 1917 – vii 1917, *A.D.E. Elmer* 17760 (BM[2], K, L, P, U); Camarines, Mt. Isaro(g), 22 iii 1997, *P. Wilkie et al.* 29142 (E); Benguet, Pauai, iv 1918 – vi 1918, *J.K. Santos* 31892 (B, K, P); Ifugao, Mt. Polis, ii 1920, *M. Ramos, G. Edano* 37622 (B, L).

■ **Notes:** Sands (2001) states that this species is confined to a specific altitudinal band, the lower limit coinciding with the bottom of the cloud base.

■ **IUCN category: LC.** Area of occupancy is c. 10,000 km². Occurs in two national parks.

Begonia oxyura Merr. & L.M. Perry [§ Petermannia], J. Arnold Arbor. 24: 49 (1943). – Type: New Guinea, Papua, Idenburg River, iii 1939, *L.J. Brass* 13457 (holo A).

■ **NEW GUINEA: Papua:** Idenburg River, iii 1939, *L.J. Brass* 13217 (para A) – Type of *Begonia oxyura* Merr. & L.M. Perry; Idenburg River, iii 1939, *L.J. Brass* 13457 (holo A) – Type of *Begonia oxyura* Merr. & L.M. Perry.

■ **IUCN category: DD.** Insufficient specimens could be georeferenced with certainty.

Begonia pachyrhachis L.B. Sm. & Wassh. [§ Sphenanthera], Phytologia 52: 442 (1983). – Type: not located.

Casparya crassicaulis A. DC., Ann. Sci. Nat. Bot., IV 11: 119 (1859) – *Begonia crassicaulis* (A. DC.) Warb., Nat. Pflanzenfam. 3(6A): 149 (1894); Smith & Wasshausen, Phytologia 52: 442 (1983). – Type: not located.

■ **Notes:** This name should be synonymised under *B. multangula* Blume according to annotations on a De Vriese specimen (Jawa, K) (Tebbitt, 1997) which I have not located.

■ **IUCN category: DD.** Taxonomic uncertainty.

Begonia padangensis Irmsch. [§ Petermannia], Webbia 9: 475 (1953). – Type: Sumatra, Mt. Singalan, 1878, *O. Beccari* PS125 (syn B[5], Fl[5], K, L).
- **SUMATRA:** Mt. Singalan, 1878, *O. Beccari* PS125 (syn B[5], Fl[5], K, L) – Type of *Begonia padangensis* Irmsch.; West Sumatra Prov., Gunung Merapi, 19 vii 2006, *D. Girmansyah et al. 760* (E); West Sumatra Prov., Gunung Merapi, 19 vii 2006, *D. Girmansyah, A. Poulsen, I. Hatta, R. Neivita 761* (BO, E).
- **Notes:** The petioles of this species have short ferrugineous curly hairs.
- **IUCN category: VU D2**. The two localities are both designated as Protection Forest. Any degradation of these areas will lead to a threat level of EN.

Begonia palawanensis Merr. [§ Petermannia], Philipp. J. Sci. 6: 380 ('1911', 1912); Merrill, Enum. Philipp. Fl. Pl. 3: 126 (1923). – Type: Philippines, Palawan, Napsahan, ix 1910, 9°43′49″N, 118°27′48″E, *E.D. Merrill 7232* (syn B, BM[2], K, L, P).
- **PHILIPPINES: Palawan:** Napsahan, ix 1910, *E.D. Merrill 7232* (syn B, BM[2], K, L, P) – Type of *Begonia palawanensis* Merr.
- **IUCN category: DD**. Insufficient specimens could be georeferenced with certainty.

Begonia paleacea Kurz [§ Monophyllon], Flora 54: 297 (1871); Clarke, Fl. Brit. Ind. 605 (1879). – Type: Burma, Martaban, Attaran-Thal, *D. Brandis 1326* (syn K).
- **BURMA:** Martaban, Attaran Valley, 20 viii 1858, *unknown s.n.* (K); Martaban, Attaran-Thal, *D. Brandis 1326* (syn K) – Type of *Begonia paleacea* Kurz.
- **Notes:** The type looks like a scrappy specimen of *B. integrifolia* (also noted by Clarke 1879, p. 650). No other material exists.
- **IUCN category: DD**. Taxonomic uncertainty.

Begonia palmata D. Don [§ Platycentrum], Prodr. Fl. Nepal. 223 (1825); Hara, Fl. E. Himalaya 215 (1966); Hara, Phot. Pl. E. Himalaya 60 (1968); Hara, Fl. E. Himalaya 2: 84 (1971). – Type: not located.
Begonia laciniata Roxb., Fl. Ind. ed. 1832 3: 649 (1832); Candolle, Prodr. 15(1): 347 (1864); Koorders, Exkurs.-Fl. Java 2: 645 (1912); Gagnepain, in Lecomte (ed.), Fl. Indo-Chine 2: 1107 (1921); Craib, Fl. Siam. 1: 775 (1931). – Type: not located.
Begonia laciniata var. *pilosa* Craib, Fl. Siam. 1: 775 (1931). – Type: Thailand, Doi Pu Ka, 26 ii 1921, *A.F.G. Kerr 4937* (holo ABD; iso BM, K).

Begonia palmata sensu lato
- **BURMA:** Valley of the Nam Tamai, 15 viii 1938, *R. Kaulback s.n.* (BM); Mahtum, 26 viii 1939, *R. Kaulback 89* (BM); Mahtum, 5 ix 1939, *R. Kaulback 380* (BM); Southern Shan States, King Jung, *R.W. MacGregor 782* (E); Nwai Valley, 15 ix 1914, *F.K. Ward 1947* (E); Kachin, Sumprabum sub-division, Ning W'Krok – Kanang, Tsuptaung – Kanang, 27 xii 1961, *J. Keenan 3012* (E[2]); Kachin, Sumprabum sub-division, Tsuptaung – Kanang, 27 xii 1961, *J. Keenan 3016* (E); Kachin, Sumprabum sub-division, Ning W'Krok – Kanang, 20 i 1962, *J. Keenan et al. 3350* (A, K) [as *Begonia cf. palmata*]; Nam Tisang, 5 i 1931, *F.K. Ward 9097* (BM); Valley of the Nam Tamai, 9 viii 1937, *F.K. Ward 12914* (BM, E); Valley of the Nam Tamai, 5 ix 1937, *F.K. Ward 13139* (BM); 4 xii 1937, *F.K. Ward 13532* (BM) [as *Begonia cf. laciniata*].
  THAILAND: Khon Kaen Prov., Wang-Kwong waterfall, 30 x 1984, *A. McAllan 3* (K) [as *Begonia cf. palmata*]; Khon Kaen Prov., Phaong Waterfall, 1 xi 1984, *A. McAllan 8* (K) [as *Begonia cf. palmata*]; Chiangmai, Doi Inthanon National Park, Warachia waterfall, 6 xii 1984, *A. McAllan 11* (K) [as *Begonia cf. palmata*]; Chiangmai, Doi Inthanon National Park, 8 xii 1984, *A. McAllan 12* (K) [as *Begonia cf. palmata*]; Chiangmai, Doi Inthanon National Park, 9 xii 1984, *A. McAllan*

13 (K) [as *Begonia cf. palmata*]; Nakawn Sawan, Mae Wong N.P., 20 viii 1997 – 2 xi 1997, *M. van de Bult* 18/46 (L); Chiangmai, Doi Sootep, 17 vii 1994, *P. Palee* 226 (A, L); Doi Angka, 16 xi 1926, *H.B.G. Garrett* 314 (K, L); Chiangmai, Doi Sootep, 6 ix 1957, *Khantelim* 669 (K) [as *Begonia cf. palmata*]; Chiangmai, Doi Nawn Ngaw, 3 xii 2001, *J.F. Maxwell* 01-697 (A, L); Chiangmai, Doi Inthanon National Park, Huai Sai Lieng waterfall, 24 i 1990, *W.S. Hoover* 722 (A) [as *Begonia cf. palmata*]; Doi Angka, 15 xii 1934, *H.B.G. Garrett* 907 (K); Chiangmai, Doi Inthanon National Park, 20 ix 1994, *P.C. Boyce* 1007 (K) [as *Begonia cf. palmata*]; Chiangmai, Doi Sootep, 24 ix 1988, *J.F. Maxwell* 88-1114 (L); Lampang Prov., Chae Son Nat. Park, 31 x 1996, *J.F. Maxwell* 96-1437 (A, L); Chiangmai, Doi Sootep, 6 x 1912, *A.F.G. Kerr* 2730 (ABD, BM, K); Doi Pu Ka, 26 ii 1921, *A.F.G. Kerr* 4937 (holo ABD; iso BM, K) – Type of *Begonia laciniata* var. *pilosa* Craib; Chieng Rai Prov., Doi Thung, 25 x 1977, *R. Geesink et al.* 8279 (L); Payap, Doi Chiengdao, 13 ix 1967, *M. Tagawa et al.* 9883 (L, P); Kao Lem, 12 i 1925, *A.F.G. Kerr* 9960 (ABD, BM, K); Phitsanulok Prov., Phu Rom Rot, 3 x 1967, *M. Tagawa et al.* 11488 (P); Doi Khun Huai Pong, 5 iii 1968, *B. Hansen, T. Smitinand* 12827 (K, L) [as *Begonia cf. palmata*]; Chiangmai, Doi Inthanon National Park, 15 x 1979, *T. Shimizu et al.* 18783 (K) [as *Begonia cf. palmata*]; Mt. Phu, 1984 – 1985, *Murata et al.* 42583 (L); Chiangmai, Doi Sootep, Queen Sirikit Bot. Gard., 19 ix 1995, *K. Larsen et al.* 46706 (L); Loei Province, Samkokwa – Langpae, 29 viii 1988, *H. Koyama* 61399 (A, L).
LAOS: Nape, 13 x 1928, *Delacour s.n.* (P); Xieng-Khouang, 28 ix 1928, *Delacour s.n.* (P); 20 ix 1920, *E. Poilane* 1938 (K); 8 x 1920, *E. Poilane* 2002 (K).
VIETNAM: Tonkin, Mt. Bavi, vii 1908, *D'Alleizette s.n.* (L); Kontum Prov., Ngoc Linh Mountain, 6 iii 1995, *L. Averyanov et al.* 541 (A).

Begonia palmata var. bowringiana (Benth.) Golding & Kareg., Phytologia 54(7): 494 (1984). – *Begonia bowringiana* Champ. ex Benth., Hooker's J. Bot. Kew Gard. Misc. 4: 120 (1852); Hooker, Bot. Mag., n.s. 86: 5182 (1860); Candolle, Prodr. 15(1): 348 (1864); Irmscher, Mitt. Inst. Allg. Bot. Hamburg 10: 533 (1939); Golding & Karegeannes, Phytologia 54(7): 494 (1984). – Type: not located.
■ **VIETNAM:** Laoke, Chapa, Song ta Van, viii 1936, *P.A. Petelot* 7116 (B).
■ **IUCN category: LC.** Widespread.

Begonia panayensis Merr. [§ Petermannia], Philipp. J. Sci. 14: 428 (1919); Merrill, Enum. Philipp. Fl. Pl. 3: 127 (1923). – Type: Philippines, Panay, Antique Province, Culasi, 8 vi 1918, *R.C. McGregor* BS32309 (holo not located).
■ **PHILIPPINES: Panay:** Antique Province, Culasi, 8 vi 1918, *R.C. McGregor* BS32309 (holo not located) – Type of *Begonia panayensis* Merr.
■ **IUCN category: DD.** Insufficient specimens could be georeferenced with certainty.

Begonia papuana Warb. in K. Schum. & Lauterb. **[§ Petermannia]**, Fl. Deutsch. Schutzgeb. Südsee 458 (1901); Irmscher, Bot. Jahrb. Syst. 50: 573 (1914). – Type: New Guinea, Papua New Guinea, Kaiser-Wilhelmsland, Gogolebene, 25 vii 1890, *C.A.G. Lauterbach* 611 (holo B; iso L).
■ **NEW GUINEA:** *O. Warburg* 20477 (B). **Papua:** Vogelkopf, Isjon River valley, 28 x 1961, *P.v. Royen, H. Sleumer* 7563 (L) [as *Begonia cf. papuana*]. **Papua New Guinea:** Milne Bay Distr., Baniara Subdistr., Bonenau, 31 x 1950, *N. Cruttewell* 218 (L); Morobe Distr., Sattelberg, 26 ix 1935, *Clemens* 233 (L); Morobe Distr., Sattelberg, 17 x 1935, *J. & M.S. Clemens* 482 (L); Kaiser-Wilhelmsland, Gogolebene, 25 vii 1890, *C.A.G. Lauterbach* 611 (holo B; iso L) – Type of *Begonia papuana* Warb.; Morobe Distr., Wareo, 29 xii 1935, *J. & M.S. Clemens* 1418 (L); Sangwep Logging Area, 13 iii 1975, *J.F. Veldkamp* 6169 (L); Central Division, Nunumai, 23 vi 1969, *R. Pullen* 7675 (A) [as *Begonia aff. papuana*]; Morobe Distr., Kajabit Mission, 8 ix 1939, *Clemens* 10669 (A); Josephstaal, *W. Takeuchi et al.* 13417 (A); Madang Prov., Kumamdeber, 22 xii 1999, *W. Takeuchi et al.* 13543 (L);

Josephstaal, 22 xii 1999, *W. Takeuchi et al.* 13545 (A, L); Josephstaal, 22 xii 1999, *W. Takeuchi et al.* 13884 (A); Yodda Valley, 23 xii 1935, *C.E. Carr* 13949 (L); *C.E. Carr* 15388 (L); Morobe Distr., Atzera Range, 17 vii 2001, *W. Takeuchi et al.* 15444 (A, L); Morobe Distr., Rawlinson Range, Tupsundu Hill, 22 ii 1963, *P.v. Royen* 16122 (L); Morobe Distr., Siboma Bay, 26 iv 2002, *W. Takeuchi et al.* 16214 (A); Kaiser-Wilhelmsland, 12 vii 1907, *F.R.R. Schlechter* 16246 (B, P); Kaiser-Wilhelmsland, Kani Range, 1 v 1908, *F.R.R. Schlechter* 17637 (B, P); Kaiser-Wilhelmsland, Kani Range, 3 v 1908, *F.R.R. Schlechter* 17637 (B, P); Busu River, 27 x 1964, *P.v. Royen* 20114 (L); Morobe Distr., Umi River, Markham Valley, 17 xi 1959, *L.J. Brass* 32582 (L).
- **IUCN category: LC**. Widespread within Papua New Guinea, and occurs within three different eco-regions.

Begonia papyraptera Sands [§ Petermannia], in Coode *et al.*, Checkl. Fl. Pl. Gymnosperms Brunei Darussalam App. 2: 432 (1997). – Type: Borneo, Brunei, Temburong, 28 iv 1992, *R.J. Johns* 7422 (holo K; iso BRUN *n.v.*).
- **BORNEO: Brunei:** Temburong, 28 iv 1992, *R.J. Johns* 7422 (holo K; iso BRUN *n.v.*) – Type of *Begonia papyraptera* Sands.
- **IUCN category: DD**. Insufficient specimens could be georeferenced with certainty.

Begonia parishii C.B. Clarke in Hook. f. **[§ Parvibegonia]**, Fl. Brit. Ind. 2: 651 (1879). – Type: Burma, Moulmein, 1862, *C.S.P. Parish s.n.* (syn K).
- **BURMA:** Moulmein, 1862, *C.S.P. Parish s.n.* (syn K) – Type of *Begonia parishii* C.B. Clarke.
- **IUCN category: DD**. Insufficient specimens could be georeferenced with certainty.

Begonia parva Merr. [§ Diploclinium], Philipp. J. Sci. 6: 402 ('1911', 1912); Merrill, Enum. Philipp. Fl. Pl. 3: 127 (1923). – Type: Philippines, Luzon, Benguet, Rio Trinidad, xii 1908, 16°21′36″N, 120°17′24″E, *M. Ramos* 5551 (syn K).
- **PHILIPPINES: Luzon:** Benguet, Rio Trinidad, xii 1908, *M. Ramos* 5551 (syn K) – Type of *Begonia parva* Merr.
- **IUCN category: DD**. Insufficient specimens could be georeferenced with certainty.

Begonia parvilimba Merr. [§ Petermannia], Philipp. J. Sci. 26: 481 (1925). – Type: Philippines, Mindanao, Zamboanga, Malangas, *M. Ramos, G. Edano* BS36936 (syn not located).
- **PHILIPPINES: Mindanao:** Zamboanga, Malangas, *M. Ramos, G. Edano* BS36933 (para not located) – Type of *Begonia parvilimba* Merr.; Zamboanga, Malangas, *M. Ramos, G. Edano* BS36936 (syn not located) – Type of *Begonia parvilimba* Merr.
- **IUCN category: DD**. Insufficient specimens could be georeferenced with certainty.

Begonia parvuliflora A. DC. [§ Diploclinium], Ann. Sci. Nat. Bot., IV 11: 136 (1859); Candolle, Prodr. 15(1): 355 (1864); Clarke, Fl. Brit. Ind. 640 (1879). – Type: Burma, Moulmein, *W. Lobb s.n.* (holo K).
  Begonia lobbiana A. DC. pro. syn. *Begonia parvuliflora* A. DC. var. *parvuliflora*, Prodr. 15(1): 355 (1864). – Type: not located.
  Begonia velutina Parish ex Kurz, J. Asiat. Soc. Bengal 42(2): 81 (1873); Clarke, Fl. Brit. Ind. 640 (1879). – Type: Burma, Moulmein, 1862, *C.S.P. Parish s.n.* (holo K).

Begonia parvuliflora var. parvuliflora
- **BURMA:** Moulmein, *C.S.P. Parish s.n.* (K, P); Moulmein, 1862, *C.S.P. Parish s.n.* (holo K) – Type of *Begonia velutina* Parish ex Kurz; Moulmein, 1861, *C.S.P. Parish s.n.* (K); Moulmein, *W. Lobb s.n.* (holo K) – Type of *Begonia parvuliflora* A. DC.; Moulmein, 1869, *C.S.P. Parish* 129 (K).

Begonia parvuliflora var. pubescens A. DC., Prodr.15(1):355 (1864). – Type:Burma,Moulmein, 1846, *W. Lobb* 380 (lecto K, here designated; isolecto BM, E, FI).

Begonia moulmeinensis C.B. Clarke *nom. superfl.* in Hook. f., Fl. Brit. Ind. 2: 643 (1879). – Type: Burma, Moulmein, 1846, *W. Lobb* 380 (lecto K, here designated; isolecto BM, E, FI).

- **BURMA:** Moulmein, *W. Lobb* 380 (lecto K; isolecto BM, E, FI) – Type of *Begonia parvuliflora* var. *pubescens* A. DC.
- **Notes:** *Begonia moulmeinensis* C.B. Clarke is a superfluous name, being based on the same type as *B. parvuliflora* var. *pubescens* A. DC.
- **IUCN category: DD.** Insufficient specimens could be georeferenced with certainty.

Begonia paupercula King [§ Platycentrum], J. Asiat. Soc. Bengal, Pt. 2, Nat. Hist. 71:64 (1902); Ridley, Fl. Malay Penins. 1: 862 (1922); Irmscher, Mitt. Inst. Allg. Bot. Hamburg 8: 122 (1929); Kiew, Begonias Penins. Malaysia 170 (2005). – Type: Peninsular Malaysia, Perak, *Kings Collector* 5952 (lecto K *n.v.*; isolecto BM, P).

- **PENINSULAR MALAYSIA:** Perak, *Kings Collector* 5952 (lecto K *n.v.*; isolecto BM, P) – Type of *Begonia paupercula* King.
- **IUCN category: VU D2.** Endemic to the Gopeng area in Perak. None of the localities are formally protected.

Begonia pavonina Ridl. [§ Platycentrum], J.Fed.Malay States Mus.4:22 (1909);Ridley,Fl.Malay Penins. 1: 863 (1922); Irmscher, Mitt. Inst. Allg. Bot. Hamburg 8: 123 (1929); Henderson, Malayan Wild Flowers – Dicotyledons 165 (1959); Kiew, Begonias Penins. Malaysia 158 (2005). – Type: Peninsular Malaysia, Pahang, Telom, *H.N. Ridley* 14125 (lecto SING *n.v.*; isolecto BM, K).

Begonia robinsonii Ridl., J. Fed. Malay States Mus. 4: 22 (1909); Kiew, Begonias Penins. Malaysia 158 (2005). – Type: Peninsular Malaysia, Pahang, Telom, *H.N. Ridley* 14125 (lecto SING *n.v.*; isolecto BM, K).

- **PENINSULAR MALAYSIA:** Pahang, Cameron Highlands, Robinsons Falls, 27 vii 2002, *S. Neale, G. Bramley* 9 (E[2]); Pahang,Cameron Highlands,Boh Tea Plantation,22 iii 1992,*Klackenberg, Lundin* 793 (L); Pahang, Cameron Highlands, Tanah Rata – Habu, 21 x 1967, *K. Iwatsuki, N. Fukuoka, M. Hutch* 13716 (L); Pahang, Telom, *H.N. Ridley* 14125 (lecto SING *n.v.*; isolecto BM, K) – Type of *Begonia pavonina* Ridl.; Pahang,Telom,*H.N. Ridley* 14125 (lecto SING *n.v.*; isolecto BM, K) – Type of *Begonia robinsonii* Ridl.; Pahang, Cameron Highlands, 3 iv 1937, *Md. Nur* SFN32874 (L) [as *Begonia robinsonii* ?]; Pahang, Cameron Highlands, Tanah Rata, 13 i 1983, *Davis* 69265 (E); Pahang,Cameron Highlands,Tanah Rata, 13 i 1983,*Davis* 69272 (E).
- **IUCN category: LC.** Described as 'locally common' in the Cameron Highlands (Kiew, 2005). Also found in secondary habitats (*Klackenberg, Lundin* 793 (L)).

Begonia pedatifida H. Lév. [§ Platycentrum], Repert. Spec. Nov. Regni Veg. 7: 21 (1909); Gagnepain, in Lecomte (ed.), Fl. Indo-Chine 2: 1104 (1921). – Type: China, Kouy-Tcheou, Pin-Fa, 21 viii 1902, *J. Cavalerie* 262 (syn not located); China, Kouy-Tcheou, Majo, 23 vii 1907, *J. Cavalerie* 3072 (syn P).

- **VIETNAM:** Tonkin, Lao-kay, *A.J.B. Chevalier* 29364 (P).
- **IUCN category: DD.** Insufficient specimens could be georeferenced with certainty.

Begonia pediophylla Merr. & L.M. Perry [§ Petermannia], J. Arnold Arbor. 24: 54 (1943). – Type: New Guinea, Papua New Guinea, Wharton Range, Murray Pass, vii 1933, *L.J. Brass* 4577 (holo A; iso B, L).

- **NEW GUINEA: Papua New Guinea:** Wharton Range, Murray Pass, vii 1933, *L.J. Brass* 4577 (holo A; iso B, L) – Type of *Begonia pediophylla* Merr. & L.M. Perry; Milne Bay Distr., Mt. Dayman, Gwariu River, 13 vi 1953, *L.J. Brass* 22935 (L); Wharton Range, Murray Pass, 10 viii 1968, *C.E. Ridsdale* 36870 (L) [as *Begonia* cf. *pediophylla*].
- **IUCN category: VU D2.** Known only from two localities.

Begonia peekelii Irmsch. [§ Petermannia], Bot. Jahrb. Syst. 50: 360 (1913); Peekel, Flora Bismarck Archipelago 391 (1984). – Type: Bismarck Archipelago, New Ireland, New Mecklenburg, 21 ix 1910, *G. Peekel* 644 (holo B).
- **BISMARCK ARCHIPELAGO: New Ireland:** New Mecklenburg, 21 ix 1910, *G. Peekel* 644 (holo B) – Type of *Begonia peekelii* Irmsch.
- **Notes:** See notes under *B. rieckei* Warb.
- **IUCN category: DD.** Taxonomic uncertainty.

Begonia pendula Ridl. [§ Petermannia], J. Straits Branch Roy. Asiat. Soc. 46: 257 (1906); Kiew & Geri, Gard. Bull. Singapore 55: 120 (2003). – Type: Borneo, Sarawak, Bau, Jambusan, *H.N. Ridley* 11772 (holo K).
- **BORNEO: Sarawak:** Bau, Bukit Tai Ton, 12 xii 1975, *P.F. Stevens* 204 (L); viii 1865, *O. Beccari* PB220 (FI); Bau, 1st Division, 18 xii 1988, *P.J.A. Kessler* 232 (L); Bau, 1st Division, 18 xii 1988, *P.J.A. Kessler* 238 (L); Gunung Skunyet, xi 1865, *O. Beccari* PB1050 (B, FI); Bau, Gunong Seburan, 9 iii 1949, *J. Sinclair* 5680 (E); Bau, vi 1957, *T. Anderson* 7799 (L); Bau, 24 iv 1955, *W.M.A. Brooke* 9874 (L); Bau, 26 iv 1955, *W.M.A. Brooke* 9891 (L); Bau, Jambusan, *H.N. Ridley* 11772 (holo K) – Type of *Begonia pendula* Ridl.; Bau, *H.N. Ridley* 11774 (K); Kuching, Tiang Bakap, 3 iv 1960, *J.A.R. Anderson* 12346 (K); Bau, Jambusan, 8 x 1977, *P.J. Martin* 39292 (L); Miri Division, Gunung Benarat, 6 ii 1995, *A. Mohtar et al.* 49665 (K); Bau, Bukit Boring, 19 xi 1985, *Yii Puan Ching* 50360 (L); Bau, Bukit Tai Ton, 18 xi 1985, *Yii Puan Ching* 51208 (L); Bau, Bukit Tai Ton, 18 xi 1985, *Yii Puan Ching* 51210 (L); Bau, Lobang Angin, 19 xi 1985, *Yii Puan Ching* 51265 (L); Bau, 1st Division, 10 x 1989, *A. Mohtar et al.* 52914 (L); Bau, Bukit Gajah, 9 ii 1999, *Jamree et al.* 82087 (L).
- **IUCN category: NT.** Based on geographic range, this species does not meet the criteria for any threatened category. It is described as 'common on most limestone hills' in the Kuching District (Kiew & Geri, 2003). However, given the threats to lowland forest and limestone areas in Sarawak, this species should be monitored carefully.

Begonia pentaphragmifolia Ridl. [§ Petermannia], Trans. Linn. Soc. London, Bot., II 9: 59 (1916). – Type: New Guinea, Papua, Camp I – Camp III, *C.B. Kloss s.n.* (syn BM); New Guinea, Papua, Camp III – Camp IV, *C.B. Kloss s.n.* (syn BM); New Guinea, Papua, Canoe Camp, *C.B. Kloss s.n.* (syn BM).
- **NEW GUINEA: Papua:** Camp I – Camp III, *C.B. Kloss s.n.* (syn BM) – Type of *Begonia pentaphragmifolia* Ridl.; Camp III – Camp IV, *C.B. Kloss s.n.* (syn BM) – Type of *Begonia pentaphragmifolia* Ridl.; Canoe Camp, *C.B. Kloss s.n.* (syn BM) – Type of *Begonia pentaphragmifolia* Ridl.; Utakwa River – Mt. Carstensz, 1912, *C.B. Kloss s.n.* (BM).
- **IUCN category: DD.** Insufficient specimens could be georeferenced with certainty.

Begonia perakensis King [§ Platycentrum], J. Asiat. Soc. Bengal, Pt. 2, Nat. Hist. 71: 64 (1902); Ridley, Fl. Malay Penins. 1: 861 (1922); Irmscher, Mitt. Inst. Allg. Bot. Hamburg 8: 129 (1929); Kiew, Begonias Penins. Malaysia 240 (2005). – Type: Peninsular Malaysia, Perak, *Kings Collector* 10338 (lecto K *n.v.*; isolecto FI, L, P); Peninsular Malaysia, Perak, *Kings Collector* 10506 (syn B, BM); Peninsular Malaysia, Perak, *Kings Collector* 10951 (syn not located).

Begonia perakensis var. perakensis
- **PENINSULAR MALAYSIA:** Perak, *Kings Collector* 10506 (syn B, BM) – Type of *Begonia perakensis* King; Perak, *Kings Collector* 10951 (syn not located) – Type of *Begonia perakensis* King; Pahang, 26 viii 1923, *Md. Nur* SFN11035 (BM); Perak, *Kings Collector* 10338 (lecto K *n.v.*; isolecto FI, L, P) – Type of *Begonia perakensis* King.

Begonia perakensis var. conjugens Irmsch., Mitt. Inst. Allg. Bot. Hamburg 8: 129 (1929); Turner, Gard. Bull. Singapore 46(2): 129 (1994); Kiew, Begonias Penins. Malaysia 244 (2005). – Type: Peninsular Malaysia, Selangor, v 1902, *C. Curtis s.n.* (lecto SING *n.v.*); Peninsular Malaysia, Selangor, Pahang Track, vii 1897, *H.N. Ridley* 8590 (syn SING *n.v.*).
- **PENINSULAR MALAYSIA:** Selangor, v 1902, *C. Curtis s.n.* (lecto SING *n.v.*) – Type of *Begonia perakensis* var. *conjugens* Irmsch.; Selangor, Pahang Track, vii 1897, *H.N. Ridley* 8590 (syn SING *n.v.*) – Type of *Begonia perakensis* var. *conjugens* Irmsch.
- **IUCN category: DD.** Insufficient specimens could be georeferenced with certainty.

Begonia perryae L.B. Sm. & Wassh. [§ Petermannia], Phytologia 52: 445 (1983). – *Begonia robinsonii* Merr., Philipp. J. Sci. 6: 375 ('1911', 1912); Merrill, Enum. Philipp. Fl. Pl. 3: 127 (1923); Smith & Wasshausen, Phytologia 52: 445 (1983). – Type: Philippines, Luzon, Camarines, Maagnas, 28 viii 1908, *C.B. Robinson* 6340 (holo not located).
- **PHILIPPINES: Luzon:** Camarines, Maagnas, 28 viii 1908, *C.B. Robinson* 6340 (holo not located) – Type of *Begonia robinsonii* Merr.
- **IUCN category: DD.** Insufficient specimens could be georeferenced with certainty.

Begonia phoeniogramma Ridl. [§ Parvibegonia], J. Straits Branch Roy. Asiat. Soc. 75: 35 (1917); Ridley, Fl. Malay Penins. 1: 858 (1922); Irmscher, Mitt. Inst. Allg. Bot. Hamburg 8: 155 (1929); Craib, Fl. Siam. 1: 777 (1931); Henderson, Malayan Wild Flowers – Dicotyledons 167 (1959); Kiew, Begonias Penins. Malaysia 85 (2005). – Type: Peninsular Malaysia, Selangor, Batu Caves, viii 1908, *H.N. Ridley* 13430 (lecto K; isolecto BM).
  Begonia paupercula auct. non King: Ridley, J. Straits Branch Roy. Asiat. Soc. 54: 42 (1909); Ridley, J. Straits Branch Roy. Asiat. Soc. 75: 35 (1917); Ridley, Fl. Malay Penins. 1: 858 (1922).
- **PENINSULAR MALAYSIA:** Selangor, Kepong, 18 vii 2002, *S. Neale, Y-Y. Sam* 4 (E); Selangor, Kepong, FRI Forest Reserve, 11 xi 1975, *P.F. Stevens* 31 (L); Selangor, Kepong, FRI Forest Reserve, 11 xi 1975, *P.F. Stevens* 31a (L); Selangor, Batu Caves, 10 vii 1906, *A. Ernst* 1115 (L); Selangor, Batu Caves, 12 viii 1937, *C.W. Franck* 1132 (P); Selangor, Batu Caves, 1908, *H.J.P. Winkler* 1801 (BM); Selangor, Batu Caves, 7 viii 1991, *R. Kiew* 3252 (L); Selangor, Batu Caves, 14 ix 1920, *I.H. Burkill* 6366 (BM); Selangor, Kepong, Bukit Lagong F.R., 17 xi 1962, *J. Sinclair* 10720 (E, L); Selangor, Batu Caves, viii 1908, *H.N. Ridley* 13430 (lecto K; isolecto BM) – Type of *Begonia phoeniogramma* Ridl.; Selangor, Batu Caves, 30 x 1967, *T. Shimizu, N. Fukuoka* 14143 (L); Selangor, Batu Caves, 4 xi 1953, *J. Sinclair* 40067 (E, L).
- **Notes:** See notes under *B. integrifolia* Dalzell.
- **IUCN category: VU B2ab(iii).** Has an area of occupancy of less than 2000 km^2, and one of the populations (Batu Caves) is threatened by disturbance (Kiew, 2005).

Begonia phuthoensis H.Q. Nguyen [§ Coelocentrum], Novon 14: 105 (2004). – Type: Vietnam, Phu Tho Province, Thanh Son Distr., Xuan Son National Park, 30 xi 2000, *V.X. Phuong* 4041 (holo HN *n.v.*; iso MO *n.v.*).
- **VIETNAM:** Thanh Hoa Prov., Ba Thuoc Distr., Co Lung, 17 iv 2001, *N.T. Hiep et al.* HAL1094 (para HAST *n.v.*, HN *n.v.*, LE *n.v.*, MO *n.v.*) – Type of *Begonia phuthoensis* H.Q. Nguyen; Phu Tho

Province, Thanh Son Distr., Xuan Son National Park, 30 xi 2000, *V.X. Phuong* 4041 (holo HN *n.v.*; iso MO *n.v.*) – Type of *Begonia phuthoensis* H.Q. Nguyen.
■ **IUCN category: LC**. Has a wide distribution in Phu Tho and Thanh Hoa provinces (Nguyen, 2004).

Begonia physandra Merr. & L.M. Perry [not placed to section], J. Arnold Arbor. 24: 41 (1943). – Type: New Guinea, Papua New Guinea, Central Division, Mafulu, x 1933, *L.J. Brass* 5199 (holo NY *n.v.*; iso B, L).
■ **NEW GUINEA: Papua New Guinea:** Central Division, Mafulu, x 1933, *L.J. Brass* 5199 (holo NY *n.v.*; iso B, L) – Type of *Begonia physandra* Merr. & L.M. Perry.
■ **IUCN category: DD**. Insufficient specimens could be georeferenced with certainty.

Begonia picta Sm. [§ Diploclinium], Exot. Bot. 2: 81 (1805); Don, Prodr. Fl. Nepal. 223 (1825); Hooker, Bot. Mag. 57: 2965 (1830); Wallich, Cat. 129: 3685 (1831); Clarke, Fl. Brit. Ind. 2: 638 (1879); Haines, Bot. Bihar Orissa 400 (1925); Mooney, Suppl. Bot. Bihar & Orissa 68 (1950); Kitamura, Fauna Fl. Nepal Himalaya 183 (1955); Kitamura, Liv. Him. Fl. 184 (1964); Hara, Fl. E. Himalaya 215 (1966); Hara, Fl. E. Himalaya 2: 84 (1971); Grierson, Fl. Bhutan 2: 242 (1991); Panda & Das, Fl. Samb. 159 (2004). – Type: not located.
■ **BURMA:** *Fide* Kress *et al.* (2003).
■ **IUCN category: LC**. Widespread and commonly collected across the main part of its range in the Himalayas.

Begonia pierrei Gagnep. [§ Reichenheimia], Bull. Mus. Hist. Nat. (Paris) 25: 276 (1919); Gagnepain, in Lecomte (ed.), Fl. Indo-Chine 2: 1100 (1921). – Type: Vietnam, Cochinchine, Poulo-condor, *F.F.J. Harmand* 689 (syn P); Vietnam, Cochinchine, China-chiang Mts., ix 1865, *J.B.L. Pierre s.n.* (syn P).
■ **VIETNAM:** Cochinchine, China-chiang Mts., ix 1865, *J.B.L. Pierre s.n.* (syn P) – Type of *Begonia pierrei* Gagnep.; Cochinchine, Poulo-condor, *F.F.J. Harmand* 689 (syn P) – Type of *Begonia pierrei* Gagnep.
■ **IUCN category: DD**. Insufficient specimens could be georeferenced with certainty.

Begonia pilosa Jack [§ Petermannia], Malayan Misc. 2(7): 13 (1822). – *Diploclinium pilosum* (Jack) Miq. – Type: not located.
■ **SUMATRA:** *Fide* Jack (1822).
■ **IUCN category: DD**. Insufficient specimens could be georeferenced with certainty.

Begonia pinamalayensis Merr. [§ Diploclinium], Philipp. J. Sci. 26: 479 (1925). – Type: Philippines, Mindoro, Pinamalayan, v 1922, *M. Ramos* 40856 (syn B, L, P).
■ **PHILIPPINES: Mindoro:** Pinamalayan, v 1922, *M. Ramos* 40856 (syn B, L, P) – Type of *Begonia pinamalayensis* Merr.
■ **IUCN category: DD**. Insufficient specimens could be georeferenced with certainty.

Begonia pinnatifida Merr. & L.M. Perry [§ Petermannia], J. Arnold Arbor. 24: 51 (1943). – Type: New Guinea, Papua New Guinea, Palmer River, vi 1936, *L.J. Brass* 7051 (holo A).
■ **NEW GUINEA: Papua New Guinea:** Palmer River, vi 1936, *L.J. Brass* 7051 (holo A) – Type of *Begonia pinnatifida* Merr. & L.M. Perry.
■ **IUCN category: DD**. Insufficient specimens could be georeferenced with certainty.

Begonia platyphylla Merr. [§ Petermannia], Philipp. J. Sci. 10: 46 (1915); Merrill, Enum. Philipp. Fl. Pl. 3: 127 (1923). – Type: Philippines, Luzon, Nueva Vizacaya, i 1912, *R.C. McGregor* 20074 (syn BM, P).
- **PHILIPPINES: Luzon:** Nueva Vizacaya, i 1912, *R.C. McGregor* 20074 (syn BM, P) – Type of *Begonia platyphylla* Merr.
- **IUCN category: DD.** Insufficient specimens could be georeferenced with certainty.

Begonia pleioclada Irmsch. [§ Petermannia], Webbia 9: 488 (1953). – Type: Borneo, Sarawak, Gunong Skunyet, xi 1865, *O. Beccari* PB1051 (holo FI).
- **BORNEO: Sarawak:** Gunong Skunyet, xi 1865, *O. Beccari* PB1051 (holo FI) – Type of *Begonia pleioclada* Irmsch.
- **IUCN category: DD.** Insufficient specimens could be georeferenced with certainty.

Begonia polilloensis Tebbitt [§ Petermannia], Edinburgh J. Bot. 61: 99 ('2004', 2005); Tebbitt, Begonias 191 (2005). – Type: Philippines, Luzon, Quezon, Real, 18 vi 1962, *Lagrimas* PNH42651 (holo L *n.v.*).
- **PHILIPPINES: Luzon:** Quezon, Real, 18 vi 1962, *Lagrimas* PNH42651 (holo L *n.v.*) – Type of *Begonia polilloensis* Tebbitt. **Negros:** Tanyas, Lake Balinsasayao, iii 1948, *G. Edano* PNH11617 (para A *n.v.*) – Type of *Begonia polilloensis* Tebbitt.
- **IUCN category: DD.** Insufficient specimens could be georeferenced with certainty.

Begonia porteri var. macrorhiza Gagnep. in Lecomte **[§ Coelocentrum]**, Fl. Indo-Chine 2: 1109 (1921). – Type: Vietnam, Tonkin, Langson, *B. Balansa s.n.* (holo P *n.v.*).
- **VIETNAM:** Tonkin, Langson, *B. Balansa s.n.* (holo P *n.v.*) – Type of *Begonia porteri* var. *macrorhiza* Gagnep.
- **IUCN category: DD.** Insufficient specimens could be georeferenced with certainty.

Begonia postarii Kiew [§ Petermannia], Gard. Bull. Singapore 50: 165 (1998). – Type: Borneo, Sabah, Kinabatangan, Bukit Panggi, *R. Kiew, S.P. Lim* 4221 (holo SAN *n.v.*; iso SING *n.v.*).
- **BORNEO: Sabah:** Kinabatangan, Bukit Panggi, *R. Kiew, S.P. Lim* 4221 (holo SAN *n.v.*; iso SING *n.v.*) – Type of *Begonia postarii* Kiew.
- **IUCN category: VU D2.** Known only from three localities.

Begonia praetermissa Kiew [§ Platycentrum], Begonias Penins. Malaysia 268 (2005). – Type: Peninsular Malaysia, Kelantan, E-W Highway, 20 ii 2003, *R. Kiew* 5275 (holo SING *n.v.*).
- **PENINSULAR MALAYSIA:** Kelantan, E-W Highway, 20 ii 2003, *R. Kiew* 5275 (holo SING *n.v.*) – Type of *Begonia praetermissa* Kiew.
- **IUCN category: DD.** Insufficient specimens could be georeferenced with certainty.

Begonia procridifolia Wall. ex A. DC. [§ Parvibegonia], Prodr. 15(1): 352 (1864); Wallich, Num. List 6292 (1831); Candolle, Prodr. 15(1): 352 (1864); Clarke, Fl. Brit. Ind. 2: 648 (1879). – Type: Burma, *N. Wallich* 6292 (holo K-W).
- **BURMA:** *N. Wallich* 6292 (holo K-W) – Type of *Begonia procridifolia* Wall. ex A. DC.
  LAOS: Pu Tat, 21 iv 1922, *A.F.G. Kerr* 21188 (K, P) [as *Begonia procridifolia* ?].
- **IUCN category: DD.** Insufficient specimens could be georeferenced with certainty.

Begonia prolifera A. DC. [§ Monophyllon], Ann. Sci. Nat. Bot., IV 11: 135 (1859); Candolle, Prodr. 15(1): 353 (1864); Golding, Phytologia 40(1): 16 (1978). – Type: Burma, Moulmein, *W. Lobb* 381 (iso BM, E, K).

- **BURMA:** Moulmein, *C.S.P. Parish s.n.* (K, P); Moulmein, 1862, *C.S.P. Parish s.n.* (K); Moulmein, *W. Lobb* 381 (iso BM, E, K) – Type of *Begonia prolifera* A. DC.
  THAILAND: Kanchanaburi, Lai Wo Subdistr., Toong Yai Naresuan Wildlife Reserve, 8 x 1993, *J.F. Maxwell* 93-1167 (A, L).
- **Notes:** De Candolle originally ascribed the type locality in error as Singapore, from which there are no native *Begonia* species known. A photograph of a vegetative plant in Kiew (2005, p. 293) could possibly be referred to this species, which would extend its distribution to Peninsular Malaysia.
- **IUCN category: DD**. Insufficient specimens could be georeferenced with certainty.

Begonia prolixa Craib [§ Platycentrum], Bull. Misc. Inform. Kew 1930: 413 (1930); Craib, Fl. Siam. 1: 777 (1931). – Type: Thailand, Pattani, Kao Kalakiri, 31 iii 1928, 6°50′N, 100°57′E, *A.F.G. Kerr* 14926 (holo ABD; iso BM, K[2]).

- **THAILAND:** Pattani, Kao Kalakiri, 31 iii 1928, *A.F.G. Kerr* 14926 (holo ABD; iso BM, K[2]) – Type of *Begonia prolixa* Craib.
- **IUCN category: DD**. Insufficient specimens could be georeferenced with certainty.

Begonia promethea Ridl. [not placed to section], J. Straits Branch Roy. Asiat. Soc. 46: 259 (1906). – Type: Borneo, Sarawak, *G.D. Haviland* 188 (syn not located); Borneo, Sarawak, *G.D. Haviland* 485 (syn not located).

- **BORNEO: Sarawak:** *G.D. Haviland* 188 (syn not located) – Type of *Begonia promethea* Ridl.; *G.D. Haviland* 485 (syn not located) – Type of *Begonia promethea* Ridl.; *O. Beccari* PB1013 (iso Fl[2], P) – Type of *Begonia beccarii* Warb.; Gunung Skunyet, xi 1865, *O. Beccari* PB1050 (Fl).
- **Notes:** See notes under *B. beccarii* Warb.
- **IUCN category: DD**. Taxonomic uncertainty.

Begonia propinqua Ridl. [§ Petermannia], J. Straits Branch Roy. Asiat. Soc. 46: 249 (1906). – Type: Borneo, Sarawak, Matang, *H.N. Ridley* 11771 (syn K); Borneo, Sarawak, Kuching, *G.D. Haviland* not known (syn not located); Borneo, Sarawak, Matang, *G.D. Haviland* not known (syn not located); Borneo, Sarawak, Matang, *R.W. Hullett* (syn not located).

- **BORNEO: Sarawak:** Kuching, *G.D. Haviland* not known (syn not located) – Type of *Begonia propinqua* Ridl.; Matang, *R.W. Hullett* (syn not located) – Type of *Begonia propinqua* Ridl.; Matang, *G.D. Haviland* not known (syn not located) – Type of *Begonia propinqua* Ridl.; Matang, i 1915, *H.N. Ridley s.n.* (BM, K); Matang, *H.N. Ridley s.n.* (BM); Matang, v 1866, *O. Beccari* PB1619 (Fl); Matang, vi 1866, *O. Beccari* PB1830 (Fl); Matang, *H.N. Ridley* 11771 (syn K) – Type of *Begonia propinqua* Ridl.
- **IUCN category: VU D2**. Known only from three localities.

Begonia pryeriana Ridl. [§ Petermannia], J. Straits Branch Roy. Asiat. Soc. 46: 252 (1906). – Type: Borneo, Sabah, Sandakan District, *H.N. Ridley s.n.* (holo not located).

- **BORNEO: Sabah:** Sandakan District, *H.N. Ridley s.n.* (holo not located) – Type of *Begonia pryeriana* Ridl.; Sandakan District, Ulu Dusun, 12 xi 1986, *P.C.v. Welzen* 872 (L); Sandakan District, ix 1920 – xii 1920, *M. Ramos* 1201 (B, K, L); Lahad Datu District, Segama River, 11 vi 1984, *J.H. Beaman et al.* 10067 (L); Lahad Datu District, Segama River, 12 vi 1984, *J.H. Beaman et al.* 10116 (L); Lahad Datu District, Segama River, 12 vi 1984, *J.H. Beaman et al.* 10121 (L); Lahad Datu

District, Bukit Blachan, 13 vi 1984, *J.H. Beaman et al.* 10128 (L); Myburgh Province, Sandakan, *A.D.E. Elmer* 20319 (BM, K, L); Kalabakan Distr., Yayasan Logged area, 9 x 1979, *Fedilis, Sumbing* 91305 (L); Kalabakan Distr., Hap Seng logged area, 4 vi 1982, *F. Kriapinus* 94859 (L).
■ **IUCN category: EN B2ab(iii).** Has a restricted distribution, and none of the known populations are in protected areas.

Begonia pseudolateralis Warb. in Perkins **[§ Petermannia]**, Fragm. Fl. Philipp. 51 (1904); Philipp. J. Sci. 285 (1907); Merrill, Philipp. J. Sci. 6: 374 ('1911', 1912); Merrill, Enum. Philipp. Fl. Pl. 3: 127 (1923); Tebbitt & Dickson, Brittonia 52(1): 115 (2000). – Type: Philippines, Luzon, Isabela Province, Malunu, *O. Warburg* 11793 (holo B; iso B).

Mezierea salaziensis var. *calleryana* A. DC., Prodr. 15(1): 408 (1864); Tebbitt & Dickson, Brittonia 52(1): 115 (2000). – *Begonia aptera* var. *calleryana* (A. DC.) Fern.-Vill., Fl. Filip. 4: 99 (1880); Merrill, Philipp. J. Sci. 6: 374 ('1911', 1912). – Type: Philippines, Luzon, Laguna, Calauan, 1840, *J.M.M. Callery s.n.* (holo P).

Begonia lateralis Elmer ex Merr. *nom. nud.*, Enum. Philipp. Fl. Pl. 3: 127 (1923). – Type: Philippines, Luzon, Sorsogon, Mt. Bulusan, vii 1915, *A.D.E. Elmer* 16661 (iso BM, BO, K, L).

■ **PHILIPPINES: Luzon:** Laguna, Calauan, 1840, *J.M.M. Callery s.n.* (holo P) – Type of *Mezierea salaziensis* var. *calleryana* A. DC.; Cagayan Province, Penablanca, Pianga Creek, v 1918 – vi 1918, *M. Adduru* 15 (K, P); Bataan, Mt. Mariveles, Lamao River, 13 i 1904, *R.S. Williams* 523 (K) [as *Begonia pseudolateralis* ?]; 1861, *A.F. Jagor* 814 (B); Albay Province, Mt. Mayon, 17 ix 1991, *Reynoso et al.* 3506 (K, L); Tayabas, Atimonan, iii 1905, *E.D. Merrill* 4009 (B, K, P); Quezon, Polillo Island, *C.B. Robinson* 6903 (BO, L, P); Ifugao, Lake Ambuaya, 29 iii 1991, *Reynoso et al.* 7283 (K); Mt. Dimaxinggay, 13 iii 1993, *Barbon et al.* 9202 (L); Isabela Province, Malunu, *O. Warburg* 11792 (B); Isabela Province, Malunu, *O. Warburg* 11793 (holo B; iso B) – Type of *Begonia pseudolateralis* Warb.; Benguet, Sablan(g), xi 1910 – xii 1910, *E. Fenix* 12661 (B, L); Tayabas, iii 1888, *O. Warburg* 13086 (B); Cavite Province, Mt. PalayPalay Nat. Park, 22 iii 1995, *Reynoso et al.* 14957 (L); Sorsogon, Mt. Bulusan, vii 1915, *A.D.E. Elmer* 16661 (iso BM, BO, K, L) – Type of *Begonia lateralis* Elmer ex Merr. *nom. nud.*; Albay Province, Mayon Volcano, 2 vi 1953, *D.R. Mendoza* 18312 (L); Cagayan Province, Claveria, 3 viii 1995, *Garcia et al.* 18313 (L); Tayabas, Mauban, i 1913, *M. Ramos* 19481 (BM, L); Isabela Province, San Jose Village, 6 iii 1997, *Reynoso et al.* 20008 (L); Albay, Taqui River – Mt. Malinao, 27 x 1995, *Reynoso et al.* 21272 (L); Albay Province, Mt. Malinao, 29 x 1995, *Reynoso et al.* 21354 (L); Camarines, Mt. Isaro(g), xi 1913 – xii 1913, *M. Ramos* 22017 (BM, L); Laguna, vi 1915 – viii 1915, *R.C. McGregor* 22801 (P); Sorsogon, vii 1915 – viii 1915, *M. Ramos* 23437 (B, P); Catanduanes, 7 xi 1996, *Reynoso, R.S. Majaducon* 24877 (L); Catanduanes, 9 xi 1996, *Reynoso, R.S. Majaducon* 24958 (L); Apayao, *E. Fenix* 28141 (BO, K, P); Apayao, v 1917, *E. Fenix* 28141 (BO, K, P); Isabela Province, 6 iii 1997, *P. Wilkie et al.* 29008 (E); Ilocos Norte Province, Bangui – Claveria, viii 1918, *M. Ramos* 33037 (L); Albay Province, Mt. Malinao, 29 i 1956, *G. Edano* 34447 (L); Sorsogon, Irosin, 26 v 1957, *G. Edano, H. Gutierrez* 38546 (BM, K, L); Sorsogon, Irosin, v 1957, *G. Edano, H. Gutierrez* 38546 (BM, K, L); Abra Prov., Poblacion Gangal, 14 xi 1996, *Fuentes* 38594 (L); Tayabas, Casiguran, v 1925 – vi 1925, *M. Ramos, G. Edano* 45733 (BM); San Jose Village, San Mariano, Bo. Disulap, 26 iv 1961, *H. Gutierrez* 78078 (L). **Mindoro:** Mt. Halcon, Lantuyan, 29 iii 1991, 440 (K, L); Sibuang River, 12 ii 1985, *C.E. Ridsdale* 831 (L); Baco River, *E.D. Merrill* 991 (para B, K, P) – Type of *Begonia pseudolateralis* Warb.; Mt. Halcon, i 1948 – ii 1948, *G. Edano* 2492 (BO[2], L); Mt. Halcon, i 1948 – ii 1948, *G. Edano* 3492 (BO[2], L); Mt. Halcon, iii 1922, *M. Ramos, G. Edano* 40679 (K). **Panay:** Capiz, Libacao, v 1919 – vi 1919, *A. Martelino* 35397 (B, BM, K, P). **Negros:** Negros Occidental, Mt. Katugasan, iii 1954, *G. Edano* 21819 (K, L); Negros Occidental, Kinabkaban River, 28 v 1954, *G. Edano* 21899 (L). **Cebu:** Cantipla, 26 iii 1971, *Anon.* 10 (L); Kantipla, i 1994, *D. Bicknell* 804 (K). **Leyte:** 30 viii 1913, *C.A. Wenzel* 517 (BM); 10 iii 1914, *C.A.*

Wenzel 634 (BM). **Samar:** Laquilacon, vi 1924, *R.C. McGregor* 437559 (BM). **Mindanao:** Agusan, Mt. Urdaneta, viii 1912, *A.D.E. Elmer* 13494 (BM, BO, E, K, L, U); Basilan, viii 1912 – ix 1912, *J. Reillo* 16142 (BM, L, P); Zamboanga, Mt. Tubuan, x 1919, *M. Ramos, G. Edano* 36654 (BM, BO); Butuan, Ojot River, 21 vi 1961, *D.R. Mendoza* 42468 (L).
SULAWESI: Sulawesi Utara, Motomboto, 23 xii 1994, *J.J. Afriastini et al.* 2872 (L).
MOLUCCAS: Halmahera: Baccan Islands, Gunung Sibela, 27 x 1974, *E.F. de Vogel* 3701 (L) [as *Begonia cf. pseudolateralis*].
- **Notes:** One of the most widespread species in the Philippines. See notes under *B. rieckei* Warb.
- **IUCN category: LC.** Widespread.

Begonia pubescens Ridl. [§ Petermannia], J. Straits Branch Roy. Asiat. Soc. 46: 254 (1906); Gibbs, J. Linn. Soc., Bot. 42: 85 (1914). – Type: Borneo, Sarawak, Matang, *R.W. Hullett* 346 (syn not located); Borneo, Sarawak, Matang, iii 1891, *G.D. Haviland* 76 (syn K).
Begonia hirsuta Brace ex Ridl. *nom. nud.*, J. Straits Branch Roy. Asiat. Soc. 46: 254 (1906). – Type: not located.
- **BORNEO: Sarawak:** Matang, viii 1905, *H.N. Ridley s.n.* (K); Matang, 3 viii 1912, *Anderson* 3 (K); Matang, Mt. Braang, iii 1891, *G.D. Haviland* 76 (syn K) – Type of *Begonia pubescens* Ridl.; Matang, *R.W. Hullett* 346 (syn not located) – Type of *Begonia pubescens* Ridl.; Matang, 25 vi 1961, *S.H. Collenette* 699 (K, L); Gunung Mulu National Park, 19 xi 1977, *G.C.G. Argent et al.* 813 (E) [as *Begonia cf. pubescens*]; *Native collector* 1656 (K); Gunung Matang, 29 v 1962, *B.L. Burtt, P.J.B. Woods* 1958 (E) [as *Begonia cf. pubescens*]; Matang, 24 iii 1955, *W.M.A. Brooke* 9755 (L); Gunung Matang, 8 iv 1966, *J.A.R. Anderson* 25108 (L); Limbang, 16 vii 1981, *R. George et al.* 42848 (E) [as *Begonia cf. pubescens*].
- **IUCN category: VU D2.** Excluding the speculative determination of the specimen from Gunung Mulu, this species has a very restricted distribution in Matang.

Begonia pulchra (Ridl.) L.L. Forrest & Hollingsw. [§ Symbegonia], Plant Syst. Evol. 241: 208 (2003). – *Symbegonia pulchra* Ridl., Trans. Linn. Soc. London, Bot., II 9: 62 (1916); Forrest & Hollingsworth, Plant Syst. Evol. 241: 208 (2003). – Type: New Guinea, Papua, Camp IX–X, i 1913, *C.B. Kloss s.n.* (syn K[3]); New Guinea, Papua, Camp XI–IX, i 1913, *C.B. Kloss s.n.* (syn BM, K[2]).
- **NEW GUINEA: Papua:** Camp XI–IX, i 1913, *C.B. Kloss s.n.* (syn BM, K[2]) – Type of *Symbegonia pulchra* Ridl.; Camp IX–X, i 1913, *C.B. Kloss s.n.* (syn K[3]) – Type of *Symbegonia pulchra* Ridl. **Papua New Guinea:** Eastern Highlands Distr., Arau, 6 x 1959, *E.H.B. Brascamp* 31094 (K) [as *Begonia aff. pulchra*]; Eastern Highlands Distr., Arau, 10 x 1959, *L.J. Brass* 31904 (K) [as *Begonia aff. pulchra*]; Eastern Highlands Distr., Arau, 10 x 1959, *L.J. Brass* 32000 (BO, K, L) [as *Begonia aff. pulchra*].
- **IUCN category: DD.** Insufficient specimens could be georeferenced with certainty.

Begonia pumila Craib [not placed to section], Bull. Misc. Inform. Kew 1930: 414 (1930); Craib, Fl. Siam. 1: 777 (1931). – Type: Thailand, Puket, Ranawng, Kao Pawta Chongdong, 9°37′N, 98°37′E, *A.F.G. Kerr* 16757 (holo ABD; iso BM, K *n.v.*).
- **THAILAND:** Puket, Ranawng, Kao Pawta Chongdong, *A.F.G. Kerr* 16757 (holo ABD; iso BM, K *n.v.*) – Type of *Begonia pumila* Craib.
- **Notes:** Tentatively assigned to § *Ridleyella* (Doorenbos *et al.*, 1998).
- **IUCN category: DD.** Insufficient specimens could be georeferenced with certainty.

Begonia pumilio Irmsch. [§ Reichenheimia], Mitt. Inst. Allg. Bot. Hamburg 8: 102 (1929). – Type: Thailand, Pulau Panji, 16 xii 1918, *M. Haniff, Md. Nur* SFN4076 (holo K ex SING).

- **THAILAND:** Pulau Panji, 16 xii 1918, *M. Haniff, Md. Nur* SFN4076 (holo K ex SING) – Type of *Begonia pumilio* Irmsch.
- **IUCN category: DD.** Insufficient specimens could be georeferenced with certainty.

Begonia punbatuensis Kiew [§ Petermannia], Gard. Bull. Singapore 53: 279 (2001). – Type: Borneo, Sabah, Pensiangan District, Pun Batu, *R. Kiew, A. Berhaman* 4260 (holo SAN *n.v.*; iso BRUN *n.v.*, K *n.v.*, L *n.v.*, SAN *n.v.*, SAR *n.v.*, SING *n.v.*).
- **BORNEO: Sabah:** Pensiangan District, Pun Batu, *R. Kiew, A. Berhaman* 4260 (holo SAN *n.v.*; iso BRUN *n.v.*, K *n.v.*, L *n.v.*, SAN *n.v.*, SAR *n.v.*, SING *n.v.*) – Type of *Begonia punbatuensis* Kiew.
- **IUCN category: VU D2.** Known only from a single hill not in a protected area.

Begonia putii Craib [§ Diploclinium], Gard. Chron., III 83: 67 (1928); Craib, Fl. Siam. 1: 778 (1931). – Type: Thailand, Payap, Doi Chiengdao, 19 x 1926, *N. Put* 403 (holo ABD; iso B, BM, K).
- **THAILAND:** Payap, Doi Chiengdao, 19 x 1926, *N. Put* 403 (holo ABD; iso B, BM, K) – Type of *Begonia putii* Craib.
- **IUCN category: DD.** Insufficient specimens could be georeferenced with certainty.

Begonia pyrrha Ridl. [not placed to section], J. Straits Branch Roy. Asiat. Soc. 46: 260 (1906). – Type: Borneo, Sarawak, Saribas, *G.D. Haviland* 1848 (syn not located); Borneo, Sarawak, Saribas, *G.D. Haviland* 2034 (syn not located).
- **BORNEO: Sarawak:** Saribas, *G.D. Haviland* 1848 (syn not located) – Type of *Begonia pyrrha* Ridl.; Saribas, *G.D. Haviland* 2034 (syn not located) – Type of *Begonia pyrrha* Ridl.; 1865 – 1868, *O. Beccari* PB3796 (FI, K); Batang Ai, Mabau Ridge, Ulu Engkari, 15 xii 1994, *Yii Puan Ching* 69739 (K) [as *Begonia cf. pyrrha*].
- **IUCN category: DD.** Insufficient specimens could be georeferenced with certainty.

Begonia quercifolia A. DC. [§ Petermannia], Ann. Sci. Nat. Bot., IV 11: 129 (1859); Candolle, Prodr. 15(1): 321 (1864); Fernández-Villar, Noviss. App. 99 (1880); Vidal, Phan. Cuming. Philipp. 116 (1885); Vidal, Rev. Pl. Vasc. Filip. 143 (1886); Merrill, Philipp. J. Sci. 6: 387 ('1911', 1912); Merrill, Philipp. J. Sci. 7: 311 (1912); Merrill, Philipp. J. Sci. 10: 277 (1915); Merrill, Enum. Philipp. Fl. Pl. 3: 127 (1923). – Type: Philippines, Samar, 1841, *H. Cuming* 1696 (syn B, FI, K, P).
    Begonia leytensis Elmer, Leafl. Philipp. Bot. 2: 739 (1910); Merrill, Philipp. J. Sci. 6: 384 ('1911', 1912); Merrill, Philipp. J. Sci. 7: 311 (1912). – Type: Philippines, Leyte, Palo, i 1906, *A.D.E. Elmer* 7255 (syn BM, E, K, L).
- **PHILIPPINES: Luzon:** Tayabas, Lucban, v 1907, *A.D.E. Elmer* 7643 (BM, E, L); Tayabas, Infanta, viii 1909, *C.B. Robinson* 9326 (B). **Leyte:** Palo, i 1906, *A.D.E. Elmer* 7255 (syn BM, E, K, L) – Type of *Begonia leytensis* Elmer. **Samar:** 1841, *H. Cuming* 1696 (syn B, FI, K, P) – Type of *Begonia quercifolia* A. DC.; Catubig River, ii 1916 – iii 1916, *M. Ramos* 24282 (K).
- **IUCN category: DD.** Insufficient specimens could be georeferenced with certainty.

Begonia rabilii Craib [not placed to section], Bull. Misc. Inform. Kew 1930: 415 (1930); Craib, Fl. Siam. 1: 778 (1931). – Type: Thailand, Nakawn Sritamarat, Kao Chem, *N. Rabil Bunnag* 127 (holo ABD; iso BM).
- **THAILAND:** Nakawn Sritamarat, Kao Chem, *N. Rabil Bunnag* 127 (holo ABD; iso BM) – Type of *Begonia rabilii* Craib.
- **Notes:** Tentatively assigned to § *Reichenheimia* (Doorenbos *et al.*, 1998).
- **IUCN category: DD.** Insufficient specimens could be georeferenced with certainty.

Begonia racemosa Jack [§ Petermannia], Malayan Misc. 2(7): 14 (1822); Candolle, Prodr. 15(1): 322 (1864). – *Petermannia racemosa* (Jack) Klotzsch, Monatsber. Kon. Preuss. Akad. Wiss. Berlin 1854: 124 (1854); Klotzsch, Abh. Kon. Akad. Wiss. Berlin 1854: 196 (1855); Klotzsch, Begoniac. 76 (1855); Miquel, Fl. Ned. Ind. 1(1): 691 (1856); Candolle, Prodr. 15(1): 322 (1864). – *Diploclinium racemosum* (Jack) Miq., Fl. Ned. Ind. 1(1): 691 (1856); Candolle, Prodr. 15(1): 322 (1864). – Type: not located.
- **SUMATRA:** *Fide* Jack (1822).
- **IUCN category: DD.** Insufficient specimens could be georeferenced with certainty.

Begonia rachmatii Tebbitt [§ Petermannia], Edinburgh J. Bot. 61: 101 ('2004', 2005). – Type: Sulawesi, Gunung Babalombang, viii 1913, *Rachmat* 475 (holo L *n.v.*).
- **SULAWESI:** Gunung Babalombang, viii 1913, *Rachmat* 475 (holo L *n.v.*) – Type of *Begonia rachmatii* Tebbitt; Masamba, 1 viii 1937, *P.J. Eyma* 1502 (L, U).
- **IUCN category: DD.** Insufficient specimens could be georeferenced with certainty.

Begonia rajah Ridl. [§ Reichenheimia], Gard. Chron., III 16: 213 (1894); Rolfe, Bull. Misc. Inform. Kew 1914: 327 (1914); Ridley, Fl. Malay Penins. 1: 855 (1922); Irmscher, Mitt. Inst. Allg. Bot. Hamburg 8: 96 (1929); Tebbitt, Begonias 198 (2005); Kiew, Begonias Penins. Malaysia 216 (2005). – Type: Peninsular Malaysia, Trengganu, 1892, *Native collector s.n.* (holo K).
Begonia peninsulae subsp. *peninsulae* Irmsch., Mitt. Inst. Allg. Bot. Hamburg 8: 98 (1929). – Type: Peninsular Malaysia, Trengganu, *H.N. Ridley s.n.* (holo K).
- **PENINSULAR MALAYSIA:** Trengganu, *H.N. Ridley s.n.* (holo K) – Type of *Begonia peninsulae* Irmsch.; Trengganu, 1892, *Native collector s.n.* (holo K) – Type of *Begonia rajah* Ridl.; Johore, Sungai Selai, *Sam et al.* FRI47082 (KEP).
- **IUCN category: DD.** Insufficient specimens could be georeferenced with certainty.

Begonia ramosii Merr. [§ Petermannia], Philipp. J. Sci. 6: 388 ('1911', 1912); Merrill, Enum. Philipp. Fl. Pl. 3: 127 (1923). – Type: Philippines, Luzon, Laguna, San Antonio, *M. Ramos* 10942 (syn B, BM).
- **PHILIPPINES: Luzon:** Laguna, ii 1913, *M. Ramos* 1485 (B, BM[2], BO, L, P); Laguna, San Antonio, 19 viii 1910, *M. Ramos* 10941 (para B, BM, K) – Type of *Begonia ramosii* Merr.; Laguna, San Antonio, *M. Ramos* 10942 (syn B, BM) – Type of *Begonia ramosii* Merr.; Laguna, San Antonio, viii 1910, *M. Ramos* 12047 (para BM, L, P) – Type of *Begonia ramosii* Merr.; Laguna, vi 1915 – viii 1915, *R.C. McGregor* 22890 (K); Laguna, Paete, iii 1917, *R.C. McGregor* 27869 (BM, BO, K, L, P).
- **IUCN category: EN B2ab(iii).** Known only from historical collections and only from two sites, in the heavily degraded Luzon rain forest eco-region.

Begonia randiana Merr. & L.M. Perry [§ Petermannia], J. Arnold Arbor. 24: 47 (1943). – Type: New Guinea, Papua New Guinea, Mt. Tafa, ix 1933, *L.J. Brass* 4989 (holo A; iso B, BO, L).
- **NEW GUINEA: Papua New Guinea:** Milne Bay Distr., Baniara Subdistr., Agaun – Bonenau, 8 viii 1969, *W.E. Fisher* 76 (L) [as *Begonia aff. randiana*]; Morobe Distr., Sattelberg, *Clemens* 151 (L); Mt. Tafa, v 1936, *L.J. Brass* 4136 (para B) – Type of *Begonia randiana* Merr. & L.M. Perry; Mt. Tafa, ix 1933, *L.J. Brass* 4989 (holo A; iso B, BO, L) – Type of *Begonia randiana* Merr. & L.M. Perry; Central Division, Mafulu, xi 1933, *L.J. Brass* 5508 (para B, BO) – Type of *Begonia randiana* Merr. & L.M. Perry; Milne Bay Distr., Mt. Dayman, 10 vii 1953, *L.J. Brass* 23380 (L) [as *Begonia aff. randiana*]; Milne Bay Distr., Mt. Dayman, 17 vii 1933, *L.J. Brass* 23484 (L) [as *Begonia cf. randiana*]; Western Distr., Goilala Subdistr., Woitape, 5 viii 1968, *C.E. Ridsdale, P.J.B. Woods* 33784 (BO, E, L) [as *Begonia cf. randiana*]; Western Distr., Goilala Subdistr., Woitape, 22 viii 1969, *A.N. Millar* 40984 (L); Milne Bay Distr., Rabaraba, 17 vi 1972, *P. Katik* 46993 (L).

- **IUCN category: LC.** Collected from several sites across the Central Range, at mid to high elevations (900–1700 m).

Begonia reginula Kiew [§ Reichenheimia], Begonias Penins. Malaysia 218 (2005). – Type: Peninsular Malaysia, Trengganu, Ulu Setui, 29 iv 1986, *R. Kiew* 2278 (holo SING *n.v.*).
   Begonia rajah auct. non Ridl.: Kiew, Begonian 56: 53 (1989); Kiew, Nat. Malaysiana 14: 66 (1989).
- **PENINSULAR MALAYSIA:** Negri Sembilan, Ulu Serting Forest Reserve, 16 xi 1996, *R. Kiew s.n.* (para SING *n.v.*) – Type of *Begonia reginula* Kiew; Trengganu, Ulu Setui, *R. Kiew* 2259 (para KEP *n.v.*) – Type of *Begonia reginula* Kiew; Trengganu, Ulu Setui, 29 iv 1986, *R. Kiew* 2278 (holo SING *n.v.*) – Type of *Begonia reginula* Kiew; Trengganu, Kuala Sungai Bok, *Mohd Shah et al.* MS3511 (para SING *n.v.*) – Type of *Begonia reginula* Kiew; Negri Sembilan, Ulu Serting Forest Reserve, *R. Kiew* 5175 (para SING *n.v.*) – Type of *Begonia reginula* Kiew; Trengganu, Ulu Setui, *Ng* FRI22013 (para KEP *n.v.*) – Type of *Begonia reginula* Kiew; Trengganu, Ulu Brang, *L. Moysey, Kiah* SFN33803 (para K, L, SING *n.v.*) – Type of *Begonia reginula* Kiew; Negri Sembilan, Pasoh Forest Reserve, *Saw, Mustafa* FRI37505 (para KEP *n.v.*) – Type of *Begonia reginula* Kiew; Negri Sembilan, Ulu Serting Forest Reserve, *Saw, Mustafa* FRI37515 (para K, KEP *n.v.*) – Type of *Begonia reginula* Kiew; Negri Sembilan, Ulu Bendol Forest Reserve, 1 xii 1922, *R.E. Holttum* 9841 (BO).
- **IUCN category: EN B2ab(iii).** The small number of populations are threatened by habitat disturbance (Kiew, 2005).

Begonia renifolia Irmsch. [§ Sphenanthera], Bot. Jahrb. Syst. 50: 379 (1913); Tebbitt, Brittonia 55(1): 27 (2003). – Type: Sulawesi, Minahassa, Bojong, *N. Wallich* 15188 (holo B).
- **SULAWESI:** Minahassa, Bojong, *N. Wallich* 15188 (holo B) – Type of *Begonia renifolia* Irmsch.
- **IUCN category: DD.** Insufficient specimens could be georeferenced with certainty.

Begonia rex Putz. [§ Platycentrum], Fl. Serres Jard. 2(8): 141 (1857); Hooker, Bot. Mag., n.s. 85: 5701 (1859); Candolle, Prodr. 15(1): 350 (1864); Gagnepain, in Lecomte (ed.), Fl. Indo-Chine 2: 1112 (1921). – Type: not located.
- **BURMA:** *Fide* Kress *et al.* (2003).
- **Notes:** This species can be confused with *B. sizemoreae* Kiew and *B. longicoliata* C.Y. Wu (see Kiew, 2004).
- **IUCN category: DD.** Insufficient specimens could be georeferenced with certainty.

Begonia rheifolia Irmsch. [§ Platycentrum], Mitt. Inst. Allg. Bot. Hamburg 8: 132 (1929); Kiew, Begonias Penins. Malaysia 280 (2005). – Type: Peninsular Malaysia, Pahang, Tahan Valley, viii 1891, *H.N. Ridley s.n.* (holo K; iso K).
   Begonia herveyana var. *robusta* Ridl., Fl. Malay Penins. 1: 861 (1922); Irmscher, Mitt. Inst. Allg. Bot. Hamburg 8: 132 (1929); Kiew, Begonias Penins. Malaysia 280 (2005). – Type: Peninsular Malaysia, Pahang, Gunong Tahan, viii 1905, *L. Wray Jr., H.C. Robinson* 5546 (lecto K; isolecto BM, SING *n.v.*).
   Begonia tiomanensis Ridl., Bull. Misc. Inform. Kew 1928: 73 (1928); Irmscher, Mitt. Inst. Allg. Bot. Hamburg 8: 133 (1929); Kiew, Begonias Penins. Malaysia 280 (2005). – Type: Peninsular Malaysia, Pahang, Pulau Tioman, 2 v 1927, *Mohd. Nur* SFN18796 (holo K; iso K).
- **PENINSULAR MALAYSIA:** Pahang, Tahan Valley, viii 1891, *H.N. Ridley s.n.* (holo K; iso K) – Type of *Begonia rheifolia* Irmsch.; Pahang, Taman Negara, Lata Berkoh, 13 viii 1990, *H. Okada et al.* 49 (L); Pahang, Gunong Tahan, viii 1905, *L. Wray Jr., H.C. Robinson* 5546 (lecto K; isolecto BM, SING *n.v.*) – Type of *Begonia herveyana* var. *robusta* Ridl.; Pahang, Pulau Tioman, 2 v 1927, *A.J.G.H. Kostermans, Md. Nur* SFN18796 (holo K; iso K) – Type of *Begonia tiomanensis* Ridl.; Pahang,

Begonian 47: 213 (1980); Tebbitt, Begonias 204 (2005); Tebbitt, Blumea 50(1): 155 (2005). – *Platycentrum robustum* (Blume) Miq., Fl. Ned. Ind. 1(1): 694 (1856); Klotzsch, Bot. Zeitung 15: 183 (1857); Candolle, Prodr. 15(1): 275 (1864). – *Casparya robusta* (Blume) A. DC., Prodr. 15(1): 275 (1864). – Type: not located.

Begonia splendida K. Koch, Berliner Allg. Gartenzeitung 10: 74 (1857); Klotzsch, Bot. Zeitung 15: 182 (1857); Henderson, Ill. Bouquet 1: sub (1857); Candolle, Prodr. 15(1): 275 (1864); Koorders, Exkurs.-Fl. Java 2: 646 (1912). – Type: not located.

Sphenanthera robusta Hassk. ex Klotzsch, Bot. Zeitung 15: 182 (1857); Candolle, Prodr. 15(1): 275 (1864). – Type: not located.

Begonia robusta var. robusta
■ **SUMATRA:** *O. Beccari* CB4513 (FI[2]); *O. Beccari* CB4513A (FI[2]).

JAVA: *P.W. Korthals s.n.* (L); *P.W. Korthals s.n.* (L); *P.W. Korthals s.n.* (L); Mt. Salak, 17 ii 1915, *H.N. Ridley s.n.* (K[2]); *C.L.v. Blume s.n.* (P); *unknown s.n.* (L); Tjibeureum, 1916, *J.G. Boerlage s.n.* (L); Preanger, Tjibodas, 12 ix 1899, *unknown s.n.* (L); *unknown s.n.* (L); Preanger, Tjibodas, *J.G. Boerlage s.n.* (L); Preanger, Tjibodas, *M.L.A. Bruggeman s.n.* (BO); Preanger, Tjibodas, *v. Woerden s.n.* (BO); Priangan, G. Pangrango, *W.S. Kurz s.n.* (K); Preanger, Gunung Gede, 14 ii 1915, *H.N. Ridley s.n.* (K); Batavia, G. Salak, 25 xi 1940, *C.N.A. de Voogd s.n.* (K); 1 ii 1915, *H.N. Ridley s.n.* (K); Preanger, Tjibodas, *W.J.C. Kooper s.n.* (U); Preanger, Tjibodas, 24 xii 1932, *W.J.C. Kooper s.n.* (U); *unknown s.n.* (B); *O. Warburg s.n.* (B); *W.B. Hillebrand s.n.* (B); West Java, Mt. Malabar, *T. Anderson s.n.* (K); Priangan, G. Pangrango, *v. Hooten* 24 (BO); Batavia, G. Salak, *B.P.G. Hochreutiner* 57 (L); Tjibeureum, 17 iv 1932, *A. Kleinhoonte* 59 (L); Preanger, Tjibodas, ii 1890, *V. Lehmann* 63 (B); Tjibeureum, 28 vi 1941, *S. Bloembergen* 80 (BO); Preanger, Tjibodas, 20 i 1895, *J.G. Hallier* 147 (BO); Preanger, Tjibodas, 17 vii 1927, *M.L.A. Bruggeman* 187 (BO); *W. Lobb* 247 (K); Gunung Malabar, vi 1930, *L.v.d. Pijl* 249 (BO); Tjibeureum, 6 vii 1941, *S. Bloembergen* 282 (BO); Banjoemas, Gunung Slamat, 15 iv 1911, *C.A.B. Backer* 311 (BO); Gunong Salak, 10 xi 1845, *H. Zollinger* 367 (P); Gunong Salak, iii 1844, *H. Zollinger* 439 (B, P); 1858, *A.F. Jagor* 463 (B[2]); Mt. Lawu, Tambak, 26 xi 1982, *J.J. Afriastini* 486 (L); Geger Bintang, 24 ix 1918, *L.G.d. Berger* 525 (BO); Tjibeureum, 26 vi 1924, *C.N.A. de Voogd* 650 (L); SE Java, *H.O. Forbes* 796a (not located); SE Java, *H.O. Forbes* 856 (B, BM); SE Java, *H.O. Forbes* 900a (not located); Priangan, G. Pangrango, 11 iv 2001, *W.S. Hoover, H. Wiriadinata* 991 (A[2]); Mt. Halimun N.P., 15 iv 2001, *W.S. Hoover, H. Wiriadinata* 998 (A); SE Java, *H.O. Forbes* 1017b (BM); Preanger, Tjibodas, 6 v 1914, *J.A. Lorzing* 1724 (BO); Tjibeureum, 2 v 1894, *V.F. Schiffner* 2265 (L); Priangan, G. Pangrango, 7 iv 1894, *V.F. Schiffner* 2267 (L); Priangan, G. Pangrango, 6 iv 1894, *V.F. Schiffner* 2271 (L); Buitenzorg, 13 iii 1923, *R.C. Bakhuizen van den Brink* 2390 (U); Preanger, Tjibodas, *Sapiin* 2441 (U); Buitenzorg, Tjirarak, i 1941, *C.N.A. de Voogd* 3152 (BO); Preanger, Gunung Gede, 15 ix 1911, *C.A.B. Backer* 3219 (BO); Salak, G. Gadjah, 13 ii 1921, *R.C. Bakhuizen van den Brink* 4146 (BO); G. Luhur, 8 viii 1982, *M.M.J.v. Balgooy, Mogea* 4279 (K, L); Pangerango Mts., *O. Beccari* CB4498 (FI); Preanger, Tjibodas, 19 i 1906, *H.G.A. Engler* 4637 (B); Preanger, Tjibodas, 17 xii 1925, *B.H. Danser* 6161 (L); West Java, Gunung Halimun, 3 x 1985, *E.F. de Vogel* 7750 (L); Preanger, Gunung Gede, 3 iii 1959, *J. Sinclair* 10075 (E[3], K); Mt. Halimun N.P., 20 v 2002, *H. Wiriadinata et al.* 10692 (A); Tjisaroes, 26 ii 1950, *S.J.v. Ooststroom* 12832 (L); Tjibeureum, 12 iii 1950, *S.J.v. Ooststroom* 12938 (L); Geger Bintang, 1 v 1950, *S.J.v. Ooststroom* 13872 (L); Preanger, Tjibodas, 8 v 1950, *S.J.v. Ooststroom* 13946 (L); Preanger, Tjibodas, 8 v 1950, *S.J.v. Ooststroom* 13947 (L); Preanger, Gunung Gede, 21 vi 1931, *Doctors v. Leeuwen-Reynvaan* 13963 (BO); Preanger, Gunung Gede, 16 vii 1931, *Doctors v. Leeuwen-Reynvaan* 14013 (BO); Preanger, Gunung Gede, 30 xii 1977, *A. Nitta* 15113 (L); Tjidadap, 13 vi 1917, *C.A.B. Backer* 22585 (BO); Batavia, G. Salak, 23 ix 1896, *S.H. Koorders* 24465B (L); Priangan, G. Pangrango, 29 x 1898, *S.H. Koorders* 31834B (BO); Preanger,

Tjibodas, 18 x 1898, *S.H. Koorders* 31835B (BO); Preanger, Tjibodas, 1 xi 1898, *S.H. Koorders* 32016 (BO).
LESSER SUNDA ISLANDS: Bali: Mt. Lesung, *McDonald, Ismail* 4851 (K).
SULAWESI: Sopu Valley – Danau Tambing, 24 v 1979, *M.M.J.v. Balgooy* 3486 (A) [as *Begonia cf. robusta*].

Begonia robusta var. hirsutior (Miq.) Golding & Kareg., Phytologia 54(7): 499 (1984). – *Platycentrum robustum* var. *hirsutior* Miq., Fl. Ned. Ind., Suppl. 1:322 (1861); Golding & Karegeannes, Phytologia 54(7): 499 (1984). – Type: Sumatra, Palembajan, *J.E. Teijsmann* (not located).
■ **SUMATRA:** Palembajan, *J.E. Teijsmann* (not located) – Type of *Platycentrum robustum* var. *hirsutior* Miq.
■ **Notes:** Var. *hirsutior* was raised by Miquel (as *Platycentrum robustum* var. *hirsutior* Miq.) based on a specimen (collected by Teijsmann from Palembajan) he originally cited under his concept of *Diploclinium areolatum*. The insufficient description given by Miquel (*foliis minoribus, paniculis longe pedunculatis, ovarii maturescentis alulis subaequalibus*) and current lack of type material means the status of this taxon is uncertain.
     Whether *Balgooy* 3486 (A) really extends the distribution of *Begonia robusta* to Sulawesi needs further investigation.
■ **IUCN category: LC.** Widespread.

Begonia rockii Irmsch. [§ Platycentrum], Mitt. Inst. Allg. Bot. Hamburg 10: 544 (1939). – Type: Burma, Sadon – Changtifang and Kambaiti, xi 1922, *J.F.C. Rock* 7461 (holo B *ex* US; iso E *n.v.*).
■ **BURMA:** Kachin, Sumprabum sub-division, Bumpha Bum, ii 1962, *J. Keenan et al.* 3500a (E); Sadon – Changtifang and Kambaiti, xi 1922, *J.F.C. Rock* 7461 (holo B *ex* US; iso E *n.v.*) – Type of *Begonia rockii* Irmsch.
■ **IUCN category: DD.** Insufficient specimens could be georeferenced with certainty.

Begonia roxburghii (Miq.) A. DC. [§ Sphenanthera], Prodr. 15(1): 398 (1864); Clarke, Fl. Brit. Ind. 2: 635 (1879); Gagnepain, in Lecomte (ed.), Fl. Indo-Chine 2: 1119 (1921); Bull. Dept. Med. Pl. Nep. 7: 93 (1976); Grierson, Fl. Bhutan 2: 243 (1991); Tebbitt, Begonias 207 (2005). – *Diploclinium roxburghii* Miq., Fl. Ned. Ind. 1(1): 692 (1856); Candolle, Prodr. 15(1): 399 (1864). – Type: not located.
     Begonia malabarica Roxb., Fl. Ind. ed. 1832 3:648 (1832); Candolle, Prodr. 15(1):399 (1864). – Type: not located.
     Casparya polycarpa A. DC., Ann. Sci. Nat. Bot., IV 11:118 (1859); Candolle, Prodr. 15(1):277 (1864); Kurz, Flora 54: 295 (1871). – Type: not located.
     Casparya oligocarpa A. DC., Ann. Sci. Nat. Bot., IV 11: 118 (1859); Candolle, Prodr. 15(1): 276 (1864); Clarke, Fl. Brit. Ind. 2:635 (1879). – Type: not located.
■ **BURMA:** Kyauk-chin, 12 i 1926, *M.P. Chin* 198 (B); Yanka, iv 1938, *F.G. Dickason* 9227 (A).
■ **IUCN category: DD.** Insufficient specimens could be georeferenced with certainty.

Begonia rubida Ridl. [§ Petermannia], J. Straits Branch Roy. Asiat. Soc. 46: 256 (1906); Kiew & Geri, Gard. Bull. Singapore 55: 121 (2003). – Type: Borneo, Sarawak, Bau, Jambusan, *H.N. Ridley* 12393 (lecto K); Borneo, Sarawak, Braang, *G.D. Haviland* 94 (syn not located).
■ **BORNEO: Sarawak:** Braang, *G.D. Haviland* 94 (syn not located) – Type of *Begonia rubida* Ridl.; Bau, Jambusan, *H.N. Ridley* 12393 (lecto K) – Type of *Begonia rubida* Ridl.
■ **IUCN category: DD.** Insufficient specimens could be georeferenced with certainty.

Begonia rubrifolia Merr. [§ Diploclinium], Philipp. J. Sci. 14: 426 (1919). – Type: Philippines, Panay, Antique Province, Culasi, *R.C. McGregor* 32430 (syn B, BM).
- **PHILIPPINES: Panay:** Antique Province, Culasi, *R.C. McGregor* 32430 (syn B, BM) – Type of *Begonia rubrifolia* Merr.
- **IUCN category: DD.** Insufficient specimens could be georeferenced with certainty.

Begonia rufipila Merr. [§ Diploclinium], Philipp. J. Sci. 6: 393 ('1911', 1912); Merrill, Enum. Philipp. Fl. Pl. 3: 127 (1923). – Type: Philippines, Luzon, Ilocos Sur Province, Dolores, 31 x 1906, *W. Klemme* FB5665 (syn not located).
- **PHILIPPINES: Luzon:** Ilocos Sur Province, Dolores, 31 x 1906, *W. Klemme* FB5665 (syn not located) – Type of *Begonia rufipila* Merr.
- **IUCN category: DD.** Insufficient specimens could be georeferenced with certainty.

Begonia rupicola Miq. [§ Parvibegonia], Pl. Jungh. 4: 418 ('1855', 1857); Candolle, Prodr. 15(1): 352 (1864); Gagnepain, in Lecomte (ed.), Fl. Indo-Chine 2: 1106 (1921). – *Platycentrum rupicolum* (Miq.) Miq., Fl. Ned. Ind. 1(1): 693 (1856); Candolle, Prodr. 15(1): 352 (1864). – Type: Java, *T. Horsfield* 309 (holo K; iso K).
- **JAVA:** Pajittaw, *T. Horsfield s.n.* (BM); *T. Horsfield* 309 (holo K; iso K) – Type of *Begonia rupicola* Miq.
- **Notes:** Gagnepain cites specimens from Cambodia and Vietnam under this name in his Fl. Indo-Chine account (1921), but these represent a different and as yet undescribed taxon (R. Kiew, pers. comm.).
- **IUCN category: DD.** Insufficient specimens could be georeferenced with certainty.

Begonia sabahensis Kiew & J.H. Tan [§ Diploclinium], Gard. Bull. Singapore 56: 73 (2004). – Type: Borneo, Sabah, Tenom, Sungai Telekoson, 11 ii 2004, *J-H. Tan* AL727/2004 (holo SAN *n.v.*).
- **BORNEO: Sarawak:** Hose Mts., 3 iv 1980, *B.L. Burtt* 12776 (E). **Sabah:** Tenom, Sungai Telekoson, 11 ii 2004, *J-H. Tan* AL727/2004 (holo SAN *n.v.*) – Type of *Begonia sabahensis* Kiew & J.H. Tan.
- **Notes:** Unique amongst currently known Bornean *Begonia* in having pure yellow flowers.
- **IUCN category: CR B2ab(iii).** A rare and very local species, in an area with active logging (Kiew, 2004).

Begonia salomonensis Merr. & L.M. Perry [§ Petermannia], J. Arnold Arbor. 24: 56 (1943). – Type: Solomon Islands, South Solomon Islands, Ulawa, x 1932, *L.J. Brass* 2950 (holo A).
- **SOLOMON ISLANDS: South Solomon Islands:** Ulawa, x 1932, *L.J. Brass* 2950 (holo A) – Type of *Begonia salomonensis* Merr. & L.M. Perry.
- **IUCN category: DD.** Insufficient specimens could be georeferenced with certainty.

Begonia samarensis Merr. [§ Petermannia], Philipp. J. Sci. 30: 413 (1926). – Type: Philippines, Samar, Loquilocon, *R.C. McGregor* BS43757 (syn not located).
- **PHILIPPINES: Samar:** Loquilocon, *R.C. McGregor* BS43757 (syn not located) – Type of *Begonia samarensis* Merr.
- **IUCN category: DD.** Insufficient specimens could be georeferenced with certainty.

Begonia sandalifolia C.B. Clarke in Hook. f. **[§ Platycentrum]**, Fl. Brit. Ind. 2: 649 (1879); Gagnepain, in Lecomte (ed.), Fl. Indo-Chine 2: 1120 (1921). – Type: Burma, *W. Griffith* 2585 (holo K; iso P).

- **BURMA:** *W. Griffith* 2585 (holo K; iso P) – Type of *Begonia sandalifolia* C.B. Clarke.
  VIETNAM: Luang Babang, 1866 – 1888, *C. Thorel* (P).
- **IUCN category: DD.** Insufficient specimens could be georeferenced with certainty.

Begonia sarasinorum Irmsch. [§ Petermannia], Bot. Jahrb. Syst. 50: 349 (1913). – Type: Sulawesi, Bada, *K.F. & P.B. Sarasin* 2114 (holo B).
- **SULAWESI:** Bada, *K.F. & P.B. Sarasin* 2114 (holo B) – Type of *Begonia sarasinorum* Irmsch.
- **IUCN category: DD.** Insufficient specimens could be georeferenced with certainty.

Begonia sarawakensis Ridl. [§ Petermannia], J. Straits Branch Roy. Asiat. Soc. 46: 250 (1906). – Type: Borneo, Sarawak, *G.D. Haviland* 76 (776?) (syn not located); Borneo, Sarawak, *G.D. Haviland* 784 (syn not located).
- **BORNEO:** *Mason* 7 (B). **Sarawak:** *G.D. Haviland* 76 (776?) (syn not located) – Type of *Begonia sarawakensis* Ridl.; Kuching, 2 i 1892, *G.D. Haviland* 187? (BM); ix 1865, *O. Beccari* PB512 (FI[2]); *G.D. Haviland* 784 (syn not located) – Type of *Begonia sarawakensis* Ridl.; xii 1865, *O. Beccari* PB1179 (FI *n.v.*); xii 1865, *O. Beccari* PB1195 (FI); Mt. Penrissen, 5 viii 1958, *M. Jacobs* 5082 (K) [as *Begonia sarawakensis* ?]; Kapit Distr., Bukit Raya, 16 xi 1963, *P. Chai* 18923 (K) [as *Begonia sarawakensis* ?]; Bukit Kemantan, Ulu Muput Kanan, 11 x 1963, *P. Chai* 19529 (K) [as *Begonia sarawakensis* ?]; 11 vii 1929, *J. & M.S. Clemens* 20880 (K) [as *Begonia sarawakensis* ?]; Kuching, 1929, *J. & M.S. Clemens* 20889 (K) [as *Begonia sarawakensis* ?]; Lundu, Gunung Gadin, x 1929, *J. & M.S. Clemens* 22266 (K) [as *Begonia sarawakensis* ?]; Kapit Distr., Ulu Mengiong, Wong Kijang, 26 x 1988, *Othman et al.* 56058 (K).
- **IUCN category: DD.** The types are not available, and the validity of the determinations needs checking.

Begonia sarcocarpa Ridl. [§ Sphenanthera], J. Fed. Malay States Mus. 8(4): 38 (1917); Tebbitt, Brittonia 55(1): 27 (2003). – Type: Sumatra, Korinchi, Barong Baru, 5 vi 1914, *H.C. Robinson, C.B. Kloss* 61 (holo BM).
- **SUMATRA:** Korinchi, Barong Baru, 5 vi 1914, *H.C. Robinson, C.B. Kloss* 61 (holo BM) – Type of *Begonia sarcocarpa* Ridl.
- **Notes:** See notes under *B. longifolia* Blume.
- **IUCN category: DD.** Insufficient specimens could be georeferenced with certainty.

Begonia sarmentosa L.B. Sm. & Wassh. [§ Petermannia], Phytologia 52: 443 (1983). – *Begonia elegans* Elmer, Leafl. Philipp. Bot. 7: 2554 (1915); Merrill, Enum. Philipp. Fl. Pl. 3: 122 (1923); Smith & Wasshausen, Phytologia 52: 443 (1983). – Type: Philippines, Mindanao, Agusan, Mt. Urdaneta, ix 1912, *A.D.E. Elmer* 13672 (syn B[2], BM, E, FI, K, P, U).
- **PHILIPPINES: Mindanao:** Agusan, Mt. Urdaneta, ix 1912, *A.D.E. Elmer* 13672 (syn B[2], BM, E, FI, K, P, U) – Type of *Begonia elegans* Elmer; Butuan, Tungao, 16 vi 1961, *D.R. Mendoza* 42402 (L).
- **Notes:** See notes under *B. aequata* A. Gray.
- **IUCN category: DD.** Taxonomically uncertain.

Begonia saxifragifolia Craib [§ Diploclinium], Bull. Misc. Inform. Kew 1930: 416 (1930); Craib, Fl. Siam. 1: 778 (1931). – Type: Thailand, Puket, Ranawng, Kao Talu, 3 ii 1927, *A.F.G. Kerr* 11804 (holo ABD; iso BM, K).
- **THAILAND:** Kin Sayok, 31 vii 1946, *A.J.G.H. Kostermans* 1424 (A); Puket, Ranawng, Kao Talu, 3 ii 1927, *A.F.G. Kerr* 11804 (holo ABD; iso BM, K) – Type of *Begonia saxifragifolia* Craib.
- **IUCN category: DD.** Insufficient specimens could be georeferenced with certainty.

Begonia scortechinii King [§ Platycentrum], J. Asiat. Soc. Bengal, Pt. 2, Nat. Hist. 71: 62 (1902); Ridley, Fl. Malay Penins. 860 (1922); Irmscher, Mitt. Inst. Allg. Bot. Hamburg 8: 133 (1929); Kiew, Begonias Penins. Malaysia 238 (2005). – Type: Peninsular Malaysia, Perak, 1895, *B. Scortechini* 1845 (lecto K *n.v.*; isolecto B, BM, L); Peninsular Malaysia, Perak, *Kings Collector* 7227 (syn P).

Begonia kunstleriana King, J. Asiat. Soc. Bengal, Pt. 2, Nat. Hist. 71: 63 (1902); Ridley, Fl. Malay Penins. 1: 860 (1922); Kiew, Begonias Penins. Malaysia 238 (2005). – *Begonia scortechinii* var. *kunstleriana* Ridl., Fl. Malay Penins. 1: 860 (1922). – Type: Peninsular Malaysia, Perak, Gunung Bujang Melaka, i 1895, *Kings Collector* 7194 (lecto K *n.v.*); Peninsular Malaysia, Perak, *H.N. Ridley* 9651 (syn not located); Peninsular Malaysia, Perak, *B. Scortechini s.n.* (syn K *n.v.*).

- **PENINSULAR MALAYSIA:** Perak, Bujang Melaka, *H.H. Kunstler s.n.* (not located); Perak, 1895, *B. Scortechini* 1845 (lecto K *n.v.*; isolecto B, BM, L) – Type of *Begonia scortechinii* King; Perak, *Kings Collector* 7227 (syn P) – Type of *Begonia scortechinii* King; Perak, Gunung Bujang Melaka, *Curtis* 3123 (K, SING).
- **IUCN category: CR B2ab(iii)**. Known only from the type locality at an altitude of 300 m; not recollected since 1895.

Begonia scottii Tebbitt [§ Sphenanthera], Blumea 50(1): 154 (2005). – Type: Sumatra, Gunung Ketambe, 3°45′N, 98°E, *de Wilde, de Wilde-Duyfies* 14309 (holo L *n.v.*; iso L *n.v.*).

- **SUMATRA:** *Bangham* 775 (para A *n.v.*) – Type of *Begonia scottii* Tebbitt; *Bangham* 776a (para A) – Type of *Begonia scottii* Tebbitt; *de Wilde, de Wilde-Duyfies* 13531 (para K *n.v.*, L *n.v.*) – Type of *Begonia scottii* Tebbitt; Gunong Batu Lopang, 8 vii 1972, *W.J.J.O. de Wilde* 13531 (para BO) – Type of *Begonia scottii* Tebbitt; Gunung Ketambe, *de Wilde, de Wilde-Duyfies* 14309 (holo L *n.v.*; iso L *n.v.*) – Type of *Begonia scottii* Tebbitt; *de Wilde, de Wilde-Duyfies* 16434 (para L *n.v.*) – Type of *Begonia scottii* Tebbitt.
- **Notes:** Ridley referred some specimens of this species to his *B. trigonocarpa*, but *B. scottii* differs in having serrate, ovate leaves (not shallowly lobed), a decumbent habit, and larger fruit. *Begonia trigonocarpa* is probably more closely allied to *B. multangula*.
- **IUCN category: DD**. Insufficient specimens could be georeferenced with certainty.

Begonia serpens Merr. [§ Diploclinium], Philipp. J. Sci. 14: 427 (1919); Merrill, Enum. Philipp. Fl. Pl. 3: 128 (1923). – Type: Philippines, Panay, Antique Province, Culasi, v 1918 – vi 1918, *R.C. McGregor* 32588 (syn not located).

- **PHILIPPINES: Panay:** Antique Province, Culasi, *R.C. McGregor* 32541 (para P) – Type of *Begonia serpens* Merr.; Antique Province, Culasi, v 1918 – vi 1918, *R.C. McGregor* 32588 (syn not located) – Type of *Begonia serpens* Merr.
- **IUCN category: DD**. Insufficient specimens could be georeferenced with certainty.

Begonia serraticauda Merr. & L.M. Perry [§ Petermannia], J. Arnold Arbor. 24: 51 (1943). – Type: New Guinea, Papua, Idenburg River, iii 1939, *L.J. Brass* 13685 (holo A).

- **NEW GUINEA: Papua:** Idenburg River, Bernhard Camp, ii 1939, *L.J. Brass* 12988 (para not located) – Type of *Begonia serraticauda* Merr. & L.M. Perry; Idenburg River, iii 1939, *L.J. Brass* 13685 (holo A) – Type of *Begonia serraticauda* Merr. & L.M. Perry.
- **IUCN category: DD**. Insufficient specimens could be georeferenced with certainty.

Begonia serratipetala Irmsch. [§ Petermannia], Bot. Jahrb. Syst. 50: 339 (1913); Sands, Bot. Mag., n.s. 183: 827 (1982). – Type: New Guinea, Papua New Guinea, Danip (near), iii 1909, *F.R.R. Schlechter* 19208 (holo B).

- **NEW GUINEA: Papua:** Ramoi, 1872, *O. Beccari* PP434 (FI[2]). **Papua New Guinea:** Sepik District, Tifalmin Valley, iii 1965 – iv 1965, *W. Steinkraus* 17 (L) [as *Begonia* cf. *serratipetala*]; Kaiser-Wilhelmsland, Saruwaged Gebirge, *C. Keysser* 37 (B); Didessa, 3 ix 1966, *W. Schiefenhovel* 108 (B); Simbu Prov., Mongoma, 24 ix 1980, *J. Sterly* 80-197 (L); Kaiser-Wilhelmsland, Torricellia-Geb., Wigote Village, 5 ix 1961, *P.J. Darbyshire* 355 (L); Enga Prov., Porgera Distr., Waimeram, vii 1983, *T.M. Reeve* 588 (E, L); Morobe Distr., Sattelberg, 15 xi 1935, *J. & M.S. Clemens* 906 (B, L); Morobe Distr., Sattelberg, 26 xi 1935, *Clemens* 981 (B, L); Southern Highlands Distr., Mendi Valley, 28 vi 1961, *R. Schodde* 1351 (L); Morobe Distr., Wareo, 6 i 1935, *J. & M.S. Clemens* 1530 (B, L); Morobe Distr., Kumbok Mt., 26 i 1963, *W. Takeuchi* 8710 (A); Mt. Bosavi, 18 x 1973, *M. Jacobs* 9313 (L); Morobe Distr., Wantoat, 1 vi 1960, *J.S. Womersley, Thorne* 11882 (L); Koitaki, 23 iv 1935, *C.E. Carr* 12022 (BM); Morobe Distr., Mumeng Subdistr., Wagau, 22 xi 1967, *A.N. Millar, A. Dockrill* 12057 (A, L); Morobe Distr., Rawlinson Range, Patek Hill, 21 ii 1963, *P.v. Royen* 16098 (L); Southern Highlands Distr., Tigibi, 10 vi 1966, *W. Vink* 16851 (L); Danip (near), iii 1909, *F.R.R. Schlechter* 19208 (holo B) – Type of *Begonia serratipetala* Irmsch.; Morobe Distr., Kipu, 13 xi 1968, *A. Gillison, A. Kairo* 25693 (A, L); Kar Kar Island, 16 vii 1968, *A.N. Millar* 37740 (L); Western Highland Province, Lake Kopiago Subdistr., Batene, 30 x 1968, *J. Vandenberg, J.S. Womersley* 39941 (L); Western Highland Province, Lake Kopiago Subdistr., Kopiago, 7 xi 1968, *J. Vandenberg, M. Galore* 42058 (A, L); Kar Kar Island, 10 vi 1969, *E. Mann, J. Vandenberg* 43197 (L); Morobe Distr., Mt. Dilmargi, Kisiingam, 16 xii 1972, *P.F. Stevens* 58016 (L); Morobe Distr., Lae Subprovince, Mt. Jasop, 4 xii 1979, *P. Katik* 74781 (L).
- **IUCN category: LC**. Widespread, and frequently cultivated suggesting a wide ecological tolerance.

Begonia sharpeana F. Muell. [§ Diploclinium], Proc. Linn. Soc., New South Wales, II 2: 420 (1888); Mueller, Descr. Not. Papuan Pl. 9: 60. – Type: New Guinea, Papua New Guinea, Aird River, *T.F. Bevan* (holo not located).
- **NEW GUINEA: Papua New Guinea:** Aird River, *T.F. Bevan* (holo not located) – Type of *Begonia sharpeana* F. Muell.; Gulf District, Omati, 15 i 1955, *J.S. Womersley, N.W. Simmonds* 5069 (A); Gulf District, Kikori River, 12 ii 1959, *K.J. White* 10712 (A); Gulf District, Baimuru, 1 vii 1974, *P. Katik, A. Kairo* 62174 (A) [as *Begonia* aff. *sharpeana*].
- **IUCN category: VU D2**. Apparently has a fairly restricted distribution. Endemic to limestone pinnacles at low altitudes (c. 20 m).

Begonia siamensis Gagnep. [§ Platycentrum], Bull. Mus. Hist. Nat. (Paris) 25: 278 (1919); Gagnepain, in Lecomte (ed.), Fl. Indo-Chine 2: 1107 (1921); Craib, Fl. Siam. 1: 779 (1931). – Type: Laos, Attopeu Plateau, iii 1877, *F.F.J. Harmand* 1387 (syn P); Thailand, Chiangmai, Doi Sootep, 31 x 1909, 18°50′N, 98°54′E, *A.F.G. Kerr* 888 (syn ABD, BM, BO, K, P).
- **THAILAND:** Chiangmai, Mae Soi Ridge, 16 i 1993, *J.F. Maxwell* 93-46 (A, L); Chiangmai, 14 xii 1904, *C.C. Hosseus* 238 (B, BM); Chiangmai, Doi Sootep, 12 viii 1987, *J.F. Maxwell* 87-781 (L); Chiangmai, Doi Sootep, Puping Palace, 29 vi 1988, *J.F. Maxwell* 88-813 (L); Chiangmai, Doi Sootep, 31 x 1909, *A.F.G. Kerr* 888 (syn ABD, BM, BO, K, P) – Type of *Begonia siamensis* Gagnep.; Chiangmai, Doi Sootep, 19 v 1912, *A.F.G. Kerr* 2606 (K); Chiangmai, Doi Sootep, 1 xii 1959, *L.B. & E.C. Abbe, T. Smitinand* 9269 (A) [as *Begonia siamensis* ?]; Chiangmai, Doi Chang, 23 x 1979, *T. Yahara, T. Santisuk* 20599 (L).
 LAOS: Attopeu Plateau, iii 1877, *F.F.J. Harmand* 1387 (syn P) – Type of *Begonia siamensis* Gagnep.
- **IUCN category: LC**. Reasonably widespread, including two localities in national parks.

Begonia sibthorpioides Ridl. [§ Heeringia], J. Fed. Malay States Mus. 7: 42 (1916); Craib, Fl. Siam. 1: 779 (1931); Kiew, Begonias Penins. Malaysia 54 (2005). – Type: Peninsular Malaysia, Kedah, Kedah Peak, xii 1915, *H.C. Robinson, C.B. Kloss* 6047 (holo K *n.v.*); Peninsular Malaysia, Kedah, Kedah Peak, xii 1915, *H.C. Robinson, C.B. Kloss* 6047 (mero B).

Begonia sibthorpioides var. sibthorpioides
- **PENINSULAR MALAYSIA:** Kedah, Kedah Peak, xii 1915, *H.C. Robinson, C.B. Kloss* 6047 (holo K *n.v.*; mero B) – Type of *Begonia sibthorpioides* Ridl.

Begonia sibthorpioides var. grandifolia Craib, Fl. Siam. 1: 779 (1931). – Type: Thailand, Puket, Pang-nga, 8 xii 1918, *M. Haniff, Md. Nur* SFN3988 (holo ABD).
- **THAILAND:** Puket, Pang-nga, 8 xii 1918, *M. Haniff, Md. Nur* SFN3988 (holo ABD) – Type of *Begonia sibthorpioides* var. *grandifolia* Craib.
- **IUCN category: CR B2ab(iii)**. Known only from two localities, and described as 'extremely rare and very local'. A population at a third locality has recently been destroyed by a telecommunications development (Kiew, 2005).

Begonia sibutensis Sands [§ Petermannia], in Coode *et al.*, Checkl. Fl. Pl. Gymnosperms Brunei Darussalam App. 2: 433 (1997). – Type: Borneo, Brunei, Batu Apoi Forest Reserve, 23 iii 1991, 4°33′N, 115°9′E, *A. Poulsen* 31 (holo K; iso AAU *n.v.*, BRUN *n.v.*).
- **BORNEO: Brunei:** Batu Apoi Forest Reserve, 23 iii 1991, *A. Poulsen* 31 (holo K; iso AAU *n.v.*, BRUN *n.v.*) – Type of *Begonia sibutensis* Sands.
- **IUCN category: LC**. A narrow endemic, but in a protected area.

Begonia siccacaudata J. Door. [§ Petermannia], Blumea 45: 400 (2000). – Type: Sulawesi, Bantimurung Waterfalls, 5°2′S, 119°41′E, *J.J. Wieringa* 1902 (holo WAG; iso L).
- **SULAWESI:** Bantimurung Waterfalls, *J. van Veldhuizen s.n.* (para WAG) – Type of *Begonia siccacaudata* J. Door.; Karaenta Nature Reserve, 16 ii 1981, *Johansson, H. Nybom, S. Riebe* 8 (L); Karaenta Nature Reserve, 17 ii 1981, *Johansson, H. Nybom, S. Riebe* 24 (L); Bantimurung Waterfalls, 20 v 1929, *A. Rant* 48 (L) [as *Begonia* cf. *siccacaudata*]; Maros, Bantimurung Park, 12 ii 1977, *E. Widjaja* 130 (BO) [as *Begonia* cf. *siccacaudata*]; Bantimurung Waterfalls, *J.J. Wieringa* 1902 (holo WAG; iso L) – Type of *Begonia siccacaudata* J. Door.; Maros, Gunung Kamaseh, 14 vi 1986, *S.C. Chin* 3538 (L) [as *Begonia siccacaudata* ?]; Maros, Bantimurung Park, 20 ii 1938, *P.J. Eyma* 3702 (BO, L); Maros, Bantimurung Park, 20 ii 1938, *P.J. Eyma* 3759 (BO, L); Maros, Bantimurung Park, 13 iv 1975, *W. Meijer* 9125 (L) [as *Begonia* cf. *siccacaudata*].
- **IUCN category: VU D2**. Although two of the sites are in reserve areas, the localities are in areas heavily used for recreation.

Begonia sikkimensis A. DC. [§ Platycentrum], Ann. Sci. Nat. Bot., IV 11: 134 (1859); Candolle, Prodr. 15(1): 349 (1864); Clarke, Fl. Brit. Ind. 2: 646 (1879); Hara, Fl. E. Himalaya 215 (1966); Grierson, Fl. Bhutan 2: 241 (1991). – Type: not located.
- **BURMA:** *Fide* Kress *et al.* (2003).
- **IUCN category: DD**. Insufficient specimens could be georeferenced with certainty.

Begonia silletensis (A. DC.) C.B. Clarke [§ Sphenanthera], Fl. Brit. Ind. 2: 636 (1879); Tebbitt & Guan, Novon 12: 134 (2002). – *Casparya silletensis* A. DC., Prodr. 15(1): 277 (1864). – Type: India, Sillet Mts., *N. Wallich* 9107 (holo G-DC *n.v.*; iso K).
 Begonia gigantea Wall. *nom. nud.*, Num. List 129: 3677B (1831).

Begonia silletensis subsp. silletensis
- **BURMA:** Myitkyina District, 7 iii 1910, *J.H. Lace* 5170 (E).
  THAILAND: Chiangmai, 11 vi 1960, *T. Smitinand, H. St. John* 6832 (K).
- **IUCN category: DD.** Insufficient specimens could be georeferenced with certainty.

Begonia simulans Merr. & L.M. Perry [§ Petermannia], J. Arnold Arbor. 24: 52 (1943). – Type: New Guinea, Papua, Balim River, xii 1933, *L.J. Brass* 11835 (holo A; iso BM, L).
- **NEW GUINEA: Papua:** Vogelkopf, Nettoti Range, Wekari River, 3 xii 1961, *P.v. Royen, H. Sleumer* 8041 (L); Vogelkopf, Arfak Mts., Beribai, 22 vi 1961, *W. Vink* 11476 (L); Balim River, xii 1933, *L.J. Brass* 11835 (holo A; iso BM, L) – Type of *Begonia simulans* Merr. & L.M. Perry. **Papua New Guinea:** Morobe Distr., Mumeng Subdistr., Wagau, 9 iii 2000, *M. Lovave* 62 (L) [as *Begonia cf. simulans*]; Morobe Distr., Mt. Kaindi, 5 xii 1985, *J.L.C.H.v. Valkenburg* 119 (L); Eastern Highlands Distr., Chimbu Valley, Goglme, 28 xi 1960, *E. Borgmann* 417 (L) [as *Begonia cf. simulans*]; Morobe Distr., Lae Subprovince, Ana Village, 29 i 1972, *H. Streimann* 24320 (L); Southern Highlands Distr., Ialibu Subdistr., 12 ix 1967, *J.S. Womersley, K. Woolliams* 24991 (L); Fergusson Island, Agamola – Ailuluai, 5 vi 1956, *L.J. Brass* 27004 (L); Fergusson Island, Agamola – Ailuluai, 16 vi 1956, *L.J. Brass* 27197 (L); Milne Bay Distr., Raba Raba Subdistr., Mayu, 2 vii 1972, *P.F. Stevens, J.F. Veldkamp* 45352 (L) [as *Begonia cf. simulans*]; Goroka Subprov., Mt. Hozeke, 29 xi 1984, *K. Kerenga, C. Baker* 56885 (L); Southern Highlands Distr., Ialibu Subdistr., Onim, 2 xii 1972, *M.H. Andrew* 57072 (L) [as *Begonia aff. simulans*].
- **IUCN category: LC.** Widespread, including sites in protected areas.

Begonia sinuata Wall. ex Meisn. [§ Parvibegonia], Ber. Verh. Naturf. Ges. Basel 2: 42 (1836); Clarke, Fl. Brit. Ind. 2: 650 (1879); King, J. Asiat. Soc. Bengal 71: 59 (1902); Gagnepain, in Lecomte (ed.), Fl. Indo-Chine 2: 1105 (1921); Ridley, Fl. Malay Penins. 1: 856 (1922); Irmscher, Mitt. Inst. Allg. Bot. Hamburg 8: 142 (1929); Henderson, Malayan Wild Flowers – Dicotyledons 163 (1959); Burtt, Notes Roy. Bot. Gard. Edinburgh 32: 274 (1973); Kiew, Begonias Penins. Malaysia 58 (2005). – Type: Peninsular Malaysia, Penang, *N. Wallich* 3680 (holo K *n.v.*; iso BM).
  Diploclinium biloculare Wight, Icon. Pl. Ind. Orient. (Wight) t. 1814 (1852). – *Begonia bilocularis* (Wight) Craib, Fl. Siam. 1: 771 (1931); Burtt, Notes Roy. Bot. Gard. Edinburgh 32: 274 (1973). – Type: not located.
  Begonia guttata Wall. *nom. nud.*, Num. List 129: 3671B (1831); Clarke, Fl. Brit. Ind. 2: 650 (1879). – Type: not located.
  Begonia subrotunda Wall. *nom. nud.*, Num. List 213: 6293 (1832); Clarke, Fl. Brit. Ind. 2: 650 (1879); Golding, Phytologia 40: 19 (1978). – Type: not located.
  Begonia elongata Wall. *nom. nud.*, Num. List 213: 6291 (1832); Clarke, Fl. Brit. Ind. 2: 650 (1879). – Type: not located.
  Begonia clivalis Ridl., J. Straits Branch Roy. Asiat. Soc. 54: 43 (1909); Irmscher, Mitt. Inst. Allg. Bot. Hamburg 8: 144 (1929); Kiew, Begonias Penins. Malaysia 62 (2005). – *Begonia sinuata* var. *clivalis* (Ridl.) Irmsch., Mitt. Inst. Allg. Bot. Hamburg 8: 144 (1929); Kiew, Begonias Penins. Malaysia 58 (2005). – Type: Peninsular Malaysia, Selangor, Klang Gates, viii 1900, *H.N. Ridley* 13523 (holo SING *n.v.*; iso B, BM).
  Begonia sinuata var. *penangensis* Irmsch., Mitt. Inst. Allg. Bot. Hamburg 8: 142 (1929); Kiew, Begonias Penins. Malaysia 58 (2005). – Type: Peninsular Malaysia, Penang, Batu Feringgi, *Forest Guard s.n.* (lecto not located); Peninsular Malaysia, Penang, Western Hill, 15 viii 1916, *I.H. Burkill* 1524 (syn SING *n.v.*); Peninsular Malaysia, Penang, *Kings Collector* 2269 (syn B, P); Peninsular Malaysia, Penang, Richmond Pool, 22 vii 1917, *I.H. Burkill* 2590 (syn SING *n.v.*); Java, Penang Hill, viii 1885, *C. Curtis* 390 (syn SING *n.v.*); Peninsular Malaysia, Penang, Polo Terojah, xi 1885, *C. Curtis* 481

(syn SING *n.v.*); Peninsular Malaysia, *A.C. Maingay* 674 (syn L); Peninsular Malaysia, Penang, Penang Hill, vi 1898, *H.N. Ridley* 9229 (syn BM, SING).

Begonia sinuata var. *monophylloides* Irmsch., Mitt. Inst. Allg. Bot. Hamburg 8: 143 (1929); Kiew, Begonias Penins. Malaysia 58 (2005). – Type: Peninsular Malaysia, Pahang, Tahan River, *H.N. Ridley* 2642 (2643?) (lecto SING *n.v.*).

Begonia sinuata var. *malacensis* Irmsch., Mitt. Inst. Allg. Bot. Hamburg 8: 143 (1929); Kiew, Begonias Penins. Malaysia 58 (2005). – Type: Peninsular Malaysia, *A.C. Maingay* 675 (holo B; iso BM *n.v.*, L).

Begonia sinuata var. *langkawiensis* Irmsch., Mitt. Inst. Allg. Bot. Hamburg 8: 142 (1929); Kiew, Begonias Penins. Malaysia 58 (2005). – Type: Peninsular Malaysia, Langkawi, ix 1890, *C. Curtis s.n.* (lecto SING *n.v.*; isolecto BM *n.v.*, P); Peninsular Malaysia, Gunong Raya, 13 xi 1921, *M. Haniff, Md. Nur* SFN7177 (syn SING *n.v.*).

Begonia sinuata var. *etamensis* Irmsch., Mitt. Inst. Allg. Bot. Hamburg 8: 142 (1929); Kiew, Begonias Penins. Malaysia 58 (2005). – Type: Peninsular Malaysia, Penang, *C. Curtis* 3098 (lecto SING *n.v.*).

Begonia sinuata sensu lato

■ **BURMA:** Moulmein, 1857, *C.S.P. Parish s.n.* (K); Tenasserim Division, Tavoy District, Paungdaw Power Station, vi 1961, *J. Keenan et al.* 510 (A, E); Tenasserim Division, Tavoy District, Paungdaw Power Station, vi 1961, *J. Keenan et al.* 536 (A, E, K); Tenasserim Division, Tavoy District, Paungdaw Power Station, vii 1961, *J. Keenan* 568 (E); Tenasserim Division, Tavoy District, Paungdaw Power Station, viii 1961, *J. Keenan et al.* 727 (A, E, K); Tenasserim Division, Tavoy District, Paungdaw Power Station, viii 1961, *J. Keenan* 825 (A, E); Tenasserim Division, Tavoy District, Paungdaw Power Station, viii 1961, *J. Keenan* 899 (E); *W. Griffith* 2595 (A, P); Trungoo District, Kyunpadawng Range, Kyankkyri, 11 viii 1911, *J.H. Lace* 5399 (E, K).

THAILAND: Chantaburi, 27 vii 1969, *J.F. Maxwell s.n.* (L); Adang Island, 22 x 1989, *C. Congdon* 89 (A); Songkhla Prov., Boripat Falls, 23 viii 1984, *J.F. Maxwell* 84-114 (A); Ranong, Pra Part Beach, 18 vii 1979, *C. Niyomdham et al.* 345 (A, K, L, P); Sisaket Prov., Kantaralak Distr., Dongrak Range, Chong Bat Lak, 15 viii 1976, *J.F. Maxwell* 76-500 (L); Kwae Noi River Basin, Brangkasi, 19 vi 1946 – 22 vi 1946, *G. Hoed, A.J.G.H. Kostermans* 683 (A, K); Kwae Noi River Basin, Brangkasi, 19 vi 1946 – 22 vi 1946, *A.J.G.H. Kostermans, G. Hoed* 683 (L); Lansagah Distr., Khao Luang National Park, 4 ii 1990, *W.S. Hoover* 744A (A); Lansagah Distr., Khao Luang National Park, 4 ii 1990, *W.S. Hoover* 744B (A); Tarutao Island, 8 vi 1980, *C. Congdon* 830 (A[2]); Satun Distr., Talaybun National Park, 6 ix 1985, *J.F. Maxwell* 85-846 (L); Adang Island, 9 ix 1980, *C. Congdon* 889 (A); Kwae Noi River Basin, 3 vi 1946, *G. Hoed* 915 (K); Adang Island, 9 ix 1980, *C. Congdon* 917 (A); Khao Lumpee N.P., 13 ix 1994, *P.C. Boyce* 925 (K); Ranong, Thungraya Nasak Wildlife Sanct., 27 viii 2002, *D.J. Middleton et al.* 1383 (A, L); Chantaburi, Khao Sabab, 21 vii 1966, *K. Larsen et al.* 1638 (K, L, P); Khao Yai National Park, 31 x 1969, *C.F. van Beusekom, C. Charoenpol* 1941 (L); Kopah, Janjan hills, 9 xii 1897, *M. Haniff, Md. Nur* SFN2093 (syn K) – Type of *Begonia sinuata* var. *longialata* Irmsch.; Kao Kuap Krat, 21 v 1890, *unknown* 2897 (BM); Kopah, Kopah – Kohkhan, 16 xii 1897, *M. Haniff, Md. Nur* SFN2988 (syn K) – Type of *Begonia sinuata* var. *longialata* Irmsch.; Terutau, Western Bay (Grotto), 27 ii 1966, *H.M. Burkill* 4015 (A, L); Chantaburi, Koh Chang Island, 3 viii 1973, *R. Geesink, C. Phengkhlai* 6275 (K, L, P); Chantaburi, Koh Chang Island, 10 v 1974, *R. Geesink et al.* 6657 (K, L); Gunong Raya, 13 xi 1921, *M. Haniff, Md. Nur* SFN7177 (syn SING *n.v.*) – Type of *Begonia sinuata* var. *langkawiensis* Irmsch.; Ranong, Hadsin Dam, 24 vi 1974, *R. Geesink et al.* 7450 (K, L); Chantaburi, Koh Chang Island, 26 ix 1924, *A.F.G. Kerr* 9172 (ABD, BM, K); Chantaburi, Koh Chang Island, 26 ix 1924, *A.F.G. Kerr* 9172a (ABD, BM, K); Terutao, 27 ii 1966, *B. Hansen, T. Smitinand* 12452 (L, P); Klawng Zai, 22 vii 1928, *A.F.G. Kerr* 15882 (B, BM, K, L); Trat Prov., Ko Chang Island, 2 viii 1973, *Murata et al.* 17403 (L); Trat Prov., Ko Chang Island, 3 viii 1973, *Murata et al.* 17516 (L); Khao Phra Mi, 7 vii 1972, *K. Larsen et al.* 30712 (B, K, L); Khao Phra Bat, Kathing Falls, 26 viii 1972, *K. Larsen et al.* 32074 (B, K, L); Chantaburi, 26 viii 1972, *K. Larsen et al.* 32075 (B); Chantaburi, 3 ix 1972, *K. Larsen et al.* 32401 (B, L); Puket, Ranawng, Ngaw waterfall forest reserve, 6 ix 1984, *N. Fukuoka* 35873 (A, L); Trang Province, Ton The Waterfall, 14 xi 1990, *K. Larsen et al.* 41339 (P); Songkhla Prov., Khao Nam Khang, 28 viii 1995, *K. Larsen et al.* 46097 (L).

VIETNAM: Cochinchine, *J.B.L. Pierre s.n.* (BM, P); Phu-quoc island, *E. Contest-Lacour* 257 (P); Phu-quoc island, *F.F.J. Harmand* 283 (P).

CAMBODIA: Baie Island, vii 1903, *Geoffray* 227 (P); Elephant Mountain, 9 viii 1919, *E. Poilane* 260 (P); Bokor Mt., 5 viii 1938, *Muller* 312 (P).

PENINSULAR MALAYSIA: Langkawi, ix 1890, *C. Curtis s.n.* (lecto SING *n.v.*; isolecto BM *n.v.*, P) – Type of *Begonia sinuata* var. *langkawiensis* Irmsch.; Penang, Batu Feringgi, *Forest Guard s.n.* (lecto not located) – Type of *Begonia sinuata* var. *penangensis* Irmsch.; Penang, West Hill, *H.N. Ridley s.n.* (BM); Selangor, Sungei Pisang, 30 vii 2002, *S. Neale et al.* 13 (E); Kelantan, Kuala Aring, 1992, *B.H. Kiew* 24 (L); Ulu Kinchin, 25 v 1986, *B.H. Kiew* 86-25 (L); Pahang, Taman Negara, Kuala Tahan – Gunung Tahan, 14 viii 1990, *H. Okada et al.* 464 (L); Penang, Polo Terojah, xi 1885, *C. Curtis* 481 (syn SING *n.v.*) – Type of *Begonia sinuata* var. *penangensis* Irmsch.; *A.C. Maingay* 674 (syn L) – Type of *Begonia sinuata* var. *penangensis* Irmsch.; Penang, Standort, viii 1885, *C. Curtis* 390 (syn SING *n.v.*) – Type of *Begonia sinuata* var. *penangensis* Irmsch.; *A.C. Maingay* 675 (holo B; iso BM *n.v.*, L) – Type of *Begonia sinuata* var. *malacensis* Irmsch.; Selangor, Klang Gates, 8 x 1983, *R. Kiew* 1254 (L); Penang, Penang Hill, 14 x 1967, *T. Shimizu et al.* 1298 (L); Penang, Penang Hill, 15 x 1967, *T. Shimizu et al.* 1298 (L); Penang, Western Hill, 15 viii 1916, *I.H. Burkill* 1524 (syn SING *n.v.*) – Type of *Begonia sinuata* var. *penangensis* Irmsch.; Penang, *Kings Collector* 2269 (syn B, P) – Type of *Begonia sinuata* var. *penangensis* Irmsch.; Trengganu, Batu Biwa, 25 x 1986, *R. Kiew* 2347 (L); Penang, Richmond Pool, 22 vii 1917, *I.H. Burkill* 2590 (syn SING *n.v.*) – Type of *Begonia sinuata* var. *penangensis* Irmsch.; Pahang, Tahan River, *H.N. Ridley* 2642 (2643?) (lecto SING *n.v.*) – Type of *Begonia sinuata* var. *monophylloides* Irmsch.; Penang, *C. Curtis* 3098 (lecto SING *n.v.*) – Type of *Begonia sinuata* var. *etamensis* Irmsch.; Penang, *N. Wallich* 3680 (holo K *n.v.*; iso BM) – Type of *Begonia sinuata* Wall. ex Meisn.; Perak, Larut, viii 1888, *Kings Collector* 4060 (BM); Pulau Jarajah, ix 1888, *Kings Collector* 4971 (L); Gunong Tahan, Kuala Teku (Kwala-Teku), *H.C. Robinson, L. Wray Jr.* 5539 (BM); Penang, Penang Hill, vi 1898, *H.N. Ridley* 9229 (syn BM, SING) – Type of *Begonia sinuata* var. *penangensis* Irmsch.; Selangor, Klang Gates, viii 1900, *H.N. Ridley* 13523 (holo SING *n.v.*; iso B, BM) – Type of *Begonia clivalis* Ridl.; Selangor, Klang Gates, viii 1900, *H.N. Ridley* 13533 (BM); Langkawi, Seven Wells, 11 xi 1979, *B.C. Stone* 14237 (L); Gunong Tahan, Kuala Teku (Kwala-Teku), 5 ix 1928, *R.E. Holttum* 20091 (BM); Kedah, Bt. Perangan F.R., 10 viii 1974, *Anon.* 21744 (L); Kota Tinggi, Polepah Kiri, 31 v 1936, *E.J.H. Corner* 31437 (L); Penang, Tiger Hill, 18 xi 1950, *J. Sinclair* 39103 (E); Selangor, Klang Gates, 12 xi 1953, *J. Sinclair* 40132 (E, L); Trengganu, Sungei Kemia, 29 viii 1996, *Chua et al.* 40560 (L).

SUMATRA: Bangka, G. Pading, 2 xii 1917, *H.A.B. Bunnemeijer* 2180 (L) [as *Begonia sinuata* ?].

Begonia sinuata var. helferi Irmsch., Mitt. Inst. Allg. Bot. Hamburg 8: 141 (1929). – Type: Burma, Tenasserim, *Helfer* 2575 (syn B, L *n.v.*, P, W *n.v.*).

■ **BURMA:** Tenasserim, *Helfer* 2575 (syn B, L *n.v.*, P, W *n.v.*) – Type of *Begonia sinuata* var. *helferi* Irmsch.

Begonia sinuata var. longialata Irmsch., Mitt. Inst. Allg. Bot. Hamburg 8: 141 (1929). – Type: Thailand, Kopah, Janjan hills, 9 xii 1917, *M. Haniff, Md. Nur* SFN2093 (syn K); Thailand, Kopah, 16 xii 1917, *M. Haniff, Md. Nur* SFN2988 (syn K).

■ **THAILAND:** Kopah, Janjan hills, 9 xii 1917, *M. Haniff, Md. Nur* SFN2093 (syn K) – Type of *Begonia sinuata* var. *longialata* Irmsch.; Kopah, 16 xii 1917, *M. Haniff, Md. Nur* SFN2988 (syn K) – Type of *Begonia sinuata* var. *longialata* Irmsch.

Begonia sinuata var. pantiensis Kiew, Begonias Penins. Malaysia 64 (2005). – Type: Peninsular Malaysia, Johore, Gunung Panti, 25 ix 2000, *R. Kiew* 5113 (holo SING *n.v.*; iso K *n.v.*, KEP *n.v.*, L *n.v.*).

■ **PENINSULAR MALAYSIA:** Johore, Gunung Panti, *Teruya* 483 (para SING *n.v.*) – Type of *Begonia sinuata* var. *pantiensis* Kiew; Johore, Gunung Panti, *R. Kiew* 3780 (iso L *n.v.*; para SING *n.v.*) – Type of *Begonia sinuata* var. *pantiensis* Kiew; Johore, Gunung Panti, 25 ix 2000, *R. Kiew* 5113 (holo SING *n.v.*; iso K *n.v.*, KEP *n.v.*, L *n.v.*) – Type of *Begonia sinuata* var. *pantiensis* Kiew; Johore, Gunung Panti, *Saw* FRI37733 (para KEP *n.v.*) – Type of *Begonia sinuata* var. *pantiensis* Kiew; Johore, Gunung Panti, *C.M. Weber* 840723-2/2 (para KEP *n.v.*) – Type of *Begonia sinuata* var. *pantiensis* Kiew.

■ **Notes:** A widespread and instantly recognisable species on account of its dense indumentum of short, stellate hairs.

■ **IUCN category: LC.** Widespread.

Begonia sizemoreae Kiew [§ Platycentrum], Gard. Bull. Singapore 56: 95 (2004). – Type: Vietnam, Ha Tay, Ba Vi National Park, *R. Kiew* 5304 (holo SING *n.v.*; iso HN *n.v.*).

■ **VIETNAM:** Tonkin, Mt. Bavi, vii 1908, *D'Alleizette s.n.* (L) [as *Begonia sizemoreae* ?]; Tonkin, Mt. Bavi, xii 1887, *B. Balansa* 3765 (B, P); Ha Tay, Ba Vi National Park, *R. Kiew* 5304 (holo SING *n.v.*; iso HN *n.v.*) – Type of *Begonia sizemoreae* Kiew.

■ **IUCN category: LC.** Described as 'locally common' and grows on roadside banks (Kiew, 2004).

Begonia smithiae Geddes [§ Platycentrum], Bull. Misc. Inform. Kew 1928: 239 (1928); Craib, Fl. Siam. 1: 779 (1931). – Type: Thailand, Nakawn Sritamarat, Kao Luang, *E. Smith* 714 (holo K; iso ABD, BM).

■ **THAILAND:** Nakawn Sritamarat, Kao Luang, *E. Smith* 714 (holo K; iso ABD, BM) – Type of *Begonia smithiae* Geddes; Nakawn Sritamarat, Kao Luang, 30 iv 1928, *A.F.G. Kerr* 15499 (ABD, B, BM, K).

■ **IUCN category: DD.** Insufficient specimens could be georeferenced with certainty.

Begonia socia Craib [§ Parvibegonia], Bull. Misc. Inform. Kew 1930: 417 (1930); Craib, Fl. Siam. 1: 779 (1931). – Type: Thailand, Payap, Pa Tup, Me Tak Me Wang – Me Na Hua Me Ing – Mt. Topapu, *H.B.G. Garrett* 196 (holo ABD; iso K).

■ **THAILAND:** Payap, Pa Tup, Me Tak Me Wang – Me Na Hua Me Ing, *H.B.G. Garrett* 196 (holo ABD; iso K) – Type of *Begonia socia* Craib.

■ **Notes:** See notes under *B. integrifolia* Dalzell.

■ **IUCN category: DD.** Taxonomic uncertainty.

Begonia sogerensis Ridl. [§ Petermannia], J. Bot. 52: 289 (1914). – Type: New Guinea, Papua New Guinea, Mt. Sogere, *H.O. Forbes* 157 (syn BM); New Guinea, Papua New Guinea, Mt. Sogere, 9°28'45"S, 147°31'37"E, *H.O. Forbes* 261 (syn BM, P); New Guinea, Papua New Guinea, Mt. Sogere, *H.O. Forbes* 444 (syn BM *n.v.*).

■ **NEW GUINEA: Papua New Guinea:** Mt. Sogere, *H.O. Forbes* 157 (syn BM) – Type of *Begonia sogerensis* Ridl.; Mt. Sogere, *H.O. Forbes* 261 (syn BM, P) – Type of *Begonia sogerensis* Ridl.; Mt. Sogere, *H.O. Forbes* 444 (syn BM *n.v.*) – Type of *Begonia sogerensis* Ridl.; 1886, *H.O. Forbes* 890 (BM); Morobe Distr., Pao River, 11 vi 1999, *M. Lovave* 82305 (L); Morobe Distr., Iye Creek, 17 vi 1999, *M. Lovave* 82433 (A, L); Morobe Distr., Pao River, 12 vi 1999, *B. Bau* 82576 (L).

- **IUCN category: VU D2**. Known from three locations at low to mid elevations, which are not in protected areas.

Begonia soluta Craib [§ Diploclinium], Bull. Misc. Inform. Kew 1930: 418 (1930); Craib, Fl. Siam. 1: 779 (1931). – Type: Thailand, Nakawn Sawan, Kampengpet, Kao Hua, 13 vi 1922, *A.F.G. Kerr* 6129 (holo ABD; iso BM, K).
- **THAILAND:** Nakawn Sawan, Kampengpet, Kao Hua, 13 vi 1922, *A.F.G. Kerr* 6129 (holo ABD; iso BM, K) – Type of *Begonia soluta* Craib.
- **IUCN category: DD**. Insufficient specimens could be georeferenced with certainty.

Begonia somervillei Hemsl. [§ Petermannia], Bull. Misc. Inform. Kew 1896: 17 (1896); Merrill & Perry, J. Arnold Arbor. 24: 57 (1943). – Type: Solomon Islands, South Solomon Islands, New Georgia, 1894 – 1895, *Officers of HMS Penguin s.n.* (holo K).
- **SOLOMON ISLANDS: South Solomon Islands:** New Georgia, 1894 – 1895, *Officers of HMS Penguin s.n.* (holo K) – Type of *Begonia somervillei* Hemsl.; Betivatu, *H.S. McKee* 1640 (K) [as *Begonia cf. somervillei*]; Guadalcanal, Umusani River, *P.F. Hunt* 2019 (A, K, L); Guadalcanal, Mt. Gallego, 3 vii 1965, *P.F. Hunt* 2092 (K[2], L) [as *Begonia cf. somervillei*]; San Cristobal, Waimamura, 11 viii 1932, *L.J. Brass* 2644 (A, B, P); Guadalcanal, Monitor Creek, 6 vii 1965, *T.C. Whitmore* 6058 (K, L); Santa Ysabel, Potoro Bay, 30 viii 1966, *W. Beers* 7679 (K) [as *Begonia cf. somervillei*]; Guadalcanal, Mt. Mambulu, 20 vii 1967, *A. Nakisi* 8022 (K) [as *Begonia cf. somervillei*]; Guadalcanal, Lungga River, 21 vi 1991, *P.I. Forster, D.J. Liddle* 507230 (L).
- **IUCN category: LC**. Widely collected, and a recent collection (1991) describes it as 'common in the area' (*P.I. Forster, D.J. Liddle* 507230 (L)).

Begonia sorsogonensis (Merr.) Elmer nom. nud. [§ Petermannia], Leafl. Philipp. Bot. 10: 3708 (1939). – *Begonia sorsogonensis* Elmer ex Merr. *nom. nud.*, Enum. Philipp. Fl. Pl. 3: 121 (1923); Elmer, Leafl. Philipp. Bot. 10: 3708 (1939). – Type: Philippines, Luzon, Sorsogon, Mt. Bulusan, *A.D.E. Elmer* 14515 (syn B, BM[3], L, P, U).
- **PHILIPPINES: Luzon:** Sorsogon, Mt. Bulusan, *A.D.E. Elmer* 14515 (syn B, BM[3], L, P, U) – Type of *Begonia sorsogonensis* (Merr.) Elmer *nom. nud.*; Catanduanes, 14 xi 1917 – 11 xii 1917, *M. Ramos* 30544 (BM, P).
- **Notes:** Strictly speaking, this is not a valid name as no Latin diagnosis was included in the description, necessary as from 1935. However, it is included here for completeness.
- **IUCN category: DD**. Insufficient specimens could be georeferenced with certainty.

Begonia speluncae Ridl. [§ Reichenheimia], J. Straits Branch Roy. Asiat. Soc. 46: 258 (1906); Kiew & Geri, Gard. Bull. Singapore 55: 122 (2003). – Type: Borneo, Sarawak, Bau, Bidi, 1893, *H.N. Ridley* 11773 (lecto K); Borneo, Sarawak, Mt. Braang, *G.D. Haviland* 873 (syn not located); Borneo, Sarawak, Tabea, *G.D. Haviland s.n.* (syn not located).
- **BORNEO: Sarawak:** Tabea, *G.D. Haviland s.n.* (syn not located) – Type of *Begonia speluncae* Ridl.; Bau, Bidi, ii 1908, *C.J. Brooks s.n.* (K); Quop, *Hewitt s.n.* (K); Bau, Bidi Cave, viii 1912, *Anderson* 75 (K); Matang, Mt. Braang, vi 1890, *G.D. Haviland* 76 (K) – Type of *Begonia pubescens* Ridl.; Bau, Bidi, 1924, *E.G. Mjoberg* 177 (BM, K); Bau, Bukit Jebong, 13 xii 1975, *P.F. Stevens* 215 (L); Mt. Braang, *G.D. Haviland* 873 (syn not located) – Type of *Begonia speluncae* Ridl.; Bau, 12 viii 1958, *M. Jacobs* 5177 (B, L); Bau, Seburan, 24 i 1960, *T. Anderson* 8983 (K, L); Bau, Bidi, 1893, *H.N. Ridley* 11773 (lecto K) – Type of *Begonia speluncae* Ridl.; Bau, Bidi Cave, 1929, *J. & M.S. Clemens* 20644 (B[2], BM, K, L, P); Padawan, Bukit Pait, 2 iii 1969, *Erwin, Paul* 27427 (L); Lobang Mawang, Bukit Selabor, 26 ix 1968, *P. Chai* 28044 (L); Bau, G. Berloban, 13 iii 1984, *Yii Puan Ching* 46142 (L); Tebedu, 8 ii 1985, *A. Mohtar* 49238 (K); Bau, Bukit Boring, 14 xi 1985, *Yii Puan Ching* 50355

(L); Bau, Bukit Boring, 14 xi 1985, *Yii Puan Ching* 50366 (L); Gunung Selebur, 20 iv 1999, *Jamree et al.* 82258 (L).
- **IUCN category: VU D2.** Although collected from most hills in the Kuching limestone area (Kiew & Geri, 2003), only one of the localities is in a national park (Gunung Gading).

Begonia sphenocarpa Irmsch. [§ Petermannia], Bot. Jahrb. Syst. 50: 369 (1913). – Type: Sulawesi, Mt. Topapu, 17 ix 1902, *K.F. & P.B. Sarasin* 2100 (holo B).
- **SULAWESI:** Mt. Topapu, 17 ix 1902, *K.F. & P.B. Sarasin* 2100 (holo B) – Type of *Begonia sphenocarpa* Irmsch.
- **IUCN category: DD.** Insufficient specimens could be georeferenced with certainty.

Begonia spilotophylla F. Muell. [§ Petermannia], Descr. Not. Papuan Pl. 4: 67 (1876); Merrill & Perry, J. Arnold Arbor. 24: 58 (1943). – Type: New Guinea, Papua New Guinea, Fly River, *C.L.M.d. Albertis s.n.* (holo MEL *n.v.*; iso FI); New Guinea, Papua New Guinea, Fly River, *C.L.M.d. Albertis s.n.* (mero B).
- **NEW GUINEA: Papua New Guinea:** Fly River, *C.L.M.d. Albertis s.n.* (holo MEL *n.v.*; iso FI; mero B) – Type of *Begonia spilotophylla* F. Muell.; Fly River, v 1936, *L.J. Brass* 6702 (A, L); Fly River, v 1936, *L.J. Brass* 6730 (BM, L).
- **IUCN category: DD.** Insufficient specimens could be georeferenced with certainty.

Begonia stenogyna Sands [§ Petermannia], in Coode *et al.*, Checkl. Fl. Pl. Gymnosperms Brunei Darussalam App. 2: 432 (1997). – Type: Borneo, Brunei, Batu Apoi Forest Reserve, 13 vii 1993, 4°33′N, 115°9′E, *M.J.S. Sands* 5782 (holo K; iso BRUN *n.v.*).
- **BORNEO: Sarawak:** Gunung Mulu, Sg. Mendalam, 17 x 1977, *J.A.R. Anderson* 39390 (L). **Brunei:** Batu Apoi Forest Reserve, 11 xi 1991, *C. Hansen* 1546 (L); Batu Apoi Forest Reserve, 13 vii 1993, *M.J.S. Sands* 5782 (holo K; iso BRUN *n.v.*) – Type of *Begonia stenogyna* Sands; Temburong, Sungei Engkiang – Kuala Belalong Field Centre, 6 iii 1991, *G.C.G. Argent, D. Mitchell* 91194 (E, L).
- **IUCN category: LC.** Occurs within the Batu Apoi Forest Reserve and is also recorded from Gunung Mulu.

Begonia stevei M. Hughes [§ Petermannia], Edinburgh J. Bot. 63: 197 (2006). – Type: Sulawesi, Loewoek distr., Sungai Hek, 25 ii 2004, 1°1′10″S, 122°10′30″E, *Hendrian, M. Newman, S. Scott, M. Nazre Saleh, D. Supriadi* 877 (holo E; iso L).
- **SULAWESI:** Loewoek distr., Sungai Hek, 25 ii 2004, *Hendrian, M. Newman, S. Scott, M. Nazre Saleh, D. Supriadi* 877 (holo E; iso L) – Type of *Begonia stevei* M. Hughes; Batui River, 19 x 1989, *M. Coode* 6013 (L[2]) [as *Begonia stevei* ?].
- **IUCN category: VU D2** (Hughes, 2006).

Begonia stichochaete K.G. Pearce [§ Petermannia], Gard. Bull. Singapore 55: 78 (2003). – Type: Borneo, Sarawak, Gunung Subis, *Ahmad* 14 (holo SAR *n.v.*; iso SING *n.v.*).
- **BORNEO: Sarawak:** Gunung Subis, *Ahmad* 14 (holo SAR *n.v.*; iso SING *n.v.*) – Type of *Begonia stichochaete* K.G. Pearce; Niah National Park, 17 vii 1981, *B. Lee* 40075 (E); Niah National Park, *B. Lee* S40075 (para SAR *n.v.*) – Type of *Begonia stichochaete* K.G. Pearce; Kuala Niah, *K.G. Pearce, Remlanto* S78539 (para SAR *n.v.*) – Type of *Begonia stichochaete* K.G. Pearce; Bukit Kasut, *K.G. Pearce, Remlanto* S78597 (para SAR *n.v.*) – Type of *Begonia stichochaete* K.G. Pearce; Sekaloh, *Jemree Sabli* S89014 (para SAR *n.v.*) – Type of *Begonia stichochaete* K.G. Pearce; Sekaloh, *K.G. Pearce* S89267 (para SAR *n.v.*) – Type of *Begonia stichochaete* K.G. Pearce.

- **IUCN category: LC**. Occurs in both primary and secondary forest, on two different soil types; also described as 'locally frequent' at the base of limestone cliffs (Pearce, 2003).

Begonia stictopoda (Miq.) A. DC. [§ Reichenheimia], Prodr. 15(1):391 (1864).– *Mitscherlichia stictopoda* Miq., Fl. Ned. Ind. 1(1): 1092 (1856); Miquel, Fl. Ned. Ind., Suppl. 1: 333 (1861); Candolle, Prodr. 15(1): 391 (1864). – Type: Sumatra, Palembajan, *J.E. Teijsmann s.n.* (syn K, L).
- **SUMATRA:** Bencoolen, Lebong Tandai, *C.J. Brooks s.n.* (K); Palembajan, *J.E. Teijsmann s.n.* (syn K, L) – Type of *Mitscherlichia stictopoda* Miq.; unknown *s.n.* (L); Mt. Singalan, 1878, *O. Beccari* PS127 (Fl[2]); Padang, Ajer Mantjoer, viii 1878, *O. Beccari* PS657 (BM, K, L); Mt. Singalan, 27 v 1918, *H.A.B. Bunnemeijer* 2626 (B, L); Barisan Range, 13 iii 1974, *E.F. de Vogel* 2956 (L) [as *Begonia cf. stictopoda*]; Pajakumbuh, Mt. Sago, 1955, *W. Meijer* 3347 (L) [as *Begonia cf. stictopoda*]; Mt. Singalan, 1878, *O. Beccari* CB4511 (Fl[2]); Mt. Singalan, 1878, *O. Beccari* CB4511A (Fl[2]).
- **IUCN category: LC**. Three of the localities are in protected areas; also grows on roadside banks (pers. obs.).

Begonia stilandra Merr. & L.M. Perry [§ Petermannia], J. Arnold Arbor. 24: 52 (1943). – Type: New Guinea, Papua, Idenburg River, i 1939, *L.J. Brass* 12301 (holo A; iso BM, L).
- **NEW GUINEA: Papua:** Wandammen Peninsula, Wondiwoi Mts., 5 iii 1962, *C. Koster* 1370 (L) [as *Begonia cf. stilandra*]; Vogelkopf, Tamrau Range, Sudjalak Village – Mt. Kusemun, 7 xi 1961, *P.v. Royen, H. Sleumer* 7772 (L); Idenburg River, i 1939, *L.J. Brass* 12301 (holo A; iso BM, L) – Type of *Begonia stilandra* Merr. & L.M. Perry; Idenburg River, ii 1939, *L.J. Brass* 12499 (para L) – Type of *Begonia stilandra* Merr. & L.M. Perry. **Papua New Guinea:** Goroka Subprov., Dunantina Valley (upper), 11 vi 1956, *R.D. Hoogland, R. Pullen* 5297 (L) [as *Begonia cf. stilandra*].
- **IUCN category: LC**. Half of the locations are in protected areas, and the distribution spans four eco-regions. Occupies mid to high elevations.

Begonia strachwitzii Warb. ex Irmsch. [§ Petermannia], Bot. Jahrb. Syst. 50: 357 (1913). – Type: Sulawesi, Minahassa, Bojong, *O. Warburg* 15192 (holo B; iso B).
- **SULAWESI:** Minahassa, Bojong, *O. Warburg* 15191 (B[2]); Minahassa, Bojong, *O. Warburg* 15192 (holo B; iso B) – Type of *Begonia strachwitzii* Warb. ex Irmsch.
- **IUCN category: DD**. Insufficient specimens could be georeferenced with certainty.

Begonia strictinervis Irmsch. [§ Petermannia], Bot. Jahrb. Syst. 50: 365 (1913). – Type: New Guinea, Papua New Guinea, Mt. Finisterre, *F.R.R. Schlechter* 18004 (holo B; iso P).
- **NEW GUINEA: Papua New Guinea:** Mt. Finisterre, *F.R.R. Schlechter* 18004 (holo B; iso P) – Type of *Begonia strictinervis* Irmsch.
- **IUCN category: DD**. Insufficient specimens could be georeferenced with certainty.

Begonia strictipetiolaris Irmsch. [§ Petermannia], Bot. Jahrb. Syst. 50: 348 (1913). – Type: Sulawesi, Minahassa, Tomohon, *K.F. & P.B. Sarasin* 400a (holo B; iso K).
- **SULAWESI:** Minahassa, Tomohon, *K.F. & P.B. Sarasin* 400a (holo B; iso K) – Type of *Begonia strictipetiolaris* Irmsch.
- **IUCN category: DD**. Insufficient specimens could be georeferenced with certainty.

Begonia strigosa (Warb.) L.L. Forrest & Hollingsw. [§ Symbegonia], Plant Syst. Evol. 241: 209 (2003). – *Symbegonia strigosa* Warb., Fl. Deutsch. Schutzgeb. Südsee 324 (1905); Forrest & Hollingsworth, Plant Syst. Evol. 241: 209 (2003). – Type: New Guinea, Papua New Guinea, Kaiser-Wilhelmsland, Bismarck-Geb., i 1902, *F.R.R. Schlechter* 14019 (holo B; iso BO, K).

- **NEW GUINEA: Papua:** Wandammen Peninsula, Wondiwoi Mts., 5 iii 1962, *F.A.W. Schram* 13362 (L) [as *Begonia aff. strigosa*]. **Papua New Guinea:** Western Highland Province, Korombi, xii 1981, *T.M. Reeve* 592 (E, K[2]) [as *cf. Symbegonia strigosa*]; Enga Prov., Porgera Distr., xii 1981, *T.M. Reeve* 593 (E) [as *cf. Symbegonia strigosa*]; Goroka Subprov., Bena Bena Valley, 24 ix 1957, *R. Pullen* 635 (L); Madang Prov., Saidor Subdistr., Moro, xi 1964, *A.C. Jermy* 4365 (BM); Morobe Distr., Finisterre Mts., Budemo-Moro track, 21 x 1964, *R. Pullen* 6009 (BM, L); Morobe Distr., Finisterre Mts., 6 xi 1964, *R. Pullen* 6050 (BM); Morobe Distr., Skindewai, 5 i 1956, *J.S. Womersley, A. Millar* NGF8344 (BO, K, L); Morobe Distr., Kasanombe, 28 viii 1973, *M. Jacobs* 8618 (L[2]); Morobe Distr., Kasanombe, 28 viii 1973, *M. Jacobs* 8619 (L); Morobe Distr., Finisterre Mts., Uva-Minga River track, 10 viii 1976, *P.v. Royen* 11687 (L) [as *Begonia aff. strigosa*]; Morobe Distr., Mumeng Subdistr., Wagau, 22 xi 1967, *A.N. Millar, A. Dockrill* NGF12081 (L); Morobe Distr., Mt. Shungol, 10 ix 1963, *T.G. Hartley* TGH12467 (K); Morobe Distr., Mt. Shungol, 10 xii 1963, *T.G. Hartley* TGH12467 (L); Eastern Highlands Distr., Waisa, 25 ix 1964, *T.G. Hartley* TGH13116 (L); Kaiser-Wilhelmsland, Bismarck-Geb., i 1902, *F.R.R. Schlechter* 14019 (holo B; iso BO, K) – Type of *Symbegonia strigosa* Warb.; Morobe Distr., Zatari, 23 ii 1963, *P.v. Royen, A. Millar* NGF15661 (BO, K) [as *Begonia aff. strigosa*]; Eastern Highlands Distr., Wonatabe, iv 1963, *J.S. Womersley* NGF17644 (BO, K, L); Morobe Distr., Mumeng Subdistr., Wagau, 3 xi 1963, *J.S. Womersley* NGF17871 (K, L); Eastern Highlands Distr., Kerowagi, 4 ix 1963, *A. Millar, R.E. Holttum* 18627 (K); Chimbu Province, Kerowagi Sub-distr., Gena, 4 ix 1963, *A. Millar, R.E. Holttum* NGF18627 (L); Morobe Distr., Bupu, 4 iii 1964, *A. Millar* NGF23330 (L); Morobe Distr., Mumeng Subdistr., Wagau, 11 vi 1964, *A. Millar* 23383 (L); Eastern Highlands Distr., Kassam, 11 i 1968, *E.E. Henty, M. Coode* NGF29253 (L); Eastern Highlands Distr., Okapa, 20 ix 1959, *L.J. Brass* 31597 (K, L); Chimbu Province, Kerowagi Sub-distr., Korongi River, 27 viii 1968, *A. Millar* 37798 (L); Southern Highlands Distr., Ialibu Subdistr., Ialibu, 22 vi 1969, *M. Coode, P. Wardle, P. Katik* NGF40320 (L) [as *Begonia aff. strigosa*]; Morobe Distr., Wau Subprov., New Yamap, 13 iii 1970, *H. Streimann, A. Kairo* NGF47542 (L); Eastern Highlands Distr., Okapa Sub-distr., Waresevieta, 26 v 1971, *E.E. Henty* NGF49160 (L); Morobe Distr., Kasanombe, 27 viii 1973, *P. Katik, K. Taho* LAE56394 (L); Morobe Distr., Mt. Dilmargi, 16 xii 1972, *P.F. Stevens* LAE58028 (K); Morobe Distr., Mt. Dilmargi, 16 xii 1972, *P.F. Stevens* 58030 (L[2]).
- **IUCN category: LC**. Widespread in the Central Range of Papua New Guinea.

Begonia subcyclophylla Irmsch. [§ Diploclinium], Bot. Jahrb. Syst. 50: 374 (1913). – Type: New Guinea, Papua New Guinea, Komlo, 22 xii 1907, *F.R.R. Schlechter* 17011 (syn B, P); New Guinea, Papua New Guinea, Kani Range, *F.R.R. Schlechter* 20371 (syn B, P).
- **NEW GUINEA: Papua New Guinea:** Komlo, 22 xii 1907, *F.R.R. Schlechter* 17011 (syn B, P) – Type of *Begonia subcyclophylla* Irmsch.; Kani Range, *F.R.R. Schlechter* 20371 (syn B, P) – Type of *Begonia subcyclophylla* Irmsch.
- **IUCN category: DD**. Insufficient specimens could be georeferenced with certainty.

Begonia subelliptica Merr. & L.M. Perry [§ Petermannia], J. Arnold Arbor. 24: 57 (1943). – Type: New Guinea, Papua, Idenburg River, iii 1939, *L.J. Brass* 13396 (holo A *n.v.*; iso L).
- **NEW GUINEA: Papua:** Idenburg River, iii 1939, *L.J. Brass* 13396 (holo A *n.v.*; iso L) – Type of *Begonia subelliptica* Merr. & L.M. Perry.
- **IUCN category: DD**. Insufficient specimens could be georeferenced with certainty.

Begonia subisensis K.G. Pearce [§ Petermannia], Gard. Bull. Singapore 55: 85 (2003). – Type: Borneo, Sarawak, Gunung Subis, *J.A.R. Anderson, S. Tan, E. Wright* S27574 (holo SAR *n.v.*; iso SING *n.v.*).

- **BORNEO: Sarawak:** Gunung Subis, *J.A.R. Anderson, S. Tan, E. Wright* S27574 (holo SAR *n.v.*; iso SING *n.v.*) – Type of *Begonia subisensis* K.G. Pearce; Great Cave, *J.A.R. Anderson* S31939 (para SAR *n.v.*) – Type of *Begonia subisensis* K.G. Pearce; Bukit Kasut, *K.G. Pearce, Remlanto* S78538 (para SAR *n.v.*) – Type of *Begonia subisensis* K.G. Pearce.
- **IUCN category: VU D2**. Although described as 'locally frequent' (Pearce, 2003) this species is a narrow endemic in the Subis limestone.

Begonia sublobata Jack [§ Diploclinium], Malayan Misc. 2(7): 10 (1822); Candolle, Prodr. 15(1): 355 (1864). – *Diploclinium sublobatum* (Jack) Miq., Fl. Ned. Ind. 1(1): 690 (1856) – Type: not located.
- **SUMATRA:** *Fide* Jack (1822).
- **IUCN category: DD**. Insufficient specimens could be georeferenced with certainty.

Begonia subnummularifolia Merr. [§ Diploclinium], Philipp. J. Sci. 29: 403 (1926). – Type: Borneo, Sabah, Banguey Island, ix 1923, *P. Castro, F. Melegrito* 1454 (syn B, K).
- **BORNEO: Sabah:** Banguey Island, ix 1923, *P. Castro, F. Melegrito* 1454 (syn B, K) – Type of *Begonia subnummularifolia* Merr.
- **IUCN category: DD**. Insufficient specimens could be georeferenced with certainty.

Begonia suborbiculata Merr. [§ Diploclinium], Philipp. J. Sci. 6: 398 ('1911', 1912); Merrill, Enum. Philipp. Fl. Pl. 3: 128 (1923). – Type: Philippines, Palawan, Malampaya Bay, ix 1910, 10°50'35"N, 119°22'14"E, *E.D. Merrill* 7229 (syn B, BM[2], K, P).
- **PHILIPPINES: Palawan:** Puerto Princesa, 20 iii 1947, *G. Edano* 563 (K); Malampaya Bay, ix 1910, *E.D. Merrill* 7229 (syn B, BM[2], K, P) – Type of *Begonia suborbiculata* Merr.; Taytay, iv 1913, *E.D. Merrill* 9187 (K, P); Lake Manguao, iv 1913, *E.D. Merrill* 9488 (BM[2], BO, K, P).
- **IUCN category: VU D2**. Known from four localities. Although all of Palawan has been designated a Fauna and Flora Watershed Reserve, much of the forest cover is degraded, especially at lower elevations.

Begonia subperfoliata Parish ex Kurz [§ Diploclinium], J. Asiat. Soc. Bengal 42(2): 81 (1873). – Type: Burma, Moulmein, 1862, *C.S.P. Parish s.n.* (syn K, P).
- **BURMA:** Moulmein, 1862, *C.S.P. Parish s.n.* (syn K, P) – Type of *Begonia subperfoliata* Parish ex Kurz.
  THAILAND: Kanchanaburi, Lai Wo Subdistr., Toong Yai Naresuan Wildlife Reserve, 7 x 1993, *J.F. Maxwell* 93-1148 (A, L) [as *Begonia cf. subperfoliata*]; Payap, Doi Chiengdao, 13 ix 1967, *M. Tagawa et al.* 9884 (L).
- **IUCN category: LC**. Two of the Thai localities are in protected areas.

Begonia subprostrata Merr. [§ Petermannia], Philipp. J. Sci. 26: 482 (1925). – Type: Philippines, Mindanao, Zamboanga, Malangas, xi 1919, *M. Ramos, G. Edano* 36977 (syn B).
- **PHILIPPINES: Mindanao:** Zamboanga, Malangas, xi 1919, *M. Ramos, G. Edano* 36977 (syn B) – Type of *Begonia subprostrata* Merr.; Zamboanga, Malangas, xi 1919, *M. Ramos, G. Edano* BS37222 (para not located) – Type of *Begonia subprostrata* Merr.
- **IUCN category: DD**. Insufficient specimens could be georeferenced with certainty.

Begonia subtruncata Merr. [§ Petermannia], Philipp. J. Sci. 6: 390 ('1911', 1912); Merrill, Enum. Philipp. Fl. Pl. 3: 128 (1923). – Type: Philippines, Luzon, Union Province, Castilla, iii 1906, *A. Loher* 6076 (syn PNH).

- **PHILIPPINES: Luzon:** Union Province, Castilla, iii 1906, *A. Loher* 6076 (syn PNH) – Type of *Begonia subtruncata* Merr.
- **IUCN category: DD.** Insufficient specimens could be georeferenced with certainty.

Begonia subviridis Craib [not placed to section], Bull. Misc. Inform. Kew 1930: 418 (1930); Craib, Fl. Siam. 1: 780 (1931). – Type: Thailand, Nakawn Sritamarat, Kao Luang, 26 v 1930, *A.F.G. Kerr s.n.* (holo ABD).
- **THAILAND:** Nakawn Sritamarat, Kao Luang, 26 v 1930, *A.F.G. Kerr s.n.* (holo ABD) – Type of *Begonia subviridis* Craib.
- **IUCN category: LC.** The type locality is in the Khao Luang National Park.

Begonia sudjanae C.-A. Jansson [§ Reichenheimia], Acta Horti Gothob. 26: 1 (1963). – Type: Sumatra, Padang, 25 ii 1962, *C-A. Jansson s.n.* (holo GB *n.v.*; iso B[3], L).
- **SUMATRA:** Padang, 25 ii 1962, *C.-A. Jansson s.n.* (holo GB *n.v.*; iso B[3], L) – Type of *Begonia sudjanae* C.-A. Jansson.
- **IUCN category: DD.** Insufficient specimens could be georeferenced with certainty.

Begonia suffrutescens Merr. & L.M. Perry [§ Petermannia], J. Arnold Arbor. 24: 49 (1943). – Type: New Guinea, Papua, Idenburg River, i 1939, *L.J. Brass* 12030 (holo A *n.v.*; iso BM, L).
- **NEW GUINEA: Papua:** Cycloop Mts., Ifar – Ormoe, 13 x 1954, *P.v. Royen* 3773 (L); Cycloop Mts., Ifar – Ormoe, 10 vi 1961, *P.v. Royen, H. Sleumer* 5769 (L); Wissel Lakes, Pekeglbaro, 26 v 1980, *W. Vink, F.A.W. Schram* 8878 (L); Idenburg River, i 1939, *L.J. Brass* 12030 (holo A *n.v.*; iso BM, L) – Type of *Begonia suffrutescens* Merr. & L.M. Perry; Idenburg River, ii 1939, *L.J. Brass* 12985 (para L) – Type of *Begonia suffrutescens* Merr. & L.M. Perry. **Papua New Guinea:** Sepik District, Sereri Creek, 26 iii 1964, *C.D. Sayers* 19553 (L).
- **IUCN category: LC.** Two of the localities are in protected areas, and the species occurs at elevations from 850 to 2000 m.

Begonia surculigera Kurz [§ Diploclinium], Flora 54: 296 (1871); Clarke, Fl. Brit. Ind. 640 (1879). – Type: Burma, Arracan, x 1869, *W.S. Kurz s.n.* (holo K; iso K).
- **BURMA:** Arracan, x 1869, *W.S. Kurz s.n.* (holo K; iso K) – Type of *Begonia surculigera* Kurz; Douzayit, 4 viii 1911, *J.H. Lace* 5390 (K) [as *Begonia surculigera* ?].
- **IUCN category: DD.** Insufficient specimens could be georeferenced with certainty.

Begonia sychnantha L.B. Sm. & Wassh. [§ Reichenheimia], Phytologia 54(7): 469 (1984). – *Begonia polyantha* Ridl., J. Fed. Malay States Mus. 8(4): 38 (1917); Smith & Wasshausen, Phytologia 54(7): 469 (1984). – *Begonia trichopoda* var. *polyantha* (Ridl.) Irmsch., Pareys Blumengartn. ed. 2: 89 (1960). – Type: Sumatra, Korinchi, Siolak Dras, 3000 feet, 16 iii 1914, *H.C. Robinson, C.B. Kloss s.n.* (BM).
- **SUMATRA:** Korinchi, Siolak Dras, 3000 feet, 16 iii 1914, *H.C. Robinson, C.B. Kloss s.n.* (BM) – Type of *Begonia polyantha* Ridl; Korinchi, Siolak Dras, 16 iii 1914, *H.C. Robinson, C.B. Kloss s.n.* (BM[2], K, L); Korinchi, Korinchi Peak – Korinchi, 30 iv 1914, *C.B. Kloss* 175 (BM); Korinchi, Mt. Korinchi, 24 vii 2006, *D. Girmansyah, A. Poulsen, I. Hatta, R. Neivita* 781 (E); Padang, Lubuk Sulasi, 30 vi 1953, *J.v. Borssum Waalkes* 2757 (BO[2]); Padang, Lubuk Sulasi, 30 vi 1953, *J.v. Borssum Waalkes* 2897 (BO[2]); Pajakumbuh, 20 viii 1957, *W. Meijer* 6717a (L) [as *Begonia cf. polyantha*]; Barisan Range, Air Sirah, 5 v 1985, *E.F. de Vogel, J.J. Vermeulen* 7447 (L); 29 i 1950, *W. Meijer* 7550 (L) [as *Begonia cf. polyantha*]; Korinchi, Batu Hampar, 1 iii 1954, *A.H.G. Alston* 14018 (BM[2]).
- **IUCN category: LC.** Occurs throughout the Kerinci-Seblat National Park.

Begonia symbeccarii L.L. Forrest & Hollingsw. [§ Symbegonia], Plant Syst. Evol. 241: 208 (2003). – *Symbegonia beccarii* Irmsch., Webbia 9: 507 (1953); Forrest & Hollingsworth, Plant Syst. Evol. 241: 208 (2003). – Type: New Guinea, Papua, Vogelkopf, Arfak Mts., 1875, *O. Beccari* CB4504 (holo B; iso FI).

■ **NEW GUINEA: Papua:** Vogelkopf, Arfak Mts., 1875, *O. Beccari* CB4504 (holo B; iso FI) – Type of *Symbegonia beccarii* Irmsch.

■ **IUCN category: DD**. Insufficient specimens could be georeferenced with certainty.

Begonia symbracteosa L.L. Forrest & Hollingsw. [§ Symbegonia], Plant Syst. Evol. 241: 208 (2003). – *Symbegonia bracteosa* Warb., Fl. Deutsch. Schutzgeb. Südsee 323 (1905); Forrest & Hollingsworth, Plant Syst. Evol. 241: 208 (2003). – Type: New Guinea, Papua New Guinea, Kaiser-Wilhelmsland, Torricellia-Geb., iv 1902, *F.R.R. Schlechter* 14369 (holo B; iso B, BM, BO, K, P).

■ **NEW GUINEA: Papua:** Vogelkopf, Nettoti Range, Wekari River, 3 xii 1961, *P.v. Royen, H. Sleumer* 8033 (BO, K); Vogelkopf, Mt. Nettoti, 5 xii 1961, *P.v. Royen, H. Sleumer* 8161 (L) [as *Begonia aff. symbracteosa*]. **Papua New Guinea:** Chimbu Province, Bomkane Village, xi 1977, *W.S. Hoover* 72 (L) [as *Begonia aff. symbracteosa*]; Mt. Hagen vicinity, xii 1977, *W.S. Hoover* 75 (B, L) [as *Begonia aff. symbracteosa*]; Kaiser-Wilhelmsland, Torricellia-Geb., Mt. Somoro, 31 viii 1961, *P.J. Darbyshire* 310 (L); Eastern Highlands Distr., Chimbu Valley, 10 x 1980, *J. Sterly* 80319 (L) [as *Begonia aff. symbracteosa*]; Simbu Prov., Gembogl, 10 x 1980, *J. Sterly* 80-320 (L); Goroka Subprov., Asaro-Mairifutica divide, 30 viii 1957, *R. Pullen* 399 (L) [as *Begonia aff. symbracteosa*]; Kaiser-Wilhelmsland, Torricellia-Geb., Wigote Village, 17 ix 1961, *P.J. Darbyshire* 462 (K, L); Eastern Highlands Distr., Daulo-Chave Road, 21 xi 1954, *H.S. McKee* 1307 (L) [as *Begonia aff. symbracteosa*]; Eastern Highlands Distr., AI River, 3 iv 1953, *J.S. Womersley* 4834 (BM, BO, L) [as *Begonia aff. symbracteosa*]; Western Highland Province, Kubor Range, Minj-Nona Divide, 26 viii 1963, *R. Pullen* 5261 (BO, L) [as *Begonia aff. symbracteosa*]; West Sepik Province, Mt. Amdutakin, 17 ix 1966, *C. Kalkman* 5307 (K, L); Eastern Highlands Distr., AI River, 7 ix 1953, *J.S. Womersley* 5334 (L) [as *Begonia aff. symbracteosa*]; Eastern Highlands Distr., Keglsugl, 20 iv 1967, *L.K. Wade* ANU7641 (BO, L) [as *Begonia aff. symbracteosa*]; Eastern Highlands Distr., Kumul Mission, 12 i 1962, *J.S. Womersley* 14134 (K, L) [as *Begonia aff. symbracteosa*]; Kaiser-Wilhelmsland, Torricellia-Geb., iv 1902, *F.R.R. Schlechter* 14369 (holo B; iso B, BM, BO, K, P) – Type of *Symbegonia bracteosa* Warb.; Eastern Highlands Distr., Kanawyroka Creek, 30 viii 1963, *A. Millar, N.G.F. van Royen* 15978 (BO, L) [as *Begonia aff. symbracteosa*]; Western Highland Province, Uinba, 9 ix 1963, *W. Vink* 16533 (K); Western Highland Province, Mazmal, 1 ix 1963, *A. Millar, N.G.F. van Royen* 18573 (L) [as *Begonia aff. symbracteosa*]; Eastern Highlands Distr., Daulo Pass, 20 v 1966, *J.S. Womersley* NGF27422 (L) [as *Begonia aff. symbracteosa*]; Southern Highlands Distr., West of Mt. Ne, 20 viii 1966, *D.G. Frodin* NGF28408 (BO, L) [as *Begonia aff. symbracteosa*]; Chimbu Province, Road to Nilguma, 19 x 1969, *A. Millar* NGF38376 (L) [as *Begonia aff. symbracteosa*]; Mt. Hagen, Mur Mur Pass – Mt. Hagen, 1 x 1963, *J. Vandenberg, P. Katik, A. Kairo* NGF39909 (L) [as *Begonia aff. symbracteosa*]; Eastern Highlands Distr., Daulo Pass, 7 xi 1968, *A. Millar* NGF40630 (L) [as *Begonia aff. symbracteosa*]; Mt. Hagen, Mur Mur Pass, 17 vi 1984, *K. Kerenga, D.E. Symon* LAE56818 (L) [as *Begonia aff. symbracteosa*].

■ **IUCN category: DD**. Insufficient specimens with firm determinations could be georeferenced with certainty.

Begonia symgeraniifolia L.L. Forrest & Hollingsw. [§ Symbegonia], Plant Syst. Evol. 241: 208 (2003). – *Symbegonia geraniifolia* Ridl., Trans. Linn. Soc. London, Bot., II 9: 61 (1916); Forrest & Hollingsworth, Plant Syst. Evol. 241: 208 (2003). – Type: New Guinea, Papua, Camp 7-8, *C.B. Kloss s.n.* (syn BM, K).

- **NEW GUINEA: Papua:** Camp 7-8, *C.B. Kloss s.n.* (syn BM, K) – Type of *Symbegonia geraniifolia* Ridl.; Mimika Regency, Utekini Valley, 3 iii 1999, *T.M.A. Utteridge* 235 (not located) [as *Begonia symgeraniifolia* ?]; Wissel Lakes, 3 x 1939 – 8 x 1939, *P.J. Eyma* 5250 (BO, K, L); Mimika Regency, Utekini Valley, 7 xii 1996, *M.J.S. Sands* 7300 (K) [as *Begonia symgeraniifolia* ?]; Mimika Regency, Utekini Valley, 18 viii 1998, *J.H. Beaman* 12264 (K) [as *Begonia symgeraniifolia* ?]. **Papua New Guinea:** St. Michael, xi 1977, *W.S. Hoover* 58 (B, E); Enga Prov., Porgera Distr., Bealo, xii 1981, *T.M. Reeve* 596 (K, L) [as *cf. Symbegonia geraniifolia*]; Eastern Highlands Distr., Mt. Gahavisuka, iv 1983, *N. Cruttewell* 2313 (L) [as *Begonia aff. symgeraniifolia*]; Eastern Highlands Distr., Kortumi, 22 xi 1954, *Floyd, J.S. Womersley* NGF6714 (L) [as *Begonia aff. symgeraniifolia*]; Eastern Highlands Distr., Kortumi, 23 xi 1954, *Floyd, J.S. Womersley* NGF6775 (L) [as *Begonia aff. symgeraniifolia*]; Eastern Highlands Distr., Mt. Michael, Kini Creek, 2 ix 1959, *J.S. Womersley* NGF11367 (BM, L) [as *Begonia aff. symgeraniifolia*]; Eastern Highlands Distr., West of Omaura, 5 vii 1963, *T.G. Hartley* 11979 (L) [as *Begonia aff. symgeraniifolia*]; Chimbu Province, Crater Mt. Wildlife Management area, 21 vii 1998, *W. Takeuchi* 12321 (L); Eastern Highlands Distr., Mt. Michael, 10 ix 1959, *L.J. Brass* 31500 (K) [as *Begonia aff. symgeraniifolia*]; Eastern Highlands Distr., Kainantu Sub-distr., Mt. Piora foothills, 12 vi 1972, *M. Coode* NGF46286 (L) [as *Begonia aff. symgeraniifolia*]; Goroka Subprov., Mt. Hozeke, 29 xi 1984, *K. Kerenga, C. Baker* 56892 (L) [as *Begonia cf. symgeraniifolia*].
- **Notes:** The specimens cited here undoubtedly cover more than one taxon; further investigation is needed.
- **IUCN category: DD.** Taxonomic uncertainty.

Begonia symhirta L.L. Forrest & Hollingsw. [§ Symbegonia], Plant Syst. Evol. 241: 208 (2003). – *Symbegonia hirta* Ridl., Trans. Linn. Soc. London, Bot., II 9: 61 (1916); Forrest & Hollingsworth, Plant Syst. Evol. 241: 208 (2003). – Type: New Guinea, Papua, i 1913, *C.B. Kloss s.n.* (syn BM, K[2]).
- **NEW GUINEA: Papua:** i 1913, *C.B. Kloss s.n.* (syn BM, K[2]) – Type of *Symbegonia hirta* Ridl.
- **IUCN category: DD.** Insufficient specimens could be georeferenced with certainty.

Begonia sympapuana L.L. Forrest & Hollingsw. [§ Symbegonia], Plant Syst. Evol. 241: 208 (2003). – *Symbegonia papuana* Merr. & L.M. Perry, J. Arnold Arbor. 24: 59 (1943); Forrest & Hollingsworth, Plant Syst. Evol. 241: 208 (2003). – Type: New Guinea, Papua, Idenburg River, i 1939, *L.J. Brass* 12161 (holo A; iso BM, BO, L).
- **NEW GUINEA: Papua:** Idenburg River, i 1939, *L.J. Brass* 12161 (holo A; iso BM, BO, L) – Type of *Symbegonia papuana* Merr. & L.M. Perry. **Papua New Guinea:** Southern Highlands Distr., Near Tari, xii 1977, *W.S. Hoover* 73 (L) [as *Begonia aff. sympapuana*]; Morobe Distr., Skindewai, 9 i 1956, *J.S. Womersley, A. Millar* NGF8474 (K, L); Chimbu Province, Crater Mt. Wildlife Management area, 17 vii 1998, *W. Takeuchi* 12178 (L); Eastern Highlands Distr., Arau, 7 x 1959, *L.J. Brass* 31911 (K, L).
- **IUCN category: DD.** Insufficient specimens could be georeferenced with certainty.

Begonia symparvifolia L.L. Forrest & Hollingsw. [§ Symbegonia], Plant Syst. Evol. 241: 208 (2003). – *Symbegonia parvifolia* Gibbs, Fl. Arfak Mts. 150 (1917); Forrest & Hollingsworth, Plant Syst. Evol. 241: 208 (2003). – Type: New Guinea, Papua, Vogelkopf, Arfak Mts., Koebre Mt., xii 1913, *L.S. Gibbs* 5644 (holo BM).
- **NEW GUINEA: Papua:** Wandammen Peninsula, Wondiwoi Mts., vii 1928, *E. Mayr* 263 (B, L) [as *Begonia aff. symparvifolia*]; Manokwari, Iraiweri, 8 vi 1991, *E. Widjaja* EAW4271 (L); Vogelkopf, Arfak Mts., Koebre Mt., xii 1913, *L.S. Gibbs* 5644 (holo BM) – Type of *Symbegonia parvifolia* Gibbs; Van Rees Mts., v 1926, *Doctors v. Leeuwen-Reynvaan* 9186 (BO, L) [as *Begonia aff. symparvifolia*]; Mamberamo, Albatros Camp, vii 1926, *Doctors v. Leeuwen-Reynvaan* 9651 (L) [as *Begonia*

aff. symparvifolia]; Wandammen Peninsula, Wondiwoi Mts., 25 ii 1962, *C. Koster* BW13565 (L) [as *Begonia aff. symparvifolia*]; Vogelkopf, Arfak Mts., Angi Lakes, 11 i 1962, *H. Sleumer, W. Vink* BW14089 (L) [as *Begonia cf. symparvifolia*]; Vogelkopf, Arfak Mts., Angi Lakes, 24 i 1962, *H. Sleumer, W. Vink* BW14310 (L) [as *Begonia cf. symparvifolia*]. **Papua New Guinea:** Western Highland Province, Jimmi valley, Tagalinga, 16 vi 1955, *J.S. Womersley, A. Millar* NGF7693 (L) [as *Begonia aff. symparvifolia*]; Aitape Subdistr., Bliri River, 12 vii 1961, *P.J. Darbyshire, R.D. Hoogland* 8116 (BO) [as *Begonia aff. symparvifolia*]; Chimbu Province, Crater Mt. Wildlife Management area, 6 iii 1997, *W. Takeuchi* 11701 (L).

■ **IUCN category: DD**. Insufficient specimens with firm determinations could be georeferenced with certainty.

Begonia sympodialis Irmsch. [§ Petermannia], Webbia 9: 495 (1953). – Type: Borneo, Sarawak, Sakarrang, x 1867, *O. Beccari* PB3867 (holo FI).

■ **BORNEO: Sarawak:** Sakarrang, x 1867, *O. Beccari* PB3867 (holo FI) – Type of *Begonia sympodialis* Irmsch.

■ **IUCN category: DD**. Insufficient specimens could be georeferenced with certainty.

Begonia symsanguinea L.L. Forrest & Hollingsw. [§ Symbegonia], Plant Syst. Evol. 241: 208 (2003); Tebbitt, Begonias 220 (2005). – *Symbegonia sanguinea* Warb., Fl. Deutsch. Schutzgeb. Südsee 323 (1905); Forrest & Hollingsworth, Plant Syst. Evol. 241: 208 (2003). – Type: Papua New Guinea, Bismarck Mts., i 1902, *F.R.R. Schlechter* 13979 (holo B; iso BO, K).

■ **NEW GUINEA: Papua New Guinea:** Bismarck Mts., i 1902, *F.R.R. Schlechter* 13979 (holo B; iso BO, K) – Type of *Symbegonia sanguinea* Warb.; Western Highland Province, Korombi, xii 1981, *T.M. Reeve* 592 (L); Enga Prov., Porgera Distr., xii 1981, *T.M. Reeve* 593 (L); Central Division, 11 viii 1968, *N. Cruttewell* 1476 (L); Southern Highlands Distr., West of Mt. Ne, 30 viii 1966, *D.G. Frodin* 32090 (L); Eastern Highlands Distr., Kassam, 31 x 1959, *L.J. Brass* 32326 (K, L); Morobe Distr., Menyama Sub. Dist., Aiwa, 18 v 1968, *H. Streimann, A. Kairo* 39020 (K, L).

■ **IUCN category: DD**. Insufficient specimens could be georeferenced with certainty.

Begonia tafaensis Merr. & L.M. Perry [§ Petermannia], J. Arnold Arbor. 24: 55 (1943). – Type: New Guinea, Papua New Guinea, Mt. Tafa, v 1933, *L.J. Brass* 4017 (holo NY *n.v.*).

■ **NEW GUINEA: Papua New Guinea:** Mt. Tafa, v 1933, *L.J. Brass* 4017 (holo NY *n.v.*) – Type of *Begonia tafaensis* Merr. & L.M. Perry; Mt. Tafa, ix 1933, *L.J. Brass* 5109 (para A) – Type of *Begonia tafaensis* Merr. & L.M. Perry; Mt. Kuni, 9 v 1957, *J.S. Womersley* 9508 (BM) [as *Begonia aff. tafaensis*]; Bismarck Mts., Mt. Oibo, 6 x 1995, *W. Takeuchi et al.* 10502 (L) [as *Begonia aff. tafaensis*]; Eastern Highlands Distr., Mt. Michael, 2 ix 1959, *J.S. Womersley* 11369 (BM) [as *Begonia aff. tafaensis*]; Wharton Range, Murray Pass, 11 viii 1968, *C.E. Ridsdale, P.J.B. Woods* 36899 (E) [as *Begonia aff. tafaensis*]; Southern Highlands Distr., Mt. Giluwe, 28 xii 1973, *J. Croft et al.* 60876 (E) [as *Begonia aff. tafaensis*].

■ **IUCN category: LC**. Area of occupancy >10,000 km², and described as 'fairly common in tall forest'. Occurs at elevations of c. 2200 m.

Begonia tambelanensis (Irmsch.) Kiew [§ Reichenheimia], Gard. Bull. Singapore 55: 63 (2003). – *Begonia peninsulae* subsp. *tambelanensis* Irmsch., Mitt. Inst. Allg. Bot. Hamburg 8: 100 (1929). – Type: Borneo, Kalimantan, Tambelan Island, 31 viii 1895, *H.N. Ridley s.n.* (holo K *n.v.*).

■ **BORNEO: Kalimantan:** Tambelan Island, 31 viii 1895, *H.N. Ridley s.n.* (holo K *n.v.*) – Type of *Begonia peninsulae* subsp. *tambelanensis* Irmsch.

■ **IUCN category: DD**. Insufficient specimens could be georeferenced with certainty.

Begonia tampinica Burkill ex Irmsch. [§ Platycentrum], Mitt. Inst. Allg. Bot. Hamburg 8: 130 (1929); Kiew, Begonias Penins. Malaysia 265 (2005). – Type: Peninsular Malaysia, Negri Sembilan, Gunung Tampin, 27 i 1918, *I.H. Burkill* SFN3108 (lecto SING *n.v.*); Peninsular Malaysia, Negri Sembilan, Gunung Tampin, 15 vii 1917, 2°30'57"N, 102°11'59"E, *I.H. Burkill* SFN2534 (syn K).

- **PENINSULAR MALAYSIA:** Negri Sembilan, Gunung Tampin, 15 vii 1917, *I.H. Burkill* SFN2534 (syn K) – Type of *Begonia tampinica* Burkill ex Irmsch.; Negri Sembilan, Gunung Tampin, 27 i 1918, *I.H. Burkill* SFN3108 (lecto SING *n.v.*) – Type of *Begonia tampinica* Burkill ex Irmsch.
- **IUCN category: CR B2ab(iii).** Known only from a single population of a few plants in a heavily populated area. 'On the brink of extinction' (Kiew, 2005).

Begonia tawaensis Merr. [§ Petermannia], Univ. Calif. Publ. Bot. 15: 211 (1929). – Type: Borneo, Sabah, Tawau, Tawau Hill Forest Reserve, *A.D.E. Elmer* 20466 (syn B, BM[2], L, P, U).

- **BORNEO: Sabah:** Tawau, Tawau Hill Forest Reserve, 16 v 1966, *D. Hou* 187 (L); Tawau, Tawau Hill Forest Reserve, *A.D.E. Elmer* 20466 (syn B, BM[2], L, P, U) – Type of *Begonia tawaensis* Merr.; Tawau, Bukit Tanaga, 16 v 1966, *H.T. Sinanggul* 56176 (L); Lahad Datu District, Mt. Silam?, 27 vii 1987, *A. Naidil* 118091 (L).
- **IUCN category: NT.** The main stronghold for this species is the Tawau reserve, and its future depends on the maintenance of this area as undisturbed forest.

Begonia tayabensis Merr. [§ Diploclinium], Philipp. J. Sci. 13: 38 (1918); Merrill, Enum. Philipp. Fl. Pl. 3: 128 (1923); Tebbitt, Begonias 223 (2005); Golding, Begonian 74: 12 (2007). – Type: Philippines, Luzon, Tayabas, Umiray River, 3 vi 1917, *M. Ramos, G. Edano* 29054 (syn P).

- **PHILIPPINES: Luzon:** Tayabas, Umiray River, 3 vi 1917, *M. Ramos, G. Edano* 29054 (syn P) – Type of *Begonia tayabensis* Merr.
- **IUCN category: DD.** Insufficient specimens could be georeferenced with certainty.

Begonia temburongensis Sands [§ Petermannia], in Coode *et al.*, Checkl. Fl. Pl. Gymnosperms Brunei Darussalam App. 2: 432 (1997). – Type: Borneo, Brunei, Temburong, 12 iii 1991, 4°22'N, 115°17'E, *M.J.S. Sands* 5399 (holo K; iso BRUN *n.v.*).

- **BORNEO: Brunei:** Temburong, 12 iii 1991, *M.J.S. Sands* 5399 (holo K; iso BRUN *n.v.*) – Type of *Begonia temburongensis* Sands.
- **IUCN category: DD.** Insufficient specimens could be georeferenced with certainty.

Begonia tenericaulis Ridl. [§ Petermannia], Bull. Misc. Inform. Kew 1925: 83 (1925). – Type: Sumatra, Bencoolen, Lebong Tandai, vi 1922, *C.J. Brooks* 7608 (holo K).

- **SUMATRA:** Pajakumbuh, 1 ix 1957, *Maradjo* 371 (L); Ulu Gadut, 10 i 1983, *M. Hotta et al.* 425 (L); Riau Province, Bukit Tiga Puluh, 31 vii 2006, *D. Girmansyah, A. Poulsen, I. Hatta, F. Antoni* 796 (E); Pajakumbuh, Mt. Sago, 3 xi 1955, *W. Meijer* 4502 (L); Pajakumbuh, Mt. Sago, 27 xii 1955, *W. Meijer* 4697 (L); Pajakumbuh, Mt. Sago, 16 ii 1956, *E. Meijer Drees* 4716 (L); Pajakumbuh, Mt. Sago, 5 xii 1956, *W. Meijer* 5381a (L); Sungai Tualang, 13 x 1939, *P. Buwalda* 7019 (P); Bencoolen, Lebong Tandai, vi 1922, *C.J. Brooks* 7608 (holo K) – Type of *Begonia tenericaulis* Ridl.; Gunung Go Lemboeh, 18 ii 1937, *C.G.G.J.v. Steenis* 8882 (L); Kloet Nature Reserve, 6 vii 1985, *de Wilde, de Wilde-Duyfies* 19731 (L) [as *Begonia cf. tenericaulis*].
- **Notes:** The elongate fruit and tiny tepals on the female flower are very distinctive of both this species and the earlier described *B. vuijckii* Koord. Whether these two species are distinct requires further investigation.
- **IUCN category: DD.** Taxonomic uncertainty.

Begonia tenuifolia Dryand. [§ Parvibegonia], Trans. Linn. Soc. 1: 162 (1791); Blume, Enum. Pl. Javae 1: 94 (1827); Candolle, Prodr. 15(1): 351 (1864); Koorders, Exkurs.-Fl. Java 2: 651 (1912); Doorenbos, Begonian 47: 215 (1980); Smith & Wasshausen, Phytologia 54(7): 468 (1984). – *Platycentrum tenuifolium* (Dryand.) Miq., Fl. Ned. Ind. 1(1): 693 (1856). – Type: Java, Princes Island, *J. Banks s.n.* (holo BM).

   Begonia varians A. DC., Ann. Sci. Nat. Bot., IV 11: 135 (1859); Candolle, Prodr. 15(1): 352 (1864); Koorders, Exkurs.-Fl. Java 2: 651 (1912). – Type: Java, *W.H. de Vriese s.n.* (holo K).

   Begonia lineata N.E. Br., Gard. Chron., II 18: 199 (1882); Koorders, Exkurs.-Fl. Java 2: 651 (1912); Backer & Brink, G. v.d., Fl. Java 1: 308 (1964); Smith & Wasshausen, Phytologia 54(7): 468 (1984). – Type: not located.

- **JAVA:** *T. Horsfield s.n.* (K[2]); *C.L.v. Blume s.n.* (P); *C.L.v. Blume s.n.* (L); *W.H. de Vriese s.n.* (holo K) – Type of *Begonia varians* A. DC.; Princes Island, *J. Banks s.n.* (holo BM) – Type of *Begonia tenuifolia* Dryand.; *J. Cook s.n.* (BM); Wynkoops Bay, iii 1920, *C.B. Kloss s.n.* (K); Pekalongan, 14 ii 1927, *E.H.B. Brascamp* 6 (BO); Preanger, Tjiandjoer, Kiara Pajoeng, 13 iv 1918, *Zwaardemaker* 29 (BO); Mt. Pajung, Udjung Kulon F.R., 4 xi 1960, *UNESCO* 38 (L); Rajamandala, 21 i 1924, *P. Dakkus* 92 (BO); Bantam, Sakahudjan, 7 ii 1954, *T.O. van Kregten* 106 (BO); Rajamandala, 16 xi 1902, *P. Dakkus* 117 (BO); Wynkoops Bay, Palaboeanratoe, 17 vi 1956, *L.L. Forman* 143 (K); Banjoemas, Noesa Kambangan, 23 xi 1938, *A.J.G.H. Kostermans* 178 (BO); Kediri, 15 i 1920, *A. Thorenaar* 327 (BO); Pangandaran Peninsula, 29 x 1958, *A.J.G.H. Kostermans* 351 (BO *n.v.*); Sempor, 31 xii 1935, *R. Brinkman* 661 (A); Tjisalak, 8 iii 1920, *R.C. Bakhuizen van den Brink* 690 (U); West Java, Pelabuanratu, 14 iv 2001, *W.S. Hoover, J.M. Hunter, H. Wiriadinata, D. Girmansyah* 995 (A); Preanger, Tjisokan, 28 xii 1916, *R.C. Bakhuizen van den Brink* 1774 (BO); Preanger, Tjadas Malang, 25 ii 1917, *G.v.d. Brink* 2607 (BO, U); Gunung Tjibodas, 13 i 1929, *C.G.G.J.v. Steenis* 2691 (BO); 22 ii 1845, *H. Zollinger* 2720 (A, B, BM, P); Preanger, Tjiareuj, *R.C. Bakhuizen van den Brink* 3023 (K, U); 17 iii 1844, *H. Zollinger* 4267 (P); Wynkoops Bay, Pelabukan, 17 iii 1951, *A.J.G.H. Kostermans* 4594 (A, K, L); Preanger, Tjibodas, 7 xii 1930, *C.G.G.J.v. Steenis* 4688 (BO); Gunung Lawoe, 14 ii 1913, *C.A.B. Backer* 6751 (BO); Gunung Ringgit, 8 iii 1940, *P. Buwalda* 7496 (BO, K); Gunung Tjibodas, 13 i 1929, *G. v.d. Brink* 7505 (BO); Gunung Tjibodas, 13 i 1929, *G. v.d. Brink* 7507 (BO); Gunung Tjibodas, 13 i 1929, *R.C. Bakhuizen van den Brink* 7508 (BO); Gunung Tjibodas, 13 i 1929, *R.C. Bakhuizen van den Brink* 7510 (BO); Nusa Kambangan, 4 iv 1962, *unknown* 7764 (BO); *Doctors v. Leeuwen-Reynvaan* 11363 (BO); Idjen Mts., Gunung Blaoe, 19 iv 1940, *C.G.G.J.v. Steenis* 12044 (L); Gunung Merapi, 9 ii 1984, *A.J.M. Leeuwenberg, Rudjiman* 13018 (L); Pangandaran Peninsula, iii 1962, *A.J.G.H. Kostermans* 18876 (A, L, P); Mt. Pajung, Udjung Kulon F.R., 9 xii 1964, *A.J.G.H. Kostermans* 21830 (A); Banjoemas, Noesa Kambangan, 14 ix 1899, *S.H. Koorders* 21926 (BO); Pulau Kangean, 28 iii 1919, *C.A.B. Backer* 27804 (BO); Besoeki, Gunung Kemiri Sanga, 12 iv 1920, *C.A.B. Backer* 30618 (BO); Wynkoops Bay, Palaboeanratoe, 14 ix 1899, *S.H. Koorders* 34483 (BO); G. Kasang, 14 ix 1899, *S.H. Koorders* 40096B (BO).
- **Notes:** See notes under *B. integrifolia* Dalzell.
- **IUCN category: LC**. Widespread.

Begonia teysmanniana (Miq.) Tebbitt [§ Platycentrum], Brittonia 52(1): 116 (2000). – *Platycentrum teysmannianum* Miq., Fl. Ned. Ind. 1(1): 1092 (1856); Miquel, Fl. Ned. Ind., Suppl. 1: 333 (1861); Candolle, Prodr. 15(1): 276 (1864); Tebbitt & Dickson, Brittonia 52(1): 116 (2000). – *Casparya teysmanniana* (Miq.) A. DC., Prodr. 15(1): 276 (1864). – Type: Sumatra, Solok, Gunung Talang, *J.E. Teijsmann* 1104 (holo L).

- **SUMATRA:** Korinchi, Kampong Baru, 23 vii 1972 – 24 vii 1972, *R.J. Morley, M.K. Kardin* 289 (L); Korinchi, Mt. Korinchi, 24 vii 2006, *D. Girmansyah, A. Poulsen, I. Hatta, R. Neivita* 782 (E); *H.A.B. Bunnemeijer* 887a (B); Solok, Gunung Talang, *J.E. Teijsmann* 1104 (holo L) – Type of *Platycentrum* *teysmannianum* Miq.; *H.A.B. Bunnemeijer* 8219 (B); Korinchi, Mt. Korinchi, 8 iv 1922, *H.A.B. Bunnemeijer* 9266 (B); Korinchi, Mt. Korinchi, *H.A.B. Bunnemeijer* 9648 (B).

■ **IUCN category: LC**. Occurs in the Kerinci-Seblat National Park at relatively high altitudes where the forest is undisturbed.

Begonia thaipingensis King [§ Parvibegonia], J. Asiat. Soc. Bengal, Pt. 2, Nat. Hist. 71: 61 (1902); Ridley, Fl. Malay Penins. 1: 857 (1922); Irmscher, Mitt. Inst. Allg. Bot. Hamburg 8: 145 (1929); Kiew, Begonias Penins. Malaysia 102 (2005). – Type: Peninsular Malaysia, Perak, *Kings Collector* 8511 (lecto K *n.v.*); Peninsular Malaysia, Perak, *B. Scortechini* 1479 (syn BM, L, P); Peninsular Malaysia, Perak, *L. Wray Jr.* 1774 (syn not located); Peninsular Malaysia, Perak, *Kings Collector* 2523 (syn B, FI).

■ **PENINSULAR MALAYSIA:** Selangor, Semangkok, 16 vii 2002, *S. Neale et al.* 1 (E); Pahang, Genting Highlands, Bunga Buah Road, 12 vii 1984, *R. Kiew* 1322 (L); Perak, *B. Scortechini* 1479 (syn BM, L, P) – Type of *Begonia thaipingensis* King; Perak, *L. Wray Jr.* 1774 (syn not located) – Type of *Begonia thaipingensis* King; Perak, *Kings Collector* 2523 (syn B, FI) – Type of *Begonia thaipingensis* King; Perak, *Kings Collector* 8511 (lecto K *n.v.*) – Type of *Begonia thaipingensis* King; Selangor, Kanching F.R., 16 i 1954, *C.G.G.J.v. Steenis* 18507 (L); Selangor, Bukit Batu Berdinding, 2 xi 1937, *Md. Nur* SFN34327 (L).

■ **IUCN category: LC**. Fairly widespread on the Main Range in Peninsular Malaysia.

Begonia thomsonii A. DC. [§ Platycentrum], Ann. Sci. Nat. Bot., IV 11: 135 (1859); Candolle, Prodr. 15(1): 349 (1864). – Type: India, Khasia, *J.D. Hooker, T. Thomson s.n.* (syn K, P).

■ **BURMA:** Kachin, Sumprabum sub-division, Sumprabum, 22 i 1953, *F.K. Ward* 20418 (A).

■ **IUCN category: DD**. Insufficient specimens could be georeferenced with certainty.

Begonia tigrina Kiew [§ Reichenheimia], Begonias Penins. Malaysia 50 (2005). – Type: Peninsular Malaysia, Kelantan, Gua Setir, 19 ii 2003, *R. Kiew* 5271 (holo SING *n.v.*; iso K *n.v.*, KEP *n.v.*).

■ **PENINSULAR MALAYSIA:** Kelantan, Gua Maka, *R. Kiew, S. Anthonysamy* 3021 (para L, SING *n.v.*) – Type of *Begonia tigrina* Kiew; Kelantan, Gua Setir, 8 viii 2002, *A. Piee s.n.* (para SING *n.v.*) – Type of *Begonia tigrina* Kiew; Kelantan, Gua Setir, 19 ii 2003, *R. Kiew* 5271 (holo SING *n.v.*; iso K *n.v.*, KEP *n.v.*) – Type of *Begonia tigrina* Kiew.

■ **IUCN category: VU D2**. Restricted to two limestone hills.

Begonia timorensis (Miq.) Golding & Kareg. [§ Petermannia], Phytologia 54(7): 493 (1984). – *Diploclinium timorense* Miq., Fl. Ned. Ind. 1(1): 692 (1856); Candolle, Prodr. 15(1): 407 (1864); Golding & Karegeannes, Phytologia 54(7): 493 (1984). – Type: not located.

   Begonia aptera Decne. Later homonym, Nouv. Ann. Mus. Hist. Nat. 3: 451 (1834); Miquel, Fl. Ned. Ind. 1(1): 692 (1856); Candolle, Prodr. 15(1): 407 (1864); Gagnepain, Bull. Mus. Hist. Nat. (Paris) 25: 282 (1919); Golding & Karegeannes, Phytologia 54(7): 493 (1984). – Type: not located.

   Begonia decaisneana Gagnep., Bull. Mus. Hist. Nat. (Paris) 25: 282 (1919); Golding & Karegeannes, Phytologia 54(7): 494 (1984). – Type: not located.

■ **Notes:** The previously applied '*aptera*' epithet raises the possibility that this could be a synonym of the widespread *B. longifolia* Blume, although I have not seen any specimens.

■ **IUCN category: DD**. Insufficient specimens could be georeferenced with certainty.

Begonia tonkinensis Gagnep. [not placed to section], Bull. Mus. Hist. Nat. (Paris) 25: 280 (1919); Gagnepain, in Lecomte (ed.), Fl. Indo-Chine 2: 1108 (1921). – Type: Vietnam, Hanoi, Vo-xa, *Bon* 3127 (holo P).

- **VIETNAM:** Hanoi, Vo-xa, *Bon* 3127 (holo P) – Type of *Begonia tonkinensis* Gagnep.; Tonkin, *Bon* 6169 (P); Tonkin, Langson, 28 iii 1941, *P.A. Petelot* 7082 (B).
- **IUCN category: DD**. Insufficient specimens could be georeferenced with certainty.

Begonia torricellensis Warb. in K. Schum. & Lauterb. **[§ Petermannia]**, Fl. Deutsch. Schutzgeb. Südsee 321 (1905). – Type: New Guinea, Papua New Guinea, Kaiser-Wilhelmsland, Torricellia-Geb., iv 1902, *F.R.R. Schlechter* 14368 (syn B[2], BM, P); New Guinea, Papua New Guinea, Kaiser-Wilhelmsland, Torricellia-Geb., iv 1902, *F.R.R. Schlechter* 14541 (syn B[2], BM, P).
- **NEW GUINEA: Papua New Guinea:** Kaiser-Wilhelmsland, Torricellia-Geb., iv 1902, *F.R.R. Schlechter* 14368 (syn B[2], BM, P) – Type of *Begonia torricellensis* Warb.; Kaiser-Wilhelmsland, Torricellia-Geb., iv 1902, *F.R.R. Schlechter* 14541 (syn B[2], BM, P) – Type of *Begonia torricellensis* Warb.
- **IUCN category: DD**. Insufficient specimens could be georeferenced with certainty.

Begonia trichocheila Warb. in Perkins **[§ Diploclinium]**, Fragm. Fl. Philipp. 53 (1904); Merrill, Philipp. J. Sci. 6: 404 ('1911', 1912); Merrill, Enum. Philipp. Fl. Pl. 3: 128 (1923). – Type: Philippines, Luzon, Rizal Province, Montalban, *O. Warburg* 13260 (holo B).
- **PHILIPPINES: Luzon:** Rizal Province, i 1910, *E.D. Merrill* 7063 (B, BM, K, P); Rizal Province, Montalban, *O. Warburg* 13260 (holo B) – Type of *Begonia trichocheila* Warb.
- **IUCN category: DD**. Insufficient specimens could be georeferenced with certainty.

Begonia trichopoda Miq. [§ Reichenheimia], Fl. Ned. Ind. 1(1): 1093 (1856); Miquel, Fl. Ned. Ind., Suppl. 1: 333 (1861); Candolle, Prodr. 15(1): 386 (1864). – *Mitscherlichia trichopoda* (Miq.) Miq., Fl. Ned. Ind., Suppl. 1: 333 (1861); Candolle, Prodr. 15(1): 386 (1864). – Type: Sumatra, *J.E. Teijsmann s.n.* (holo not located).
- **SUMATRA:** *J.E. Teijsmann s.n.* (holo not located) – Type of *Begonia trichopoda* Miq.
- **IUCN category: DD**. Insufficient specimens could be georeferenced with certainty.

Begonia tricuspidata C.B. Clarke in Hook. f. **[§ Alicida]**, Fl. Brit. Ind. 2: 637 (1879). – Type: Burma, Moulmein, 1862, *C.S.P. Parish s.n.* (holo K).
- **BURMA:** Moulmein, 1862, *C.S.P. Parish s.n.* (holo K) – Type of *Begonia tricuspidata* C.B. Clarke.
- **Notes:** Resembles *B. demissa* in habit slightly, but lacks an asymmetric androecium.
- **IUCN category: DD**. Insufficient specimens could be georeferenced with certainty.

Begonia trigonocarpa Ridl. [§ Sphenanthera], J. Fed. Malay States Mus. 8(4): 38 (1917). – Type: Sumatra, Korinchi, Sungai Kumbang, 1 iv 1914, *H.C. Robinson, C.B. Kloss s.n.* (holo K; iso BM[2]).
- **SUMATRA:** Korinchi, Sungai Kumbang, 1 iv 1914, *H.C. Robinson, C.B. Kloss s.n.* (holo K; iso BM[2]) – Type of *Begonia trigonocarpa* Ridl.; Korinchi, Barong Baru, 8 vi 1914, *H.C. Robinson, C.B. Kloss s.n.* (BM).
- **Notes:** See notes under *B. scottii* Tebbitt.
- **IUCN category: DD**. Taxonomic uncertainty; the limits between this species and *B. scottii* Tebbitt and *B. multangula* Blume need defining.

Begonia triradiata C.B. Clarke in Hook. f. **[§ Alicida]**, Fl. Brit. Ind. 2: 637 (1879). – Type: Burma, Moulmein, *C.S.P. Parish s.n.* (holo K).
- **BURMA:** Moulmein, *C.S.P. Parish s.n.* (holo K) – Type of *Begonia triradiata* C.B. Clarke.
- **IUCN category: DD**. Insufficient specimens could be georeferenced with certainty.

Begonia turbinata Ridl. [§ Sphenanthera], J. Fed. Malay States Mus. 8(4): 37 (1917); Tebbitt, Brittonia 55(1): 28 (2003). – Type: Sumatra, Korinchi, Siolak Dras, 15 iii 1914, *H.C. Robinson, C.B. Kloss s.n.* (lecto BM; isolecto BM *n.v.*, K); Sumatra, Korinchi, Sungei Kumbang, 3 iv 1914, *H.C. Robinson, C.B. Kloss s.n.* (syn BM *n.v.*).
- **SUMATRA:** Korinchi, Siolak Dras, 15 iii 1914, *H.C. Robinson, C.B. Kloss s.n.* (lecto BM; isolecto BM *n.v.*, K) – Type of *Begonia turbinata* Ridl.; Berastagi Woods, 8 ii 1921, *H.N. Ridley s.n.* (BM, K); Berastagi Woods, West Hill, 14 ii 1921, *H.N. Ridley s.n.* (K); Korinchi, Sungei Kumbang, 3 iv 1914, *H.C. Robinson, C.B. Kloss s.n.* (syn BM *n.v.*) – Type of *Begonia turbinata* Ridl.; Rintis, 11 ix 1941, *H. Surbeck* 536 (L); Sibayak Volcano, 15 ii 1932, *Bangham* 1018 (K); Padang, Lubuk Sulasi, 30 vi 1953, *J.v. Borssum Waalkes* 2765 (BO); Asahan, Dolok Si Manoek-manoek, 5 x 1936 – 20 xi 1936, *Rhamat Si Boeea* 10246 (A, K, L); Korinchi, Sungei Karing, 2 iii 1954, *A.H.G. Alston* 14040 (BM, L); Korinchi, Sungei Penoh – Indrapura, 8 iii 1954, *A.H.G. Alston* 14312 (BM, L).
- **Notes:** Very close to *B. longifolia* Blume, with duplicates sometimes carrying conflicting determinations. Tebbitt (2003b) keeps this species apart from *B. longifolia* due to its turbinate fruit and slender red-tinged stems.
- **IUCN category: DD.** Whether the specimens from Berastagi are conspecific with those from the type locality needs further investigation.

Begonia urdanetensis Elmer [§ Petermannia], Leafl. Philipp. Bot. 7: 2559 (1915); Merrill, Enum. Philipp. Fl. Pl. 3: 128 (1923). – Type: Philippines, Mindanao, Agusan, Mt. Urdaneta, *A.D.E. Elmer* 13819 (syn BM, E, FI, L).
- **PHILIPPINES: Mindanao:** Agusan, Mt. Urdaneta, *A.D.E. Elmer* 13819 (syn BM, E, FI, L) – Type of *Begonia urdanetensis* Elmer.
- **IUCN category: DD.** Insufficient specimens could be georeferenced with certainty.

Begonia urunensis Kiew [§ Petermannia], Gard. Bull. Singapore 53: 283 (2001). – Type: Borneo, Sabah, Pensiangan District, Batu Urun, *R. Kiew, S. Anthonysamy* 4473 (holo SAN *n.v.*; iso SING *n.v.*).
- **BORNEO: Sabah:** Pensiangan District, Batu Urun, *R. Kiew, S. Anthonysamy* 4473 (holo SAN *n.v.*; iso SING *n.v.*) – Type of *Begonia urunensis* Kiew.
- **IUCN category: VU D2.** Known only from the type locality, which is not inside a national park.

Begonia vaccinioides Sands [§ Petermannia], in Beaman *et al.*, Pl. Mt. Kinabalu 161 (2001). – Type: Borneo, Sabah, Mt. Kinabalu, Gunung Tambuyukon, 12 iii 1980, *G.C.G. Argent, P. Walpole* 1478 (holo E *n.v.*; iso K).
- **BORNEO: Sabah:** Mt. Kinabalu, Mt. Tembuyuken, *Phillips, Lamb* 189 (para E *n.v.*, L *n.v.*) – Type of *Begonia vaccinioides* Sands; Mt. Kinabalu, Gunung Tambuyukon, 12 iii 1980, *G.C.G. Argent, P. Walpole* 1478 (holo E *n.v.*; iso K) – Type of *Begonia vaccinioides* Sands.
- **IUCN category: NT.** Known only from one locality near the summit of Mt. Tembuyuken in the Kinabalu National Park. Could possibly become vulnerable through the effects of climate change.

Begonia vagans Craib [§ Parvibegonia], Bull. Misc. Inform. Kew 1930: 419 (1930); Craib, Fl. Siam. 1: 780 (1931). – Type: Thailand, Nakawn Sawan, Kampengpet, Kao Hua, 13 vi 1922, *N. Put* 1777 (holo ABD; iso BM, K).
- **THAILAND:** Payao Prov., Doi Luang N.P., 26 vii 1997, *O. Petrmitr* 53 (L) [as *Begonia vagans* ?]; Mae Hong Sawn Prov., San Ban Dan Wildlife Sanct., 6 viii 1999, *J.F. Maxwell* 99-98 (A, L); Kanchanaburi, Lai Wo Subdistr., Toong Yai Naresuan Wildlife Reserve, 16 vi 1993, *J.F. Maxwell* 93-624 (A, L) [as *Begonia cf. vagans*]; Nakawn Sawan, Rachaburi, Kampengpet, Kanburi, Kao Hua, Sai Yok, 30 vii

1928, *N. Put* 1777 (holo ABD; iso BM, K) – Type of *Begonia vagans* Craib; Rachaburi, Kanburi, Sai Yok, 30 vii 1928, *A. Marcan* 2344 (ABD, BM, K).

- **Notes:** The epithet derives from the habit of this species of forming plantlets at the leaf tip. This is also found in the later described *B. elisabethae* Kiew, and hence the two are initially easily confused. *Begonia vagans* has a more elongate inflorescence, which appears different in structure to the cyme found in *B. elisabethae* and resembles that of *B. alicida* C.B. Clarke, but the type material is rather fragmentary. The fruits of *B. vagans* also differ from those found on other species in § *Parvibegonia*, having sub-equal wings and possibly being 3-locular. It may be that this species is better placed in § *Alicida*.
- **IUCN category: LC.** Quite widely distributed in Thailand, including two localities in national parks.

Begonia vallicola Kiew [§ Platycentrum], Begonias Penins. Malaysia 145 (2005). – Type: Peninsular Malaysia, Perak, Bujang Melaka, 15 ii 2000, *R. Kiew* 4899 (holo SING *n.v.*).

- **PENINSULAR MALAYSIA:** Perak, Gunung Bujang Melaka, *C. Curtis* 3323 (para SING *n.v.*) – Type of *Begonia vallicola* Kiew; Perak, Gunung Bujang Melaka, *C. Curtis* 3325 (para SING *n.v.*) – Type of *Begonia vallicola* Kiew; Perak, Gunung Bujang Melaka, *R. Kiew* 4891 (para SING *n.v.*) – Type of *Begonia vallicola* Kiew; Perak, Bujang Melaka, 15 ii 2000, *R. Kiew* 4899 (holo SING *n.v.*) – Type of *Begonia vallicola* Kiew.
- **IUCN category: VU D2.** Known only from the type locality.

Begonia vanderwateri Ridl. [§ Petermannia], Trans. Linn. Soc. London, Bot., II 9: 58 (1916). – Type: New Guinea, Papua, Camp VII – Camp X, *C.B. Kloss* (not located).

- **NEW GUINEA: Papua:** Camp VII – Camp X, *C.B. Kloss* (not located) – Type of *Begonia vanderwateri* Ridl.
- **IUCN category: DD.** Insufficient specimens could be georeferenced with certainty.

Begonia vanoverberghii Merr. [§ Diploclinium], Philipp. J. Sci. 6: 403 ('1911', 1912); Merrill, Enum. Philipp. Fl. Pl. 3: 128 (1923). – Type: Philippines, Luzon, Bontoc (sub-province), Bauco, ix 1910, *M. Vanovergergh* 831 (syn B[2], FI, U).

- **PHILIPPINES: Luzon:** Bontoc (sub-province), Bauco, ix 1910, *M. Vanovergergh* 831 (syn B[2], FI, U) – Type of *Begonia vanoverberghii* Merr.; Bontoc (sub-province), xi 1912, *M. Vanovergergh* 1920 (P).
- **IUCN category: DD.** Insufficient specimens could be georeferenced with certainty.

Begonia variabilis Ridl. [§ Parvibegonia], J. Straits Branch Roy. Asiat. Soc. 57: 50 (1911); Ridley, Fl. Malay Penins. 1: 858 (1922); Irmscher, Mitt. Inst. Allg. Bot. Hamburg 8: 147 (1929); Craib, Fl. Siam. 1: 780 (1931); Tebbitt, Begonias 230 (2005); Kiew, Begonias Penins. Malaysia 92 (2005). – Type: Peninsular Malaysia, Ulu Temango, vii 1904, *H.N. Ridley* 14730 (lecto SING *n.v.*; isolecto BM *n.v.*).

- **THAILAND:** Songkhla Prov., Dton Nga Chang Reserve, 9 xi 1984, *J.F. Maxwell* 84-413 (A); Nakawn Sritamarat, Kiriwong, Tap Chang, 26 vii 1951, *T. Smitinand* 705 (L); Yala Province, Kue Long waterfalls, 22 x 1970, *K. Larsen et al.* 4157 (L); Kao Nawng, 9 viii 1927, *A.F.G. Kerr* 13240 (ABD, BM, K); Takuapah, 14 vii 1972, *K. Larsen et al.* 30942 (B).
  PENINSULAR MALAYSIA: Ulu Temango, vii 1904, *H.N. Ridley* 14730 (lecto SING *n.v.*; isolecto BM *n.v.*) – Type of *Begonia variabilis* Ridl.
- **IUCN category: NT.** Described as 'not common' by Kiew (2005), and although this species occurs from Kelantan to southern Thailand, it prefers low altitude sites which are more prone to disturbance.

Begonia venusta King [§ Platycentrum], J. Asiat. Soc. Bengal, Pt. 2, Nat. Hist. 71: 65 (1902); Ridley, Fl. Malay Penins. 1: 862 (1922); Irmscher, Mitt. Inst. Allg. Bot. Hamburg 8: 126 (1929); Kiew, Gard. Bull. Singapore 51:117 (1999); Kiew, Begonias Penins. Malaysia 179 (2005). – Type: Peninsular Malaysia, Perak, Ulu Batang Padang, *L. Wray Jr.* 1598 (lecto K *n.v.*).

   Begonia megapteroidea King, J. Asiat. Soc. Bengal, Pt. 2, Nat. Hist. 71:65 (1902); Ridley, Fl. Malay Penins.1:862 (1922); Kiew, Begonias Penins. Malaysia 179 (2005). – Type: Peninsular Malaysia, Perak, *L. Wray Jr.* 1450 (syn not located); Peninsular Malaysia, Perak, *L. Wray Jr.* 1573 (syn not located).

- **PENINSULAR MALAYSIA:** Pahang, Cameron Highlands, Gunong Brinchang, 26 vii 2002, *S. Neale, G. Bramley* 7 (E); Perak, *L. Wray Jr.* 1450 (syn not located) – Type of *Begonia megapteroidea* King; Perak, *L. Wray Jr.* 1573 (syn not located) – Type of *Begonia megapteroidea* King; Perak, Ulu Batang Padang, *L. Wray Jr.* 1598 (lecto K *n.v.*) – Type of *Begonia venusta* King; Pahang, Cameron Highlands, Robinsons Falls, 6 iii 1980, *B.C. Stone* 14407 (L).
- **IUCN category: LC**. Described as 'the most rampant of all begonias in Peninsular Malaysia' and is capable of invading the banks on the edges of vegetable plots (Kiew, 2005).

Begonia versicolor Irmsch. [§ Platycentrum], Mitt. Inst. Allg. Bot. Hamburg 10:546 (1939); Shui & Huang, Acta Bot. Yunnan. 21(1): 18 (1999); Tebbitt, Begonias 237 (2005). – Type: not located.
- **VIETNAM:** Cha-pa, *E. Poilane* 12881 (P).
- **IUCN category: DD**. Insufficient specimens could be georeferenced with certainty.

Begonia villifolia Irmsch. [§ Platycentrum], Notes Roy. Bot. Gard. Edinburgh 21: 43 (1951); Huang & Shui, Acta Bot. Yunnan. 21(1): 17 (1999). – Type: China, Yunnan, xii 1939, *C.W. Wang* 83071 (mero B).
- **BURMA:** Mountains E of Fort Hertz, 3 ix 1926, *F.K. Ward* 7354 (B, K[2]); Ti Hka River, Fort Hertz, 9 xii 1931, *F.K. Ward* 10211 (BM).
   VIETNAM: Laoke, Chapa, Song ta Van, viii 1930, *P.A. Petelot* 7119 (B[2]); Laoke, Chapa, 14 viii 1926, *E. Poilane* 12922 (P); Tonkin, Lao-kay, 1 xii 1937, *E. Poilane* 26661 (P).
- **IUCN category: DD**. Insufficient specimens could be georeferenced with certainty.

Begonia vitiensis A.C. Sm. [§ Diploclinium], Bernice P. Bishop Mus. Bull. 141:99 (1936). – Type: Fiji, Vanua Levu, Thakaundrove, Mt. Mariko, 14 xi 1933, *A.C. Smith* 466 (holo BISH *n.v.*; iso BO, K, P).
- **VANUATU: Santo Island:** Tauratu, 12 xii 1994, *F. Kamasteia* 62 (K, L); Apouna Valley, 25 viii 1971, *J. Raynal* 16283 (K, L); Apouna Valley, 3 ix 1971, *H.S. McKee* 24163 (K[2]).
   FIJI: Vanua Levu: Thakaundrove, Mt. Mariko, 14 xi 1933, *A.C. Smith* 466 (holo BISH *n.v.*; iso BO, K, P) – Type of *Begonia vitiensis* A.C. Sm.; Waiwai – Lomaloma, *Horne* 618 (para K *n.v.*) – Type of *Begonia vitiensis* A.C. Sm.
- **IUCN category: VU D2**. Neither of the localities are protected, and the forests at altitudes of 600–900 m are disturbed.

Begonia vuijckii Koord. [§ Petermannia], Exkurs.-Fl. Java 2:647 (1912). – Type: Gunong Salak, *C.L.v. Blume* 6088 (mero B).
- **SUMATRA:** *Cramer* 60 (B); Riau Province, Tigapulu Mts., 6 xi 1988, *J.S. Burley et al.* 1205 (A); Riau Province, Tigapulu Mts., 25 xi 1988, *J.S. Burley* 2148 (A); Asahan, Loemban Ria, 5 ii 1934 – 12 iv 1934, *Rahmat Si Boeea* 7566 (A).
   JAVA: Gunong Salak, *C.L.v. Blume* 6088 (mero B) – Type of *Begonia vuijckii* Koord.; G. Soenarari, 1 i 1913, *C.A.B. Backer* 6335 (B).
- **Notes:** See notes under *B. tenericaulis* Ridl.; assigned here to § *Petermannia*.
- **IUCN category: LC**. Reasonably widespread.

Begonia wadei Merr. & Quisumb. [§ Diploclinium], Addisonia 17: 57 (1932). – Type: Philippines, Palawan, Coron Island, 24 xii 1929, *W.H. Brown* BS78801 (holo PNH *n.v.*; iso NY).

- ■ **PHILIPPINES: Palawan:** vii 1912, *E. Fenix* 15540 (B, BM); Miniloc Island, 24 iv 1997, *D.A. Madulid et al.* 27564 (L); Coron Island, 24 xii 1929, *W.H. Brown* BS78801 (holo PNH *n.v.*; iso NY) – Type of *Begonia wadei* Merr. & Quisumb.
- ■ **Notes:** An ecologically unusual species of *Begonia*, as it grows near the coast on exposed coralline limestone where it may be exposed to salt spray, and where it must endure a dry season of several months (Merrill & Quisumbing, 1932).
- ■ **IUCN category: DD.** Insufficient specimens could be georeferenced with certainty.

Begonia walteriana Irmsch. [§ Petermannia], Beitr. Phytol. 30: 14 (1964). – Type: Borneo, Kalimantan, Bukit Raja, 19 xii 1924, *H.J.P. Winkler* 953 (holo HBG *n.v.*).

- ■ **BORNEO: Kalimantan:** Bukit Raja, 19 xii 1924, *H.J.P. Winkler* 953 (holo HBG *n.v.*) – Type of *Begonia walteriana* Irmsch.
- ■ **IUCN category: DD.** Insufficient specimens could be georeferenced with certainty.

Begonia warburgii K. Schum. & Lauterb. [not placed to section], Fl. Deutsch. Schutzgeb. Südsee 459 (1901). – Type: New Guinea, Papua New Guinea, Bismarck Mts., 5 vii 1899, *H. Rodatz, T.F. Bevan* 223 (holo B).

- ■ **NEW GUINEA: Papua:** Cycloop Mts., Anafre River, 15 iii 1957, *C. Koster* 4283 (P). **Papua New Guinea:** Bismarck Mts., 5 vii 1899, *H. Rodatz, T.F. Bevan* 223 (holo B) – Type of *Begonia warburgii* K. Schum. & Lauterb.; Kaiser-Wilhelmsland, 7 ix 1907, *F.R.R. Schlechter* 16509 (B, P); Kaiser-Wilhelmsland, Rani-Gebirge, 16 v 1908, *F.R.R. Schlechter* 17722 (B, P).
- ■ **Notes:** See notes under *B. bipinnatifida* J.J. Sm. Likely to belong in § *Petermannia*.
- ■ **IUCN category: DD.** Insufficient specimens could be georeferenced with certainty.

Begonia wariana Irmsch. [§ Petermannia], Bot. Jahrb. Syst. 50: 352 (1913); Irmscher, Bot. Jahrb. Syst. 50: 569 (1914). – Type: New Guinea, Papua New Guinea, Jaduna, 14 iv 1909, *F.R.R. Schlechter* 19240 (holo B; iso P).

- ■ **NEW GUINEA: Papua New Guinea:** Jaduna, 14 iv 1909, *F.R.R. Schlechter* 19240 (holo B; iso P) – Type of *Begonia wariana* Irmsch.
- ■ **IUCN category: DD.** Insufficient specimens could be georeferenced with certainty.

Begonia weberi Merr. [§ Petermannia], Philipp. J. Sci. 6: 381 ('1911', 1912); Merrill, Enum. Philipp. Fl. Pl. 3: 128 (1923). – Type: Philippines, Mindanao, Butuan, Mt. Hilong-hilong, iii 1911, *C.M. Weber* 1210 (syn PNH).

- ■ **PHILIPPINES: Mindanao:** Butuan, Mt. Hilong-hilong, iii 1911, *C.M. Weber* 1210 (syn PNH) – Type of *Begonia weberi* Merr.
- ■ **IUCN category: DD.** Insufficient specimens could be georeferenced with certainty.

Begonia weigallii Hemsl. [§ Petermannia], Bull. Misc. Inform. Kew 1896: 17 (1896). – Type: Solomon Islands, South Solomon Islands, New Georgia, 1894 – 1895, *Officers of HMS Penguin s.n.* (holo K).

- ■ **NEW GUINEA: Papua New Guinea:** Milne Bay Distr., Gwariu River, Biniguni Camp, 6 viii 1953, *L.J. Brass* 23863 (L); Milne Bay Distr., Kwagira River, Peria Creek, 31 viii 1953, *L.J. Brass* 24263 (L); Fergusson Island, Agamola – Ailuluai, 18 vi 1956, *L.J. Brass* 27203 (L); Misima Island, Narian, 3 viii 1956, *L.J. Brass* 27576 (L); Rossel Island, Abaleti, 1 x 1956, *L.J. Brass* 28278 (L); Misima Island, Bwagaoia, 31 iii 1969, *E.B. Mann* 43270 (L).

BISMARCK ARCHIPELAGO: New Britain: Mt. Kavangi, 12 vi 1995, *G. Weiblen* 458 (L).
SOLOMON ISLANDS: North Solomon Islands: Bougainville Island, Crown Prince Range, i 1936, *A.H. Voyce* B4 (K) [as *Begonia cf. weigallii*]; Bougainville Island, Barapinang Camp (nr), 12 iv 1965, *M.M. Cole* 21 (K) [as *Begonia weigallii* ?]; Bougainville Island, *Kajewski* 1630 (B); Bougainville Island, Kieta Distr., Panguna Creek, 21 vi 1974, *J.S. Womersley* 48643 (L). **South Solomon Islands:** New Georgia, 1894 – 1895, *Officers of HMS Penguin s.n.* (holo K) – Type of *Begonia weigallii* Hemsl.; Ulawa, iv 1898, *R.B. Comins* 331 (K); Guadalcanal, Tina River, 12 xi 1962, *T.C. Whitmore, J.S. Womersley* 1101 (L); Baga Island, 9 i 1963, *T.C. Whitmore* 1315 (L); Rendova Island, Zaimane River, 12 ix 1963, *T.C. Whitmore* 1865 (L); San Cristobal, Pegato River, 23 vii 1965, *P.F. Hunt* 2215 (A, K[3], L) [as *Begonia cf. weigallii*]; San Cristobal, 26 vii 1965, *P.F. Hunt* 2245 (K); San Cristobal, Warahito River, 26 vii 1965, *P.F. Hunt* 2445 (L); Kolombangara, *P.F. Hunt* 2447 (K[2], L); Kolombangara, 17 viii 1965, *P.F. Hunt* 2447 (K[2], L); Malaita Island, Mt. Alasaa, 22 xi 1965, *P.F. Hunt* 3049 (K[2], L) [as *Begonia cf. weigallii*]; Malaita Island, Mt. Alasaa, 22 xi 1965, *P.F. Hunt* 3051 (A, K[2], L); Guadalcanal, Rere River, 20 xi 1963, *Z. Lipaqeto* 3370 (L); San Cristobal, Wairaha River, 12 v 1964, *T.C. Whitmore* 4314 (L); Guadalcanal, Mt. Gallego, 19 vi 1966, *G.F.C. Dennis* 4618 (L); Guadalcanal, Mt. Gallego, 19 vi 1966, *G.F.C. Dennis* 4633 (L); Ulawa Island, Arosha, 4 ii 1965, *R. Teona* 6221 (L); San Cristobal, Warahito River, 1 viii 1965, *T.C. Whitmore* 6270 (K); Santa Ysabel, Allardyce Harbour, 9 xi 1967, *A. Kinifu* 8299 (K) [as *Begonia cf. weigallii*]; Kolombangara, Kokove, 12 i 1968, *R. Mauriasi* 8602 (K) [as *Begonia cf. weigallii*]; Kolombangara, 29 v 1968, *R. Mauriasi* 9592 (L); Malaita Island, Rokera, 27 vii 1968, *P. Runikera* 10443 (L); Malaita Island, Dala, 17 viii 1968, *I. Gafui* 10587 (L); San Cristobal, Anganawai, 11 x 1968, *P. Runikera* 10890 (L); San Cristobal, Marogu, 18 x 1968, *P. Runikera* 11098 (L); Guadalcanal, Duidui, 4 x 1968, *R. Mauriasi* 11756 (L); Guadalcanal, Wanderer Bay, 21 x 1968, *H. Fa'arodo* 12156 (L); Savo Island, 14 iv 1969, *I. Gafui* 12970 (L); Shortlands, 13 iii 1969, *R. Mauriasi* 13297 (L); Ovau Island, Savaava, 26 iii 1969, *R. Mauriasi* 13385 (L); Malaita Island, Harumou, 7 xii 1968, *H. Fa'arodo* 13715 (L); Mono Island, *R. Mauriasi* 14171 (not located); Treasury Island, Maleaini, 3 vi 1969, *R. Mauriasi* 14275 (L); Ranongga Isl., Dae, 10 vi 1969, *R. Mauriasi* 14397 (L); Ranongga Isl., 19 vi 1969, *R. Mauriasi* 14484 (L); Savo Island, Pokilo, 10 iv 1969, *I. Gafui* 14578 (L); Big Nggela, Ghairavu, 31 vii 1969, *I. Gafui* 14771 (L); Ranongga Isl., 23 vi 1969, *R. Mauriasi* 15589 (L); Ranongga Isl., Palaina, 28 vi 1969, *R. Mauriasi* 15703 (L); Ranongga Isl., 1 vii 1969, *R. Mauriasi* 15756 (L).
- **IUCN category: LC**. Quite widespread, and shows tolerance of secondary forest.

Begonia wenzelii Merr. [§ Petermannia], Philipp. J. Sci. 10: 277 (1915); Merrill, Enum. Philipp. Fl. Pl. 3: 128 (1923). – *Begonia leytensis* Merr., Philipp. J. Sci. 9: 379 (1914); Merrill, Philipp. J. Sci. 10: 277 (1915). – Type: not located.
- **PHILIPPINES: Luzon:** Kalinga Sub Prov., Lubuagan, Mt. Masingit, ii 1920, *M. Ramos, G. Edano* 37545 (B, P) [as *Begonia cf. wenzelii*]. **Leyte:** Jaro, 11 ii 1914, *C.A. Wenzel* 580 (syn BM) – Type of *Begonia leytensis* Merr.; 12 xii 1914, *C.A. Wenzel* 1219 (BM); 30 iii 1916, *C.A. Wenzel* 1695 (BM); Biliran, vi 1914, *R.C. McGregor* 18754 (K, P).
- **Notes:** See notes under *B. aequata* A. Gray.
- **IUCN category: DD**. Taxonomic uncertainty.

Begonia woodii Merr. [§ Diploclinium], Philipp. J. Sci. 26: 478 (1925). – Type: Philippines, Palawan, Malampaya Bay, 8 x 1922, *E.D. Merrill* 11589 (syn B, L, P).
- **PHILIPPINES: Palawan:** Malampaya Bay, 8 x 1922, *E.D. Merrill* 11589 (syn B, L, P) – Type of *Begonia woodii* Merr.
- **IUCN category: DD**. Insufficient specimens could be georeferenced with certainty.

Begonia wrayi Hemsl. [§ Petermannia], J. Bot. 25: 203 (1887); Ridley, Fl. Malay Penins. 1: 855 (1922); Kiew, Malayan Nat. J. 47: 312 (1994); Kiew, Begonias Penins. Malaysia 118 (2005). – Type: Peninsular Malaysia, Perak, Ulu Kenering, *L. Wray Jr.* 55 (holo K; iso SING *n.v.*); Peninsular Malaysia, Perak, 1885, *L. Wray Jr.* 3367 (syn L).

    Begonia pseudoisoptera Irmsch., Mitt. Inst. Allg. Bot. Hamburg 8: 112 (1929); Kiew, Malayan Nat. J. 47: 312 (1994). – Type: Peninsular Malaysia, Pahang, Tahan River, 1891, *H.N. Ridley* 2246 (lecto SING *n.v.*; isolecto BM, K); Peninsular Malaysia, Pahang, Gua Kechapi, 1 ii 1924, *Mohd. Nur, Foxworthy* 11911 (syn not located); Peninsular Malaysia, Kelantan, Batu Panjang, Sungai Keteh, 7 ii 1924, *Md. Nur, F.W. Foxworthy* 12116 (syn K[3]); Peninsular Malaysia, Selangor, Semangkok, *H.N. Ridley* 12997 (syn not located); Peninsular Malaysia, Selangor, Klang Gates, viii 1908, 3°13′N, 101°45′E, *H.N. Ridley* 13431 (syn B, BM, K); Peninsular Malaysia, Perak, Grik, 19 vi 1924, *I.H. Burkill, M. Haniff* 13636 (syn not located); Peninsular Malaysia, Perak, Grik, vii 1904, *H.N. Ridley* 14607 (syn BM); Peninsular Malaysia, Ulu Temango, vii 1909, *H.N. Ridley* 14607A (syn BO); Peninsular Malaysia, Selangor, Ginting Simpah, iv 1911, *unknown* 15579 (syn K); Peninsular Malaysia, Selangor, Dusun Tua, *Campbell* 1909 (syn not located); Peninsular Malaysia, Selangor, Ginting Bedai, v 1896, *H.N. Ridley* 7826 (syn not located); Peninsular Malaysia, Selangor, Pahang Track, 1897, *Ridley* 8589 (syn not located); Peninsular Malaysia, Selangor, Ginting Peras, v 1896, *H.N. Ridley s.n.* (syn not located).

    Begonia isoptera auct. non Dryand. ex Sm.: King, J. Asiat. Soc. Bengal 71 (1902); Ridley, Fl. Malay Penins. 1: 855 (1922), *pro parte*; Craib, Fl. Siam. 1: 775 (1931).

■ **THAILAND:** Narathiwat, Waeng, iv 1968, *S. Phusomsaeng* 442 (L, P); Satun Distr., Talaybun National Park, 2 vii 1985, *J.F. Maxwell* 85-664 (A); Kao Re Chaw, 20 iv 1931, *M.C. Lakshnakara* 726 (BM); Yala Province, Bang Lang Nat. Park, 8 ii 1990, *W.S. Hoover* 753 (A); Yala Province, 21 x 1970, *C. Charoenpol et al.* 4120 (L); 28 vii 1923, *A.F.G. Kerr* 7382 (ABD, B, BM, K); 28 vii 1923, *A.F.G. Kerr* 15873 (ABD, B, BM, K); Yala Province, Than To waterfall, 27 xi 1990, *K. Larsen et al.* 41736 (P).

PENINSULAR MALAYSIA: Selangor, Ginting Peras, v 1896, *H.N. Ridley s.n.* (syn not located) – Type of *Begonia pseudoisoptera* Irmsch.; Selangor, Pahang Track, v 1903, *Machado s.n.* (syn not located) – Type of *Begonia pseudoisoptera* Irmsch.; Selangor, Ulu Gombak, 11 iii 1915, *H.N. Ridley s.n.* (BM); Selangor, Ginting Sempah, 7 ix 1917, *H.N. Ridley, H.C. Robinson, C.B. Kloss s.n.* (K); Selangor, Ulu Gombak, 14 iii 1915, *H.N. Ridley s.n.* (BM); Selangor, Awana Eco Park – Genting Highlands, 29 vii 2002, *S. Neale et al.* 12 (E[2]); Perak, Ulu Kenering, *L. Wray Jr.* 55 (holo K; iso SING *n.v.*) – Type of *Begonia wrayi* Hemsl.; Legeh, 23 vii 1899, *D.T. Gwynne-Vaughan* 634 (K); Perak, Slim Hills F.R., 8 ix 1966, *T.C. Whitmore* 827 (L); Frasers Hill, 12 xii 1981, *R. Kiew* 1102 (L); Selangor, Dusun Tua, *Campbell* 1909 (syn not located) – Type of *Begonia pseudoisoptera* Irmsch.; Pahang, Tahan River, 1891, *H.N. Ridley* 2246 (lecto SING *n.v.*; isolecto BM, K) – Type of *Begonia pseudoisoptera* Irmsch.; Pahang, Taman Negara, 22 iv 1975, *M.M.J.v. Balgooy* 2490 (L); Perak, 1885, *L. Wray Jr.* 3367 (syn L) – Type of *Begonia wrayi* Hemsl.; Perak, Grik, Temengor Dam, 31 viii 1993, *A. Latiff et al.* 3979 (not located); Selangor, Ginting Bedai, v 1896, *H.N. Ridley* 7286 (BM, K); Selangor, Ginting Bedai, v 1896, *H.N. Ridley* 7826 (syn not located) – Type of *Begonia pseudoisoptera* Irmsch.; Selangor, Pahang Track, 1897, *Ridley* 8589 (syn not located) – Type of *Begonia pseudoisoptera* Irmsch.; Pahang, Gua Kechapi, 1 ii 1924, *Mohd. Nur, Foxworthy* 11911 (syn not located) – Type of *Begonia pseudoisoptera* Irmsch.; Selangor, Semangkok, viii 1904, *H.N. Ridley* 12097 (K); Kelantan, Batu Panjang, Sungai Keteh, 7 ii 1924, *Md. Nur, F.W. Foxworthy* 12116 (syn K[3]) – Type of *Begonia pseudoisoptera* Irmsch.; Selangor, Semangkok, *H.N. Ridley* 12997 (syn not located) – Type of *Begonia pseudoisoptera* Irmsch.; Selangor, Klang Gates, viii 1908, *H.N. Ridley* 13430 (B, BM, K); Selangor, Klang Gates, viii 1908, *H.N. Ridley* 13431 (syn B, BM, K) – Type of *Begonia pseudoisoptera* Irmsch.; Perak, Grik, 19 vi 1924, *I.H. Burkill, M. Haniff* 13636 (syn not located) – Type of *Begonia pseudoisoptera* Irmsch.; Selangor, Ulu Sungei Gombak, 28

x 1967, *T. Shimizu et al.* 14052 (L); Perak, viii 1909, *H.N. Ridley* 14604 (B); Perak, Grik, vii 1904, *H.N. Ridley* 14607 (syn BM) – Type of *Begonia pseudoisoptera* Irmsch.; Ulu Temango, vii 1909, *H.N. Ridley* 14607A (syn BO) – Type of *Begonia pseudoisoptera* Irmsch.; Selangor, Ginting Simpai, iv 1911, *unknown* 15579 (syn K) – Type of *Begonia pseudoisoptera* Irmsch.; Pahang, Gua Peningat, 16 vii 1970, *L.H. Shing* 17253 (L); Perak, Grik, Temengor Dam, 21 ix 1993, *FRIM* 26870 (K, L); Perak, Grik, 26 vii 1936, *M.R. Henderson* 31629 (BM, L); Kedah, Gunong Lang, 25 iii 1938, *Kiah* 35046 (K); Kedah, Baling, 10 v 1938, *Kiah* 35387 (BM, K[2]); Kedah, Baling – Kroh, 5 viii 1941, *J.C. Nauen* 38041 (K); Perak, Grik, Temengor Dam, Sungei Singor, 8 xi 1993, *Saw* 39940 (K, L); Perak, Grik, Kg. Tera, 15 xi 1966, *Ismail* 95025 (L); Selangor, Chadangan F.R., 28 ix 1966, *S. Chelliah* 98214 (L); Negeri Sembilan, 11 x 1969, *B. Everett* KEP104940 (L).
- **IUCN category: LC.** Widespread in Peninsular Malaysian lowland forest.

Begonia wyepingiana Kiew [§ Platycentrum], Begonias Penins. Malaysia 140 (2005). – Type: Peninsular Malaysia, Perak, Ulu Selim, 30 ix 2000, *R. Kiew* 5115 (holo SING *n.v.*; iso K *n.v.*, KEP *n.v.*).
- **PENINSULAR MALAYSIA:** Perak, Ulu Selim, 30 ix 2000, *R. Kiew* 5115 (holo SING *n.v.*; iso K *n.v.*, KEP *n.v.*) – Type of *Begonia wyepingiana* Kiew; Perak, Lipas-Slim Pass, *E.J. Strugnell* 20396 (para KEP *n.v.*) – Type of *Begonia wyepingiana* Kiew.
- **IUCN category: VU D2.** Has a very narrow distribution, although the area is described as inaccessible by Kiew (2005).

Begonia xiphophylla Irmsch. [§ Petermannia], Bot. Jahrb. Syst. 76: 100 (1953). – *Begonia polygonoides* Ridl., J. Straits Branch Roy. Asiat. Soc. 46: 254 (1906); Irmscher, Bot. Jahrb. Syst. 76: 100 (1953). – Type: Matang, *G.D. Haviland* 1906 (syn not located); Matang, *H.N. Ridley* 11770 (syn not located).
- **BORNEO: Sarawak:** Matang, 25 vi 1961, *S.H. Collenette* 698 (K, L); *Native collector* 719 (B, K, L); Matang, v 1866, *O. Beccari* PB1728 (Fl); Matang, *G.D. Haviland* 1906 (syn not located) – Type of *Begonia polygonoides* Ridl.; Matang, *J. & M.S. Clemens* 7734 (K); Matang, *H.N. Ridley* 11770 (syn not located) – Type of *Begonia polygonoides* Ridl.; Kuching, Bukit Tanduk, 23 iv 1987, *B.L.M. Hock* 54032 (L); 14 xii 1994, *Yii Puan Ching* 69578 (K) [as *Begonia cf. xiphophylla*].
- **IUCN category: DD.** Insufficient specimens could be georeferenced with certainty.

Begonia yappii Ridl. [§ Reichenheimia], Bull. Misc. Inform. Kew 1929: 258 (1929); Kiew, Begonias Penins. Malaysia 223 (2005). – Type: Peninsular Malaysia, Kelantan, Kuala Aring, *R.H. Yapp* 25 (holo not located).
- **PENINSULAR MALAYSIA:** Kelantan, Kuala Aring, *R.H. Yapp* 25 (holo not located) – Type of *Begonia yappii* Ridl.; Kelantan, Kuala Aring, *R.H. Yapp* 36 (K *n.v.*); Kelantan, Kuala Aring, *R. Kiew* RK5198 (K *n.v.*, KEP *n.v.*, L *n.v.*, SING *n.v.*).
- **IUCN category: VU D2.** No specific threats have been identified for the type and only locality; Kiew (2005) describes it as 'not common'.

Begonia zamboangensis Merr. [§ Petermannia], Philipp. J. Sci. 26: 481 (1925). – Type: Philippines, Mindanao, Zamboanga, Malangas, xi 1919, *M. Ramos, G. Edano* 37371 (syn B).
- **PHILIPPINES: Mindanao:** Zamboanga, Malangas, xi 1919, *M. Ramos, G. Edano* 37271 (para not located) – Type of *Begonia zamboangensis* Merr.; Zamboanga, Malangas, xi 1919, *M. Ramos, G. Edano* 37371 (syn B) – Type of *Begonia zamboangensis* Merr.
- **IUCN category: DD.** Insufficient specimens could be georeferenced with certainty.

Begonia zollingeriana A. DC. [§ Parvibegonia], Ann. Sci. Nat. Bot., IV 11 : 135 (1859); Candolle, Prodr. 15(1) : 351 (1864). – Type: Java, *H. Zollinger* 2630 (syn B, BM, FI[2], P).

   Platycentrum zollingerianum Klotzsch, Abh. Kon. Akad. Wiss. Berlin 1854: 245 (1855); Klotzsch, Begoniac. 125 (1855); Candolle, Prodr. 15(1) : 351 (1864). – Type: Java, *H. Zollinger* 2630 (syn B, BM, FI[2], P).

- **JAVA:** *H. Zollinger* 2630 (syn B, BM, FI[2], P) – Type of *Begonia zollingeriana* A. DC.; 1845, *H. Zollinger* 6937 (P).
- **IUCN category: DD**. Insufficient specimens could be georeferenced with certainty.

Geographic List

BURMA

Begonia acetosella Craib var. *acetosella*
Begonia acetosella var. *hirtifolia* Irmsch.
Begonia adenopoda Lem.
Begonia adscendens C.B. Clarke
Begonia alicida C.B. Clarke
Begonia annulata K. Koch
Begonia brandisiana Kurz
Begonia burkillii Dunn
Begonia burmensis L.B. Sm. & Wassh.
Begonia cathcartii Hook. f.
Begonia crenata Dryand.
Begonia delicatula Parish ex C.B. Clarke
Begonia demissa Craib
Begonia discrepans Irmsch.
Begonia dux C.B. Clarke
Begonia fibrosa C.B. Clarke
Begonia flaccidissima Kurz
Begonia flaviflora H. Hara var. *flaviflora*
Begonia flaviflora var. *vivida* (Irmsch.) Golding
    & Kareg.
Begonia forrestii Irmsch.
Begonia goniotis C.B. Clarke
Begonia griffithiana (A. DC.) Warb.
Begonia handelii Irmsch. var. *handelii*
Begonia handelii var. *prostrata* (Irmsch.)
    Tebbitt
Begonia hatacoa Buch.-Ham. ex D. Don
Begonia hemsleyana Hook. f.
Begonia hymenophylloides Kingdon-Ward ex
    L.B. Sm. & Wassh.
Begonia integrifolia Dalzell
Begonia iridescens Dunn
Begonia josephii A. DC.
Begonia kingdon-wardii Tebbitt
Begonia labordei H. Lév.
Begonia longifolia Blume
Begonia martabanica A. DC.

Begonia megaptera A. DC.
Begonia modestiflora Kurz
Begonia nepalensis (A. DC.) Warb.
Begonia nivea Parish ex Kurz
Begonia obovoidea Craib
Begonia oreodoxa Chun & F. Chun ex C.Y. Wu
    & T.C. Ku
Begonia paleacea Kurz
Begonia palmata D. Don var. *palmata*
Begonia palmata var. *bowringiana* (Benth.)
    Golding & Kareg.
Begonia parishii C.B. Clarke
Begonia parvuliflora A. DC. var. *parvuliflora*
Begonia parvuliflora var. *pubescens* A. DC.
Begonia picta Sm.
Begonia procridifolia Wall. ex A. DC.
Begonia prolifera A. DC.
Begonia rex Putz.
Begonia rockii Irmsch.
Begonia roxburghii (Miq.) A. DC.
Begonia sandalifolia C.B. Clarke
Begonia sikkimensis A. DC.
Begonia silletensis (A. DC.) C.B. Clarke subsp.
    silletensis
Begonia sinuata Wall. ex Meisn. var. *sinuata*
Begonia sinuata var. *helferi* Irmsch.
Begonia subperfoliata Parish ex Kurz
Begonia surculigera Kurz
Begonia thomsonii A. DC.
Begonia tricuspidata C.B. Clarke
Begonia triradiata C.B. Clarke
Begonia villifolia Irmsch.

Excluded names:
Begonia andamensis Parish ex C.B. Clarke
    Endemic to the Andaman Islands; this
    record from Kress *et al.* (2003) may refer to
    the similar *B. sinuata*.
Begonia burkei hort.

Begonia herveyana King
  Listed by Kress *et al.* (2003), but Kiew (2005) regards this species as endemic to Peninsular Malaysia.
Begonia lacei C.Y. Wu ined.
Begonia moulmeinensis C.B. Clarke (see notes under *Begonia parvuliflora* var. *pubescens* A. DC.)
Begonia rhoephila Ridl. (see notes under that species)

Notes: A flora of 57 species, with § *Platycentrum* containing the most (20 spp.). The remainder are split between § *Diploclinium* (9 spp.), § *Parvibegonia* (8 spp.), § *Sphenanthera* (8 spp.), § *Reichenheimia* (3 spp.), § *Alicida* (2 spp.), § *Lauchea* (2 spp.), § *Monopteron* (2 spp.) and § *Apterobegonia* (1 sp.). Two species are unplaced to section.

THAILAND

Begonia aceroides Irmsch.
Begonia acetosella Craib var. *acetosella*
Begonia alicida C.B. Clarke
Begonia brandisiana Kurz
Begonia cardiophora Irmsch.
Begonia cathcartii Hook. f.
Begonia cf. *furfuracea* Hook. f.
Begonia demissa Craib
Begonia discreta Craib
Begonia elisabethae Kiew
Begonia festiva Craib
Begonia fibrosa C.B. Clarke
Begonia garrettii Craib
Begonia grantiana Craib
Begonia grata Geddes ex Craib
Begonia handelii Irmsch. var. *handelii*
Begonia handelii var. *prostrata* (Irmsch.) Tebbitt
Begonia hatacoa Buch.-Ham. ex D. Don
Begonia incerta Craib
Begonia incondita Craib
Begonia integrifolia Dalzell
Begonia intermixta Irmsch.
Begonia kerrii Craib
Begonia longifolia Blume
Begonia martabanica A. DC.
Begonia modestiflora Kurz

Begonia modestiflora var. *sootepensis* (Craib) Z. Badcock ex M. Hughes
Begonia murina Craib
Begonia notata Craib
Begonia obovoidea Craib
Begonia palmata D. Don
Begonia prolifera A. DC.
Begonia prolixa Craib
Begonia pumila Craib
Begonia pumilio Irmsch.
Begonia putii Craib
Begonia rabilii Craib
Begonia rimarum Craib
Begonia saxifragifolia Craib
Begonia siamensis Gagnep.
Begonia sibthorpioides var. *grandifolia* Craib
Begonia silletensis (A. DC.) C.B. Clarke subsp. *silletensis*
Begonia sinuata Wall. ex Meisn. var. *sinuata*
Begonia sinuata var. *longialata* Irmsch.
Begonia smithiae Geddes
Begonia socia Craib
Begonia soluta Craib
Begonia subperfoliata Parish ex Kurz
Begonia subviridis Craib
Begonia vagans Craib
Begonia variabilis Ridl.
Begonia wrayi Hemsl.

Excluded names:
Begonia finlaysoniana Wall. *nom. nud.*
  Possibly a synonym of *B. prolifera* A. DC., but further investigation is needed (Golding, 1970).

Notes: A flora of 49 species, with § *Diploclinium* having the most species (14). The remainder are split between § *Parvibegonia* (10 spp.), § *Platycentrum* (6 spp.), § *Reichenheimia* (5 spp.), § *Sphenanthera* (5 spp.), § *Alicida* (1 sp.), § *Heeringia* (1 sp.), § *Monophyllon* (1 sp.), § *Petermannia* (1 sp.) and § *Tetraphila* (1 sp.; see notes under *B. furfuracea* Hook. f.). There are four species not placed to section.

LAOS

Begonia acetosella Craib var. *acetosella*
Begonia adscendens C.B. Clarke

Begonia brandisiana Kurz
Begonia cladotricha M. Hughes
Begonia cf. *furfuracea* Hook. f.
Begonia handelii Irmsch. var. *handelii*
Begonia handelii var. *prostrata* (Irmsch.)
    Tebbitt
Begonia hymenophylla Gagnep.
Begonia integrifolia Dalzell
Begonia martabanica A. DC.
Begonia modestiflora Kurz
Begonia palmata D. Don
Begonia procridifolia Wall. ex A. DC.
Begonia siamensis Gagnep.

Notes: A flora of 13 species, each known
from very few specimens. It is likely that
Laos has a far richer *Begonia* flora than
this. The species are distributed between
§ *Diploclinium* (3 spp.), § *Parvibegonia* (3 spp.),
§ *Platycentrum* (2 spp.), § *Reichenheimia* (2
spp.), § *Sphenanthera* (2 spp.) and § *Tetraphila*
(1 sp.; see notes under *B. furfuracea* Hook. f.).

VIETNAM

Begonia acetosella Craib. var. *acetosella*
Begonia alveolata T.T. Yu
Begonia anceps Irmsch.
Begonia annulata K. Koch
Begonia balansana Gagnep.
Begonia bataiensis Kiew
Begonia baviensis Gagnep.
Begonia boisiana Gagnep.
Begonia bonii Gagnep.
Begonia brevipedunculata Y.M. Shui
Begonia cathayana Hemsl.
Begonia cavaleriei H. Lév.
Begonia cucphuongensis H.Q. Nguyen &
    Tebbitt
Begonia eberhardtii Gagnep.
Begonia hahiepiana H.Q. Nguyen & Tebbitt
Begonia handelii Irmsch. var. *handelii*
Begonia handelii var. *prostrata* (Irmsch.)
    Tebbitt
Begonia harmandii Gagnep.
Begonia hatacoa Buch.-Ham. ex D. Don
Begonia integrifolia Dalzell
Begonia kui C.-I Peng
Begonia labordei H. Lév.

Begonia langbianensis Baker f.
Begonia longicarpa K.-Y Guan & D.K. Tian
Begonia longifolia Blume
Begonia macrotoma Irmsch.
Begonia palmata D. Don var. *palmata*
Begonia palmata var. *bowringiana* (Benth.)
    Golding & Kareg.
Begonia pedatifida H. Lév.
Begonia phuthoensis H.Q. Nguyen
Begonia pierrei Gagnep.
Begonia porteri var. *macrorhiza* Gagnep.
Begonia sandalifolia C.B. Clarke
Begonia sinuata Wall. ex Meisn.
Begonia sizemoreae Kiew
Begonia tonkinensis Gagnep.
Begonia versicolor Irmsch.
Begonia villifolia Irmsch.

Excluded names:
Begonia chapaensis Irmsch. *nom. nud.*
Begonia doliifera Irmsch. *nom. nud.*
Begonia semicava Irmsch. *nom. nud.*
Begonia specicola Irmsch. *nom. nud.*
    These names are cited by Ho (1999), based
    on unpublished annotations on specimens
    by Irmscher.

Notes: A flora of 36 species, with the largest
section being § *Platycentrum* (13 spp.). The
remaining species are distributed between
§ *Diploclinium* (4 spp.), § *Sphenanthera* (4
spp.), § *Coelocentrum* (3 spp.), § *Reichenheimia*
(3 spp.), § *Leprosae* (2 spp.), § *Parvibegonia* (2
spp.) and § *Petermannia* (1 sp.). There are four
species unplaced to section.

CAMBODIA

Begonia geoffrayi Gagnep.
Begonia hymenophylla Gagnep.
Begonia sinuata Wall. ex Meisn.

Excluded names:
Begonia rupicola Miq. (see notes under that
    species)

Notes: Although this list is undoubtedly
an underestimate, it is unlikely there is
a large number of species waiting to be
recorded in Cambodia, as the majority of

the country lies below 100 m altitude. The three species represent two sections, namely § *Parvibegonia* (2 spp.) and § *Reichenheimia* (1 sp.).

PENINSULAR MALAYSIA

Begonia abdullahpieei Kiew
Begonia aequilateralis Irmsch.
Begonia alpina L.B. Sm. & Wassh.
Begonia barbellata Ridl.
Begonia carnosula Ridl.
Begonia corneri Kiew
Begonia decora Stapf
Begonia eiromischa Ridl.
Begonia elisabethae Kiew
Begonia forbesii King
Begonia foxworthyi Burkill ex Ridl.
Begonia fraseri Kiew
Begonia herveyana King var. *herveyana*
Begonia herveyana var. *barnesii* Irmsch.
Begonia holttumii Irmsch.
Begonia ignorata Irmsch.
Begonia integrifolia Dalzell
Begonia isopteroidea King
Begonia jayaensis Kiew
Begonia jiewhoei Kiew
Begonia kingiana Irmsch.
Begonia klossii Ridl.
Begonia koksunii Kiew
Begonia lengguanii Kiew
Begonia longicaulis Ridl.
Begonia longifolia Blume
Begonia lowiana King
Begonia martabanica var. *pseudoclivalis* Irmsch.
Begonia maxwelliana King
Begonia nurii Irmsch.
Begonia oreophila Kiew
Begonia paupercula King
Begonia pavonina Ridl.
Begonia perakensis King var. *perakensis*
Begonia perakensis var. *conjugens* Irmsch.
Begonia phoeniogramma Ridl.
Begonia praetermissa Kiew
Begonia rajah Ridl.
Begonia reginula Kiew
Begonia rheifolia Irmsch.

Begonia rhoephila Ridl.
Begonia rhyacophila Kiew
Begonia scortechinii King
Begonia sibthorpioides Ridl.
Begonia sinuata Wall. ex Meisn. var. *sinuata*
Begonia sinuata var. *pantiensis* Kiew
Begonia tampinica Burkill ex Irmsch.
Begonia thaipingensis King
Begonia tigrina Kiew
Begonia vallicola Kiew
Begonia variabilis Ridl.
Begonia venusta King
Begonia wrayi Hemsl.
Begonia wyepingiana Kiew
Begonia yappii Ridl.

Notes: A flora of 52 species, the largest section being § *Platycentrum* (23 spp.). The remaining species are distributed between § *Reichenheimia* (10 spp.), § *Parvibegonia* (8 spp.), § *Petermannia* (5 spp.), § *Diploclinium* (2 spp.), § *Ridleyella* (2 spp.), § *Heeringia* (1 sp.) and § *Sphenanthera* (1 sp.).

SUMATRA

Begonia aberrans Irmsch.
Begonia altissima Ridl.
Begonia areolata Miq.
Begonia atricha (Miq.) A. DC.
Begonia axillaris Ridl.
Begonia barbellata Ridl.
Begonia beccariana Ridl.
Begonia bifolia Ridl.
Begonia bracteata Jack
Begonia caespitosa Jack
Begonia coriacea Hassk.
Begonia divaricata Irmsch. forma *divaricata*
Begonia divaricata forma *minor* Irmsch.
Begonia fasciculata Jack
Begonia flexula Ridl.
Begonia goegoensis N.E. Br.
Begonia hasskarliana (Miq.) A. DC.
Begonia holttumii Irmsch.
Begonia horsfieldii Miq. ex A. DC.
Begonia inversa Irmsch.
Begonia inversa forma *nana* Irmsch.
Begonia ionophylla Irmsch.
Begonia isoptera Dryand. ex Sm.

Begonia laevis Ridl.
Begonia lepida Blume
Begonia lepidella Ridl.
Begonia longifolia Blume
Begonia longipedunculata Golding & Kareg.
Begonia mollis A. DC.
Begonia muricata Blume
Begonia orbiculata Jack
Begonia padangensis Irmsch.
Begonia pilosa Jack
Begonia racemosa Jack
Begonia robusta Blume var. *robusta*
Begonia robusta var. *hirsutior* (Miq.) Golding & Kareg.
Begonia sarcocarpa Ridl.
Begonia scottii Tebbitt
Begonia sinuata Wall. ex Meisn.
Begonia stictopoda (Miq.) A. DC.
Begonia sudjanae C.-A. Jansson
Begonia sychnantha L.B. Sm. & Wassh.
Begonia tenericaulis Ridl.
Begonia teysmanniana (Miq.) Tebbitt
Begonia trichopoda Miq.
Begonia trigonocarpa Ridl.
Begonia turbinata Ridl.
Begonia vuijckii Koord.

Excluded names:
Begonia coronata Irmsch. ined.
Begonia forbesii King (see notes under that species)
Begonia gracilicyma Irmsch. ined.
Begonia korthalsiana Miq. ined.
Begonia maxwelliana King (see notes under that species)
Begonia meysselliana Linden
Locality not certain, known only from the protologue.
Begonia stictopoda var. *elongata* Miq. ined.
Begonia villipes Miq. ined.

Notes: A flora of 45 species with § *Petermannia* representing the most (15 spp). The remainder are distributed between § *Reichenheimia* (9 spp.), § *Platycentrum* (6 spp.), § *Sphenanthera* (6 spp.), § *Diploclinium* (3 spp.), § *Bracteibegonia* (3 spp.) and § *Parvibegonia* (1 sp.). Two species are unplaced to section. The § *Petermannia*

element to the flora is likely to contain a considerable number of undescribed species.

JAVA

Begonia areolata Miq.
Begonia atricha (Miq.) A. DC.
Begonia bracteata Jack
Begonia coriacea Hassk.
Begonia erosa Blume
Begonia isoptera Dryand. ex Sm.
Begonia lepida Blume
Begonia lobbii A. DC.
Begonia longifolia Blume
Begonia mollis A. DC.
Begonia multangula Blume var. *multangula*
Begonia multangula var. *glabrata* Miq.
Begonia muricata Blume
Begonia robusta Blume var. *robusta*
Begonia rupicola Miq.
Begonia tenuifolia Dryand.
Begonia vuijckii Koord.
Begonia zollingeriana A. DC.

Excluded names:
Begonia glanduloso-ciliata Irmsch. ined.
Begonia groenewegensis hort. ex K. Koch & G.A. Fintelmann
Of questionable provenance.
Begonia hochbaumii hort. ex E. Otto
Of questionable provenance.
Begonia ignorata Irmsch. (see notes under that species)
Begonia madurensis Irmsch. ined.
Begonia pachyrhachis L.B. Sm. & Wassh. (see notes under that species)
Begonia oligocarpa
Considered by Koorders (1912) to be of uncertain provenance.
Begonia palmata D. Don
Koorders (1912) cites a specimen from east Java, but it is unlikely that this species is native.
Begonia resecta Miq. ined.
Begonia strigulosa A. DC.
Considered by de Candolle (1864) and Koorders (1912) to be of uncertain provenance.

Notes: A relatively small flora of 17 species, 11 of which also occur on Sumatra. This number is unlikely to increase much further. § *Parvibegonia* and § *Reichenheimia* each contain four species, with the remainder being in § *Petermannia* (3 spp.), § *Sphenanthera* (3 spp.), § *Bracteibegonia* (2 spp.) and § *Platycentrum* (1 sp.).

LESSER SUNDA ISLANDS

Begonia brachybotrys Merr. & L.M. Perry
Begonia isoptera Dryand. ex Sm.
Begonia longifolia Blume
Begonia multangula Blume
Begonia muricata Blume
Begonia pseudolateralis Warb.
Begonia robusta Blume
Begonia tenuifolia Dryand.
Begonia timorensis (Miq.) Golding & Kareg.

Excluded names:
Begonia sumbaensis Irmsch. ined.

Notes: A flora of nine species. With the exception of *B. timorensis* (though see notes under that species), all of the species listed here are quite widespread and mostly represent an attenuated version of the Javanese flora. See also notes under *B. rieckei* Warb. The sections represented are § *Petermannia* (4 spp.), § *Sphenanthera* (3 spp.), § *Reichenheimia* (1 sp.) and § *Parvibegonia* (1 sp.).

BORNEO

Begonia adenodes Irmsch.
Begonia adenostegia Stapf
Begonia amphioxus Sands
Begonia angustilimba Merr.
Begonia anthonyi Kiew
Begonia articulata Irmsch.
Begonia artior Irmsch.
Begonia awongii Sands
Begonia bahakensis Sands
Begonia baramensis Merr.
Begonia baturongensis Kiew
Begonia beccarii Warb.

Begonia berhamanii Kiew
Begonia beryllae Ridl.
Begonia borneensis A. DC.
Begonia bruneiana Sands subsp. *bruneiana*
Begonia bruneiana subsp. *angustifolia* Sands
Begonia bruneiana subsp. *labiensis* Sands
Begonia bruneiana subsp. *retakensis* Sands
Begonia burbidgei Stapf
Begonia calcarea Ridl.
Begonia cauliflora Sands
Begonia chlorandra Sands
Begonia chlorocarpa Irmsch. ex Sands
Begonia chlorosticta Sands
Begonia chongii Sands
Begonia cincinnifera Irmsch.
Begonia cognata Irmsch.
Begonia congesta Ridl.
Begonia conipila Irmsch. ex Kiew
Begonia consanguinea Merr.
Begonia cyanescens Sands
Begonia densiretis Irmsch.
Begonia diwolii Kiew
Begonia elatostemma Ridl.
Begonia erythrogyna Sands
Begonia eutricha Sands
Begonia fuscisetosa Sands
Begonia gibbsiae Irmsch. ex Sands
Begonia gomantongensis Kiew
Begonia gueritziana Gibbs
Begonia havilandii Ridl.
Begonia heliostrophe Kiew
Begonia hexaptera Sands
Begonia hullettii Ridl.
Begonia humericola Sands
Begonia imbricata Sands
Begonia inostegia Stapf
Begonia kachak K.G. Pearce
Begonia kasutensis K.G. Pearce
Begonia keeana Kiew
Begonia keithii Kiew
Begonia kinabaluensis Sands
Begonia laccophora Sands
Begonia lailana Kiew & Geri
Begonia lambii Kiew
Begonia layang-layang Kiew
Begonia lazat Kiew & Reza Azmi
Begonia leucochlora Sands
Begonia leucotricha Sands

Begonia longiseta Irmsch.
Begonia lunatistyla Irmsch.
Begonia madaiensis Kiew
Begonia malachosticta Sands
Begonia mamutensis Sands
Begonia melikopia Kiew
Begonia microptera Hook.
Begonia minutiflora Sands
Begonia murudensis Merr.
Begonia niahensis K.G. Pearce
Begonia oblongifolia Stapf
Begonia papyraptera Sands
Begonia pendula Ridl.
Begonia pleioclada Irmsch.
Begonia postarii Kiew
Begonia promethea Ridl.
Begonia propinqua Ridl.
Begonia pryeriana Ridl.
Begonia pubescens Ridl.
Begonia punbatuensis Kiew
Begonia pyrrha Ridl.
Begonia rubida Ridl.
Begonia sabahensis Kiew & J.H. Tan
Begonia sarawakensis Ridl.
Begonia sibutensis Sands
Begonia speluncae Ridl.
Begonia stenogyna Sands
Begonia stichochaete K.G. Pearce
Begonia subisensis K.G. Pearce
Begonia subnummularifolia Merr.
Begonia sympodialis Irmsch.
Begonia tambelanensis (Irmsch.) Kiew
Begonia tawaensis Merr.
Begonia temburongensis Sands
Begonia urunensis Kiew
Begonia vaccinioides Sands
Begonia walteriana Irmsch.
Begonia xiphophylla Irmsch.

Excluded names:
Begonia aphanantha Miq. ined.
Begonia cuspidata Miq. ined.
Begonia decumbens Irmsch. ined.
Begonia densinervis Irmsch. ined.
Begonia diadema Linden ex Rodigas
    A horticultural name validated by
    Rodigas, stated to come from Borneo but
    of doubtful provenance. May even be a
    hybrid.

Begonia latiovata Irmsch. ined.
Begonia mollis A. DC.
    A record from de Candolle (1864), but
    stated to be 'ex. syn. dub. Miq.'.
Begonia transparens Miq. ined.

Notes: A flora of 95 species, the vast majority
of which are in § *Petermannia* (83 spp.). The
remainder are either in § *Diploclinium* (6 spp.),
§ *Platycentrum* (1 sp., but see notes under
B. adenostegia), § *Reichenheimia* (2 spp.),
§ *Sphenanthera* (1 sp.) or unplaced to section
(2 spp.). Only five of the species listed are
recorded from Kalimantan, which is certainly
a vast underestimate.

PHILIPPINES

Begonia acuminatissima Merr.
Begonia aequata A. Gray
Begonia affinis Merr.
Begonia agusanensis Merr.
Begonia alba Merr.
Begonia alvarezii Merr.
Begonia angilogensis Merr.
Begonia anisoptera Merr.
Begonia apayaoensis Merr.
Begonia biliranensis Merr.
Begonia binuangensis Merr.
Begonia bolsteri Merr.
Begonia brevipes Merr.
Begonia calcicola Merr.
Begonia casiguranensis Quisumb. & Merr.
Begonia castilloi Merr.
Begonia caudata Merr.
Begonia chloroneura P. Wilkie & Sands
Begonia ciliifera Merr.
Begonia collisiae Merr.
Begonia colorata Warb.
Begonia contracta Warb.
Begonia copelandii Merr.
Begonia coronensis Merr.
Begonia crispipila Elmer
Begonia cumingiana (Klotzsch) A. DC.
Begonia cumingii A. Gray
Begonia dolichotricha Merr.
Begonia edanoi Merr.
Begonia elatostematoides Merr.
Begonia elmeri Merr.

Begonia esculenta Merr.
Begonia everettii Merr.
Begonia fasciculiflora Merr.
Begonia fenicis Merr.
Begonia gitingensis Elmer
Begonia gracilipes Merr.
Begonia halconensis Merr.
Begonia hernandioides Merr.
Begonia incisa A. DC.
Begonia isabelensis Quisumb. & Merr.
Begonia jagorii Warb.
Begonia klemmei Merr.
Begonia lacera Merr.
Begonia lagunensis Elmer
Begonia lancifolia Merr.
Begonia lancilimba Merr.
Begonia latistipula Merr.
Begonia leptantha C.B. Rob.
Begonia leucosticta Warb.
Begonia littleri Merr.
Begonia loheri Merr.
Begonia longibractea Merr.
Begonia longinoda Merr.
Begonia longiscapa Warb.
Begonia longistipula Merr.
Begonia longovillosa A. DC.
Begonia luzonensis Warb.
Begonia macgregorii Merr.
Begonia malindangensis Merr.
Begonia manillensis A. DC.
Begonia mearnsii Merr.
Begonia megacarpa Merr.
Begonia megalantha Merr.
Begonia merrittii Merr.
Begonia mindanaensis Warb.
Begonia mindorensis Merr.
Begonia negrosensis Elmer
Begonia neopurpurea L.B. Sm. & Wassh. *nom. nud.*
Begonia nigritarum (Kamel) Steud.
Begonia oblongata Merr.
Begonia obtusifolia Merr.
Begonia oligantha Merr.
Begonia oxysperma A. DC.
Begonia palawanensis Merr.
Begonia panayensis Merr.
Begonia parva Merr.
Begonia parvilimba Merr.

Begonia perryae L.B. Sm. & Wassh.
Begonia pinamalayensis Merr.
Begonia platyphylla Merr.
Begonia polilloensis Tebbitt
Begonia pseudolateralis Warb.
Begonia quercifolia A. DC.
Begonia ramosii Merr.
Begonia rizalensis Merr.
Begonia rubrifolia Merr.
Begonia rufipila Merr.
Begonia samarensis Merr.
Begonia sarmentosa L.B. Sm. & Wassh.
Begonia serpens Merr.
Begonia sorsogonensis (Merr.) Elmer *nom. nud.*
Begonia suborbiculata Merr.
Begonia subprostrata Merr.
Begonia subtruncata Merr.
Begonia tayabensis Merr.
Begonia trichocheila Warb.
Begonia urdanetensis Elmer
Begonia vanoverberghii Merr.
Begonia wadei Merr. & Quisumb.
Begonia weberi Merr.
Begonia wenzelii Merr.
Begonia woodii Merr.
Begonia zamboangensis Merr.

Excluded names:
Begonia camarinensis Merr. ined.
Begonia edulis Merr. ined.
Begonia lancibracteata Merr. ined.
Begonia longicapsularis A. DC. ined.

Notes: A flora of 104 species, all except three of which are placed in two sections, § *Petermannia* (63 spp.) and § *Diploclinium* (38 spp.). Of the remaining three, two are unplaced and one is in the endemic § *Baryandra*. In the Philippines, § *Petermannia* encompasses relatively less morphological variation than on the neighbouring islands of Borneo and Sulawesi. The species in § *Diploclinium* are also more morphologically similar to each other than to others on the Asian mainland; the fact that this section is scarcely represented on the other nearby Malesian islands possibly indicates a largely endemic radiation. *Begonia neopurpurea* and *B. sorsogonensis* are included here for completeness, but strictly speaking these species need validating with a Latin diagnosis.

SULAWESI

Begonia aptera Blume
Begonia bonthainensis Hemsl.
Begonia capituliformis Irmsch.
Begonia carnosa (Teijsm. & Binn.) Teijsm. & Binn.
Begonia celebica Irmsch.
Begonia chiasmogyna M. Hughes
Begonia cuneatifolia Irmsch.
Begonia flacca Irmsch.
Begonia gemella Warb. ex L.B. Sm. & Wassh.
Begonia grandipetala Irmsch.
Begonia heteroclinis Miq. ex Koord.
Begonia hispidissima Zipp. ex Koord.
Begonia humilicaulis Irmsch.
Begonia imperfecta Irmsch.
Begonia insularum Irmsch.
Begonia koordersii Warb. ex L.B. Sm. & Wassh.
Begonia longifolia Blume
Begonia macintyreana M. Hughes
Begonia masarangensis Irmsch.
Begonia mendumae M. Hughes
Begonia pseudolateralis Warb.
Begonia rachmatii Tebbitt
Begonia renifolia Irmsch.
Begonia rieckei Warb.
Begonia sarasinorum Irmsch.
Begonia siccacaudata J. Door.
Begonia sphenocarpa Irmsch.
Begonia stevei M. Hughes
Begonia strachwitzii Warb. ex Irmsch.
Begonia strictipetiolaris Irmsch.

Excluded names:
Begonia cincinnosa Irmsch. ined.
Begonia isoptera Dryand. ex Sm. (see notes under that species)
Begonia lamii Irmsch. ined.
Begonia mollis A. DC.
From a Beccari specimen in FI (HB4499), determined by Irmscher. However, the specimen has no flowers or fruit, and whether this Javanese species really occurs on Sulawesi needs further investigation.

Begonia robusta Blume (see notes under that species)
Begonia tenuifolia Dryand.
A few specimens from Sulawesi are determined as 'aff. *B. tenuifolia*', but it would seem they are not in § *Parvibegonia*.

Notes: A flora of 30 species, dominated by § *Petermannia* (27 spp.), with the remaining three being in § *Sphenanthera*. It is likely that a complete account of Sulawesi *Begonia* will more than double this total.

MOLUCCAS

Begonia brachybotrys Merr. & L.M. Perry
Begonia holosericea (Teijsm. & Binn.) Teijsm. & Binn.
Begonia longifolia Blume
Begonia pseudolateralis Warb.
Begonia rieckei Warb.

Excluded names:
Begonia isoptera Dryand. ex Sm. (see notes under that species)
Begonia mollis A. DC. (see notes under that species)
Begonia muricata Blume (see notes under that species)

Notes: Although most of the species listed here are quite widespread, it is evident from herbarium collections that a number of endemic species remain to be described. Three of the species are in § *Petermannia* (see notes under *B. rieckei* Warb.), one is in § *Sphenanthera* and one is unplaced.

NEW GUINEA

Begonia acaulis Merr. & L.M. Perry
Begonia albobracteata Ridl.
Begonia archboldiana Merr. & L.M. Perry
Begonia arfakensis (Gibbs) L.L. Forrest & Hollingsw.
Begonia argenteomarginata Tebbitt
Begonia augustae Irmsch.
Begonia axillipara Ridl.
Begonia bartlettiana Merr. & L.M. Perry

Begonia bipinnatifida J.J. Sm.

Begonia brachybotrys Merr. & L.M. Perry

Begonia brassii Merr. & L.M. Perry

Begonia brevirimosa Irmsch. subsp.
brevirimosa

Begonia brevirimosa subsp. *exotica* Tebbitt

Begonia calliantha Merr. & L.M. Perry

Begonia clemensiae Merr. & L.M. Perry

Begonia diffusiflora Merr. & L.M. Perry

Begonia djamuensis Irmsch.

Begonia dosedlae Gilli

Begonia eliasii Warb.

Begonia filibracteosa Irmsch.

Begonia flexicaulis Ridl.

Begonia fruticella Ridl.

Begonia fulvo-villosa Warb.

Begonia gilgiana Irmsch.

Begonia glabricaulis Irmsch. var. *glabricaulis*

Begonia glabricaulis var. *brachyphylla* Irmsch.

Begonia hirsuticaulis Irmsch.

Begonia humboldtiana Gibbs

Begonia kaniensis Irmsch.

Begonia kelliana Irmsch.

Begonia kerstingii Irmsch.

Begonia lauterbachii Warb.

Begonia ledermannii Irmsch.

Begonia malmquistiana Irmsch.

Begonia media Merr. & L.M. Perry

Begonia minjemensis Irmsch.

Begonia monantha Warb.

Begonia montis-bismarckii Warb.

Begonia mooreana (Irmsch.) L.L. Forrest &
Hollingsw.

Begonia moszkowskii Irmsch.

Begonia multidentata Warb.

Begonia mystacina L.B. Sm. & Wassh.

Begonia naumoniensis Irmsch.

Begonia novoguineensis Merr. & L.M. Perry

Begonia oligandra Merr. & L.M. Perry

Begonia otophora Merr. & L.M. Perry

Begonia oxyura Merr. & L.M. Perry

Begonia papuana Warb.

Begonia pediophylla Merr. & L.M. Perry

Begonia pentaphragmifolia Ridl.

Begonia physandra Merr. & L.M. Perry

Begonia pinnatifida Merr. & L.M. Perry

Begonia pulchra (Ridl.) L.L. Forrest & Hollingsw.

Begonia randiana Merr. & L.M. Perry

Begonia rhodantha Ridl.

Begonia rieckei Warb.

Begonia serraticauda Merr. & L.M. Perry

Begonia serratipetala Irmsch.

Begonia sharpeana F. Muell.

Begonia simulans Merr. & L.M. Perry

Begonia sogerensis Ridl.

Begonia spilotophylla F. Muell.

Begonia stilandra Merr. & L.M. Perry

Begonia strictinervis Irmsch.

Begonia strigosa (Warb.) L.L. Forrest &
Hollingsw.

Begonia subcyclophylla Irmsch.

Begonia subelliptica Merr. & L.M. Perry

Begonia suffrutescens Merr. & L.M. Perry

Begonia symbeccarii L.L. Forrest & Hollingsw.

Begonia symbracteosa L.L. Forrest & Hollingsw.

Begonia symgeraniifolia L.L. Forrest &
Hollingsw.

Begonia symhirta L.L. Forrest & Hollingsw.

Begonia sympapuana L.L. Forrest & Hollingsw.

Begonia symparvifolia L.L. Forrest & Hollingsw.

Begonia symsanguinea L.L. Forrest & Hollingsw.

Begonia tafaensis Merr. & L.M. Perry

Begonia torricellensis Warb.

Begonia vanderwateri Ridl.

Begonia warburgii K. Schum. & Lauterb.

Begonia wariana Irmsch.

Begonia weigallii Hemsl.

Excluded names:

Begonia hombronensis A. DC. ined.

Begonia keysseri Irmsch. ined.

Begonia lauterbachii forma *monopoda* Irmsch.
nom. inval.
No holotype designated, necessary for a
valid publication in 1959.

Notes: A flora of 79 species, dominated by
§ *Petermannia* (55 spp.) and the endemic
§ *Symbegonia* (13 spp.). The remaining
species are either in § *Diploclinium* (7 spp.) or
unplaced to section (4 spp.).

BISMARCK ARCHIPELAGO

Begonia brachybotrys Merr. & L.M. Perry

Begonia brevirimosa Irmsch. subsp.
brevirimosa

Begonia peekelii Irmsch.
Begonia rieckei Warb.
Begonia weigallii Hemsl.

Excluded names:
Begonia lacustris Irmsch. ined.
  Although this name was published by
  Peekel (1984), no Latin description was
  given and no type was cited.
Begonia lamekotensis Irmsch. ined.

Notes: All these species are in § *Petermannia*.
See notes under *B. rieckei* Warb.

SOLOMON ISLANDS

Begonia salomonensis Merr. & L.M. Perry
Begonia somervillei Hemsl.
Begonia weigallii Hemsl.

Notes: All these species are in § *Petermannia*.

VANUATU AND FIJI

Begonia vitiensis A.C. Sm.

Notes: Why the only native (and endemic)
species in Vanuatu and Fiji should belong
to § *Diploclinium* and not § *Petermannia*
which dominates the neighbouring floras is
a mystery.

References

BADCOCK, Z. (1998). *A phylogenetic investigation of* Begonia *section* Knesebeckia *(Klotzsch) A. D.C.* PhD thesis, University of Glasgow.

BRIDSON, G. D. R. & SMITH, E. R. (1991). *Botanico-Periodicum-Huntianum/Supplementum*. Hunt Institute for Botanical Documentation, Pittsburgh.

BRUMMITT, R. K. (2001). *World Geographical Scheme for Recording Plant Distributions.* Hunt Institute for Botanical Documentation, Pittsburgh.

BRUMMITT, R. K. & POWELL, C. E. (eds.) (1992). *Authors of Plant Names.* Royal Botanic Gardens, Kew.

CANDOLLE, A. DE (1864). Begoniaceae. *Prodr.* 15: 266–408.

COLLAR, N. J., ALDRIN, N., MALLARI, D. & TABARANZA, B. R. JR. III. (1999). *Threatened birds of the Philippines: the Haribon Foundation/BirdLife International Red Data Book.* Bookmark, Inc., Makati City, Philippines.

COODE, M. J. E. *et al.* (eds.) (1997). *Checklist of the Flowering Plants and Gymnosperms of Brunei Darussalam* App. 2.

DOORENBOS, J., SOSEF, M. S. M. & DE WILDE, J. J. F. E. (1998). The sections of *Begonia* including descriptions, keys and species lists (Studies in Begoniaceae VI). *Wageningen Agricultural University Papers* 98(2): 1–266.

FUCHS, H. P. (1958). Nomenclature proposals for the Montreal Congress. *Taxon* 7(8): 218–223.

GOLDING, J. (1970). *Begonia* nomenclature notes, 2. The *Begonia* in Wallich's numerical list. *Phytologia* 40(1): 7–20.

HO, P. H. (1999). Begoniaceae. *Cay Co Viet Nam* 1: 577–588. Tre Publishers, Vietnam.

HUGHES, M. (2006). Four new species of *Begonia* (*Begoniaceae*) from Sulawesi. *Edinburgh Journal of Botany* 63(2&3): 191–199.

HUGHES, M. (2007). *Begonia cladotricha* (*Begoniaceae*): a new species from Laos. *Edinburgh Journal of Botany* 64(1): 101–105.

KIEW, R. (2004). *Begonia sizemoreae* Kiew (Begoniaceae), a handsome new begonia from Vietnam. *Gardens' Bulletin Singapore* 56: 95–100.

KIEW, R. (2005). *Begonias of Peninsular Malaysia.* Natural History Publications (Borneo).

KOORDERS, S. H. (1912). *Exkursionsflora von Java* 2: 640–653.

KRESS, W. J., DE FILIPPS, R. A., FARR, E. & KYI, D. Y. Y. (2003). A checklist of the trees, shrubs, herbs and climbers of Myanmar. *Contributions from the United States National Herbarium, Smithsonian Institution* 45: 1–590.

MERRILL, E. D. (1923). *An enumeration of Philippine flowering plants (Begoniaceae)* 3: 119–129. Bureau of Science, Manila.

MOAT, J. (2007). *Conservation assessment tools extension for ArcView 3.x, version 1.0.* GIS Unit, Royal Botanic Gardens, Kew.

PEEKEL, P. G. (1984). *Flora of the Bismarck Archipelago for Naturalists.* Office of Forests, Division of Botany, Lae, Papua New Guinea.

PROCTOR, J. & ARGENT, G. (1998). The submontane forest of Mt. Giting Giting, Sibuyan Island, Philippines. *Edinburgh Journal of Botany* 55(2): 295–316.

SANDS, M. J. S. (2001). Begoniaceae in the Flora Malesiana region. In: SAW, L. G., CHUA, L. S. L. & KHOO, K. C. (eds.), *Taxonomy, the cornerstone of biodiversity. Proceedings of the fourth international Flora Malesiana symposium*. Forest Research Institute Malaysia. Pp. 161–168.

SHUI, Y-M. (2006). A new species of *Begonia* section *Platycentrum* (Begoniaceae) from Vietnam. *Novon* 16: 269–271.

SHUI, Y-M., PENG, C.-I & WU, C-Y. (2002). Synopsis of the Chinese species of *Begonia* (Begoniaceae), with a reappraisal of sectional delimitation. *Botanical Bulletin of Academia Sinica* 43: 313–327.

STANDARDS AND PETITIONS WORKING GROUP. (2006). *Guidelines for Using the IUCN Red List Categories and Criteria. Version 6.1.* Prepared by the Standards and Petitions Working Group of the IUCN SSC Biodiversity Assessments Sub-Committee.

TEBBITT, M. C. (1997). *A systematic investigation of* Begonia *section* Sphenanthera *(Hassk.) Benth. & Hook. f.* PhD thesis, University of Glasgow.

TEBBITT, M. C. (2003a). Notes on South Asian *Begonia* (Begoniaceae). *Edinburgh Journal of Botany* 60(1): 1–9.

TEBBITT, M. C. (2003b). Taxonomy of *Begonia longifolia* Blume (Begoniaceae) and related species. *Brittonia* 55(1): 19–29.

TEBBITT, M. C. (2007). *Begonia kingdon-wardii* (Begoniaceae), a new species from Myanmar. *Kew Bulletin* 62: 143–146.

TEBBITT, M. C. & DICKSON, J. H. (2000). Amended descriptions and revised sectional assignment of some Asian begonias (Begoniaceae). *Brittonia* 52(1): 112–117.

TEBBITT, M. C., FORREST, L. L., SANTORIELLO, A., CLEMENT, W. L. & SWENSEN, S. M. (2006). Phylogenetic relationships of Asian *Begonia*, with an emphasis on the evolution of rain-ballist and animal dispersal mechanisms in sections *Platycentrum*, *Sphenanthera* and *Leprosae*. *Systematic Botany* 31(2): 327–336.

WIKRAMANAYAKE, E., DINERSTEIN, E., LOUCKS, C., OLSON, D., MORRISON, J., LAMOREUX, J., MCKNIGHT, M. & HEDAO, P. (2001). *Terrestrial Ecoregions of the Indo-Pacific: A Conservation Assessment*. Island Press, Washington, DC.

WILLIS, F., MOAT, J. & PATON, A. (2003). Defining a role for herbarium data in Red List assessments: a case study of *Plectranthus* from eastern and southern tropical Africa. *Biodiversity and Conservation* 12(7): 1537–1552.

Index of Names

Accepted names are in bold font. Homotypic synonyms are indicated by '≡'; heterotypic ones by '='.

Acetosa nigritarum Kamel
  ≡ **Begonia nigritarum (Kamel) Steud.**
Begonia abdullahpieei Kiew
Begonia aberrans Irmsch.
Begonia acaulis Merr. & L.M. Perry
Begonia aceroides Irmsch.
Begonia acetosella Craib
Begonia acetosella var. hirtifolia Irmsch.
Begonia acuminatissima Merr.
Begonia adenodes Irmsch.
Begonia adenopoda Lem.
Begonia adenostegia Stapf
Begonia adscendens C.B. Clarke
Begonia aequata A. Gray
Begonia aequilateralis Irmsch.
Begonia affinis Merr.
Begonia agusanensis Merr.
Begonia alba Merr.
Begonia albobracteata Ridl.
Begonia alicida C.B. Clarke
Begonia alpina L.B. Sm. & Wassh.
Begonia altissima Ridl.
Begonia alvarezii Merr.
Begonia alveolata T.T. Yu
Begonia amphioxus Sands
Begonia anceps Irmsch.
Begonia angilogensis Merr.
Begonia angustifolia Blume
  = **Begonia isoptera Dryand. ex Sm.**
Begonia angustilimba Merr.
Begonia anisoptera Merr.
Begonia annulata K. Koch
Begonia anthonyi Kiew
Begonia apayaoensis Merr.
Begonia aptera Blume
Begonia aptera Decne.
  = **Begonia timorensis (Miq.) Golding & Kareg.**
Begonia aptera var. calleryana (A. DC.) Fern.-Vill.
  = **Begonia pseudolateralis Warb.**
Begonia archboldiana Merr. & L.M. Perry
Begonia areolata Miq.
Begonia areolata var. minor Miq.
Begonia arfakensis (Gibbs) L.L. Forrest & Hollingsw.
Begonia argenteomarginata Tebbitt
Begonia articulata Irmsch.
Begonia artior Irmsch.
Begonia atricha (Miq.) A. DC.
Begonia augustae Irmsch.
Begonia awongii Sands
Begonia axillaris Ridl.
Begonia axillipara Ridl.
Begonia bahakensis Sands
Begonia bakeri Elmer nom. nud.
  = **Begonia luzonensis Warb.**
Begonia balansana Gagnep.
Begonia baramensis Merr.
Begonia barbellata Ridl.
Begonia bartlettiana Merr. & L.M. Perry
Begonia bataiensis Kiew
Begonia baturongensis Kiew
Begonia baviensis Gagnep.
Begonia beccariana Ridl.
Begonia beccarii Warb.
Begonia bellii H. Lév.
  = **Begonia porteri H. Lév. & Vaniot**
Begonia berhamanii Kiew
Begonia beryllae Ridl.
Begonia bifolia Ridl.
Begonia biliranensis Merr.
Begonia bilocularis (Wight) Craib
  = **Begonia sinuata Wall. ex Meisn.**
Begonia binuangensis Merr.
Begonia bipinnatifida J.J. Sm.

Begonia boisiana Gagnep.
Begonia bolsteri Merr.
Begonia bombycina Blume
   = **Begonia isoptera Dryand. ex Sm.**
Begonia bonii Gagnep.
Begonia bonthainensis Hemsl.
Begonia borneensis A. DC.
Begonia bowringiana Champ. ex Benth.
   = **Begonia palmata var. bowringiana**
     (Champ. ex Benth.) Golding & Kareg.
Begonia brachybotrys Merr. & L.M. Perry
Begonia brachyptera Merr. & L.M. Perry
   = **Begonia brachybotrys Merr. & L.M.**
     Perry
Begonia bracteata Jack
Begonia bracteata var. gedeana A. DC.
Begonia brandisiana Kurz
Begonia brassii Merr. & L.M. Perry
Begonia brevipedunculata Y.M. Shui
Begonia brevipes Merr.
Begonia brevirimosa Irmsch.
Begonia brevirimosa subsp. exotica Tebbitt
Begonia bruneiana Sands
Begonia bruneiana subsp. angustifolia
   Sands
Begonia bruneiana subsp. labiensis Sands
Begonia bruneiana subsp. retakensis Sands
Begonia bulusanensis Elmer ex Merr. nom. nud.
   ≡ **Begonia binuangensis Merr.**
Begonia burbidgei Stapf
Begonia burkillii Dunn
Begonia burkillii Irmsch.
   ≡ **Begonia aceroides Irmsch.**
Begonia burmensis L.B. Sm. & Wassh.
Begonia caespitosa Jack
Begonia calcarea Ridl.
Begonia calcicola Merr.
Begonia calliantha Merr. & L.M. Perry
Begonia camiguinensis Elmer
   = **Begonia acuminatissima Merr.**
Begonia capensis Blanco
   = **Begonia nigritarum (Kamel) Steud.**
Begonia capituliformis Irmsch.
Begonia cardiophora Irmsch.
Begonia carnosa (Teijsm. & Binn.) Teijsm. &
   Binn.
Begonia carnosula Ridl.
Begonia casiguranensis Quisumb. & Merr.
Begonia castilloi Merr.
Begonia cathayana Hemsl.
Begonia cathcartii Hook. f.

Begonia caudata Merr.
Begonia cauliflora Sands
Begonia cavaleriei H. Lév.
Begonia celebica Irmsch.
Begonia chiasmogyna M. Hughes
Begonia chlorandra Sands
Begonia chlorocarpa Irmsch. ex Sands
Begonia chloroneura P. Wilkie & Sands
Begonia chlorosticta Sands
Begonia chongii Sands
Begonia ciliifera Merr.
Begonia cincinnifera Irmsch.
Begonia cladotricha M. Hughes
Begonia clemensiae Merr. & L.M. Perry
Begonia clivalis Ridl.
   = **Begonia sinuata Wall. ex Meisn.**
Begonia cognata Irmsch.
Begonia collina Irmsch.
   = **Begonia rhoephila Ridl.**
Begonia collisiae Merr.
Begonia colorata Warb.
Begonia congesta Ridl.
Begonia conipila Irmsch. ex Kiew
Begonia consanguinea Merr.
Begonia contracta Warb.
Begonia copelandii Merr.
Begonia coriacea Hassk.
Begonia corneri Kiew
Begonia coronensis Merr.
Begonia crassicaulis (A. DC.) Warb.
   = **Begonia pachyrhachis L.B. Sm. & Wassh.**
Begonia crassirostris Irmsch.
   = **Begonia longifolia Blume**
Begonia crenata Dryand.
Begonia crispipila Elmer
Begonia cristata Warb. ex L.B. Sm. & Wassh.
   = **Begonia aptera Blume**
Begonia cucphuongensis H.Q. Nguyen &
   Tebbitt
Begonia cumingiana (Klotzsch ex Klotzsch)
   A. DC.
Begonia cumingii A. Gray
Begonia cuneatifolia Irmsch.
Begonia curtisii Ridl.
   = **Begonia integrifolia Dalzell**
Begonia cyanescens Sands
Begonia debilis King
   = **Begonia integrifolia Dalzell**
Begonia debilis var. punicea Craib
   = **Begonia integrifolia Dalzell**
Begonia decaisneana Gagnep.

Begonia hatacoa Buch.-Ham. ex D. Don
Begonia havilandii Ridl.
Begonia heliostrophe Kiew
Begonia hemsleyana Hook. f.
Begonia hernandifolia Hook.
  = **Begonia coriacea Hassk.**
Begonia hernandioides Merr.
Begonia herveyana King
Begonia herveyana var. barnesii Irmsch.
Begonia herveyana var. robusta Ridl.
  = **Begonia rheifolia Irmsch.**
Begonia heteroclinis Miq. ex Koord.
Begonia hexaptera Sands
Begonia hirsuticaulis Irmsch.
Begonia hispidissima Zipp. ex Koord.
Begonia holosericea (Teijsm. & Binn.)
    Teijsm. & Binn.
Begonia holttumii Irmsch.
Begonia horsfieldii Miq. ex A. DC.
Begonia hullettii Ridl.
Begonia humboldtiana Gibbs
Begonia humericola Sands
Begonia humilicaulis Irmsch.
Begonia hymenophylla Gagnep.
Begonia hymenophylloides Kingdon-Ward
    ex L.B. Sm. & Wassh.
Begonia ignorata Irmsch.
Begonia imbricata Sands
Begonia imperfecta Irmsch.
Begonia incerta Craib
Begonia incisa A. DC.
Begonia incondita Craib
Begonia inflata C.B. Clarke
  = **Begonia longifolia Blume**
Begonia inostegia Stapf
Begonia insularum Irmsch.
Begonia integrifolia Dalzell
Begonia integrifolia var. guttata (Wall. ex A. DC.)
    Gagnep.
  = **Begonia integrifolia Dalzell**
Begonia intermixta Irmsch.
Begonia inversa Irmsch.
Begonia inversa forma nana Irmsch.
Begonia ionophylla Irmsch.
Begonia iridescens Dunn
Begonia isabelensis Quisumb. & Merr.
Begonia isoptera Dryand. ex Sm.
Begonia isopteroidea King
Begonia jagorii Warb.
Begonia jayaensis Kiew
Begonia jiewhoei Kiew

Begonia josephii A. DC.
Begonia junghuhniana Miq.
  = **Begonia coriacea Hassk.**
Begonia junghuhniana forma acutifolia Miq. ex
    Koord.
  = **Begonia coriacea Hassk.**
Begonia kachak K.G. Pearce
Begonia kaniensis Irmsch.
Begonia kasutensis K.G. Pearce
Begonia keeana Kiew
Begonia keithii Kiew
Begonia kelliana Irmsch.
Begonia kerrii Craib
Begonia kerstingii Irmsch.
Begonia kinabaluensis Sands
Begonia kingdon-wardii Tebbitt
Begonia kingiana Irmsch.
Begonia klemmei Merr.
Begonia klossii Ridl.
Begonia koksunii Kiew
Begonia koordersii Warb. ex L.B. Sm. &
    Wassh.
Begonia kui C.-I Peng
Begonia kunstleriana King
  = **Begonia scortechinii King**
Begonia labordei H. Lév.
Begonia laccophora Sands
Begonia lacera Merr.
Begonia laciniata Roxb.
  = **Begonia palmata D. Don**
Begonia laciniata subsp. flaviflora Irmsch.
  = **Begonia flaviflora var. vivida (Irmsch.)
    Golding & Kareg.**
Begonia laciniata var. pilosa Craib
  = **Begonia palmata D. Don**
Begonia laevis Ridl.
Begonia lagunensis Elmer
Begonia lailana Kiew & Geri
Begonia lambii Kiew
Begonia lancifolia Merr.
Begonia lancilimba Merr.
Begonia langbianensis Baker f.
Begonia latistipula Merr.
Begonia lauterbachii Warb.
Begonia layang-layang Kiew
Begonia lazat Kiew & Reza Azmi
Begonia lecomtei Gagnep.
  = **Begonia handelii Irmsch.**
Begonia ledermannii Irmsch.
Begonia lengguanii Kiew
Begonia lepida Blume

≡ **Begonia oreophila Kiew**
Begonia nurii Irmsch.
Begonia oblongata Merr.
Begonia oblongifolia Stapf
Begonia obovoidea Craib
Begonia obtusifolia Merr.
Begonia oligandra Merr. & L.M. Perry
Begonia oligantha Merr.
Begonia orbiculata Jack
Begonia oreodoxa Chun & F. Chun ex C.Y. Wu & T.C. Ku
Begonia oreophila Kiew
Begonia otophora Merr. & L.M. Perry
Begonia oxysperma A. DC.
Begonia oxyura Merr. & L.M. Perry
Begonia pachyrhachis L.B. Sm. & Wassh.
Begonia padangensis Irmsch.
Begonia palawanensis Merr.
Begonia paleacea Kurz
Begonia palmata D. Don
Begonia palmata var. bowringiana (Champ. ex Benth.) Golding & Kareg.
Begonia panayensis Merr.
Begonia papuana Warb.
Begonia papyraptera Sands
Begonia parishii C.B. Clarke
Begonia parva Merr.
Begonia parvilimba Merr.
Begonia parvuliflora A. DC.
Begonia parvuliflora var. pubescens A. DC.
Begonia paupercula King
Begonia pavonina Ridl.
Begonia pedatifida H. Lév.
Begonia pediophylla Merr. & L.M. Perry
Begonia peekelii Irmsch.
Begonia peltata Elmer
   ≡ **Begonia elmeri Merr.**
Begonia peltata Hassk.
   ≡ **Begonia coriacea Hassk.**
Begonia pendula Ridl.
Begonia peninsulae Irmsch.
   = **Begonia rajah Ridl.**
Begonia peninsulae subsp. tambelanensis Irmsch.
   ≡ **Begonia tambelanensis (Irmsch.) Kiew**
Begonia pentaphragmifolia Ridl.
Begonia perakensis King
Begonia perakensis var. conjugens Irmsch.
Begonia perakensis var. rotunda Irmsch.
   = **Begonia perakensis var. perakensis**

Begonia perryae L.B. Sm. & Wassh.
Begonia philippinensis A. DC.
   ≡ **Begonia cumingii A. Gray**
Begonia phoeniogramma Ridl.
Begonia phuthoensis H.Q. Nguyen
Begonia physandra Merr. & L.M. Perry
Begonia picta Sm.
Begonia pierrei Gagnep.
Begonia pilosa Jack
Begonia pinamalayensis Merr.
Begonia pingbiensis C.Y. Wu
   = **Begonia alveolata T.T. Yu**
Begonia pinnatifida Merr. & L.M. Perry
Begonia platyphylla Merr.
Begonia pleioclada Irmsch.
Begonia polilloensis Tebbitt
Begonia polyantha Ridl.
   ≡ **Begonia sychnantha L.B. Sm. & Wassh.**
Begonia polygonoides Ridl.
   ≡ **Begonia xiphophylla Irmsch.**
Begonia porteri H. Lév. & Vaniot
Begonia porteri var. macrorhiza Gagnep.
Begonia postarii Kiew
Begonia praeclara King
   = **Begonia decora Stapf**
Begonia praetermissa Kiew
Begonia procridifolia Wall. ex A. DC.
Begonia prolifera A. DC.
Begonia prolixa Craib
Begonia promethea Ridl.
Begonia propinqua Ridl.
Begonia prostrata Irmsch.
   ≡ **Begonia handelii var. prostrata (Irmsch.) Tebbitt**
Begonia pryeriana Ridl.
Begonia pseudoisoptera Irmsch.
   = **Begonia wrayi Hemsl.**
Begonia pseudolateralis Warb.
Begonia pubescens Ridl.
Begonia pulchra (Ridl.) L.L. Forrest & Hollingsw.
Begonia pumila Craib
Begonia pumilio Irmsch.
Begonia punbatuensis Kiew
Begonia putii Craib
Begonia pyrrha Ridl.
Begonia quercifolia A. DC.
Begonia rabilii Craib
Begonia racemosa Jack
Begonia rachmatii Tebbitt
Begonia rajah Ridl.

Begonia ramosii Merr.

Begonia randiana Merr. & L.M. Perry

Begonia reginula Kiew

Begonia renifolia Irmsch.

Begonia repanda Blume
    = **Begonia isoptera Dryand. ex Sm.**

Begonia repens Blume
    = **Begonia mollis A. DC.**

Begonia rex Putz.

Begonia rheifolia Irmsch.

Begonia rhodantha Ridl.

Begonia rhoephila Ridl.

Begonia rhombicarpa A. DC.
    = **Begonia nigritarum (Kamel) Steud.**

Begonia rhombicarpa var. lobbii A. DC.
    = **Begonia nigritarum (Kamel) Steud.**

Begonia rhyacophila Kiew

Begonia richardsoniana Merr. & L.M. Perry
    ≡ **Begonia mystacina L.B. Sm. & Wassh.**

Begonia rieckei Warb.

Begonia rimarum Craib

Begonia rizalensis Merr.

Begonia robinsonii Merr.
    ≡ **Begonia perryae L.B. Sm. & Wassh.**

Begonia robinsonii Ridl.
    = **Begonia pavonina Ridl.**

Begonia robusta Blume

Begonia robusta var. hirsutior (Miq.)
    Golding & Kareg.

Begonia rockii Irmsch.

Begonia roxburghii (Miq.) A. DC.

Begonia rubida Ridl.

Begonia rubra Blume
    = **Begonia muricata Blume**

Begonia rubrifolia Merr.

Begonia rubrovenia Hook.
    = **Begonia hatacoa Buch.-Ham. ex D. Don**

Begonia rufipila Merr.

Begonia rupicola Miq.

Begonia sabahensis Kiew & J.H. Tan

Begonia salomonensis Merr. & L.M. Perry

Begonia samarensis Merr.

Begonia sandalifolia C.B. Clarke

Begonia sarasinorum Irmsch.

Begonia sarawakensis Ridl.

Begonia sarcocarpa Ridl.

Begonia sarmentosa L.B. Sm. & Wassh.

Begonia saxatilis Blume
    = **Begonia muricata Blume**

Begonia saxifragifolia Craib

Begonia scortechinii King

Begonia scortechinii var. kunstleriana Ridl.
    = **Begonia scortechinii King**

Begonia scottii Tebbitt

Begonia serpens Merr.

Begonia serraticauda Merr. & L.M. Perry

Begonia serratipetala Irmsch.

Begonia sharpeana F. Muell.

Begonia siamensis Gagnep.

Begonia sibthorpioides Ridl.

Begonia sibthorpioides var. grandifolia
    Craib

Begonia sibutensis Sands

Begonia siccacaudata J. Door.

Begonia sikkimensis A. DC.

Begonia silletensis (A. DC.) C.B. Clarke

Begonia simulans Merr. & L.M. Perry

Begonia sinuata Wall. ex Meisn.

Begonia sinuata var. clivalis (Ridl.) Irmsch.
    = **Begonia sinuata Wall. ex Meisn.**

Begonia sinuata var. etamensis Irmsch.
    = **Begonia sinuata Wall. ex Meisn.**

Begonia sinuata var. helferi Irmsch.

Begonia sinuata var. langkawiensis Irmsch.
    = **Begonia sinuata Wall. ex Meisn.**

Begonia sinuata var. longialata Irmsch.

Begonia sinuata var. malacensis Irmsch.
    = **Begonia sinuata Wall. ex Meisn.**

Begonia sinuata var. monophylloides Irmsch.
    = **Begonia sinuata Wall. ex Meisn.**

Begonia sinuata var. pantiensis Kiew

Begonia sinuata var. penangensis Irmsch.
    = **Begonia sinuata Wall. ex Meisn.**

Begonia sizemoreae Kiew

Begonia smithiae Geddes

Begonia socia Craib

Begonia sogerensis Ridl.

Begonia soluta Craib

Begonia somervillei Hemsl.

Begonia sootepensis Craib
    ≡ **Begonia modestiflora var. sootepensis**
    (Craib) Z. Badcock ex M. Hughes

Begonia sootepensis var. thorelii Gagnep.
    = **Begonia modestiflora Kurz**

Begonia sordidissima Elmer
    = **Begonia mindorensis Elmer**

Begonia sorsogonensis (Merr.) Elmer

Begonia speluncae Ridl.

Begonia sphenocarpa Irmsch.

Begonia spilotophylla F. Muell.

Begonia stenogyna Sands

Begonia stevei M. Hughes

Begonia stichochaete K.G. Pearce
Begonia stictopoda (Miq.) A. DC.
Begonia stilandra Merr. & L.M. Perry
Begonia strachwitzii Warb. ex Irmsch.
Begonia strictinervis Irmsch.
Begonia strictipetiolaris Irmsch.
Begonia strigosa (Warb.) L.L. Forrest & Hollingsw.
Begonia subcyclophylla Irmsch.
Begonia subelliptica Merr. & L.M. Perry
Begonia subisensis K.G. Pearce
Begonia sublobata Jack
Begonia subnummularifolia Merr.
Begonia suborbiculata Merr.
Begonia subperfoliata Parish ex Kurz
Begonia subprostrata Merr.
Begonia subtruncata Merr.
Begonia subviridis Craib
Begonia sudjanae C.-A. Jansson
Begonia suffrutescens Merr. & L.M. Perry
Begonia surculigera Kurz
Begonia sychnantha L.B. Sm. & Wassh.
Begonia symbeccarii L.L. Forrest & Hollingsw.
Begonia symbracteosa L.L. Forrest & Hollingsw.
Begonia symgeraniifolia L.L. Forrest & Hollingsw.
Begonia symhirta L.L. Forrest & Hollingsw.
Begonia sympapuana L.L. Forrest & Hollingsw.
Begonia symparvifolia L.L. Forrest & Hollingsw.
Begonia sympodialis Irmsch.
Begonia symsanguinea L.L. Forrest & Hollingsw.
Begonia tafaensis Merr. & L.M. Perry
Begonia tambelanensis (Irmsch.) Kiew
Begonia tampinica Burkill ex Irmsch.
Begonia tawaensis Merr.
Begonia tayabensis Merr.
Begonia temburongensis Sands
Begonia tenericaulis Ridl.
Begonia tenuicaulis Irmsch.
    ≡ Begonia discrepans Irmsch.
Begonia tenuifolia Dryand.
Begonia tetragona Irmsch.
    = Begonia acetosella Craib
Begonia teysmanniana (Miq.) Tebbitt
Begonia thaipingensis King
Begonia thomsonii A. DC.

Begonia tigrina Kiew
Begonia timorensis (Miq.) Golding & Kareg.
Begonia tiomanensis Ridl.
    = Begonia rheifolia Irmsch.
Begonia tonkinensis Gagnep.
Begonia torricellensis Warb.
Begonia trichocheila Warb.
Begonia trichopoda Miq.
Begonia trichopoda var. polyantha (Ridl.) Irmsch.
    ≡ Begonia sychnantha L.B. Sm. & Wassh.
Begonia tricornis Ridl.
    = Begonia longifolia Blume
Begonia tricuspidata C.B. Clarke
Begonia trigonocarpa Ridl.
Begonia triradiata C.B. Clarke
Begonia trisulcata (A. DC.) Warb.
    = Begonia longifolia Blume
Begonia turbinata Ridl.
Begonia urdanetensis Elmer
Begonia urunensis Kiew
Begonia vaccinioides Sands
Begonia vagans Craib
Begonia vallicola Kiew
Begonia vanderwateri Ridl.
Begonia vanoverberghii Merr.
Begonia variabilis Ridl.
Begonia varians A. DC.
    = Begonia tenuifolia Dryand.
Begonia velutina Parish ex Kurz
    = Begonia parvuliflora A. DC.
Begonia venusta King
Begonia versicolor Irmsch.
Begonia verticillata Hook.
    = Begonia adenopoda Lem.
Begonia villifolia Irmsch.
Begonia vitiensis A.C. Sm.
Begonia vuijckii Koord.
Begonia wadei Merr. & Quisumb.
Begonia walteriana Irmsch.
Begonia warburgii K. Schum. & Lauterb.
Begonia wariana Irmsch.
Begonia weberi Merr.
Begonia weigallii Hemsl.
Begonia wenzelii Merr.
Begonia woodii Merr.
Begonia wrayi Hemsl.
Begonia wyepingiana Kiew
Begonia xiphophylla Irmsch.
Begonia yappii Ridl.

= **Begonia coriacea Hassk.**
Mitscherlichia lobbii Hassk.
≡ **Begonia lobbii A. DC.**
Mitscherlichia repens Miq.
≡ **Begonia mollis A. DC.**
Mitscherlichia stictopoda Miq.
= **Begonia stictopoda (Miq.) A. DC.**
Mitscherlichia trichopoda (Miq.) Miq.
≡ **Begonia trichopoda Miq.**
Petermannia cumingiana Klotzsch ex Klotzsch
≡ **Begonia cumingiana (Klotzsch ex
Klotzsch) A. DC.**
Petermannia fasciculata (Jack) Klotzsch
≡ **Begonia fasciculata Jack**
Petermannia geniculata (Jack) Klotzsch
= **Begonia isoptera Dryand. ex Sm.**
Petermannia racemosa (Jack) Klotzsch
≡ **Begonia racemosa Jack**
Platycentrum cathcartii (Hook. f.) Klotzsch
≡ **Begonia cathcartii Hook. f.**
Platycentrum erosum (Blume) Miq.
≡ **Begonia erosa Blume**
Platycentrum multangulum (Blume) Miq.
≡ **Begonia multangula Blume**
Platycentrum multangulum var. glabrata (Miq.)
Miq.
= **Begonia multangula Blume**
Platycentrum robustum (Blume) Miq.
≡ **Begonia robusta Blume**
Platycentrum robustum var. hirsutior Miq.
≡ **Begonia robusta var. hirsutior (Miq.)
Golding & Kareg.**
Platycentrum rubrovenium (Hook.) Klotzsch
= **Begonia hatacoa Buch.-Ham. ex D. Don**
Platycentrum rupicolum (Miq.) Miq.
≡ **Begonia rupicola Miq.**
Platycentrum tenuifolium (Dryand.) Miq.
≡ **Begonia tenuifolia Dryand.**
Platycentrum teysmannianum Miq.
≡ **Begonia teysmanniana (Miq.) Tebbitt**
Platycentrum zollingerianum Klotzsch
≡ **Begonia zollingeriana A. DC.**
Scheidweileria repens Hassk.
= **Begonia mollis A. DC.**
Sphenanthera erosa Hassk. ex Klotzsch
= **Begonia erosa Blume**

Sphenanthera multangula (Blume) Klotzsch
≡ **Begonia multangula Blume**
Sphenanthera robusta Hassk. ex Klotzsch
= **Begonia robusta Blume**
Sphenanthera robusta var. rubra Hassk.
= **Begonia muricata Bluma**
Sphenanthera robusta var. viridis Hassk.
= **Begonia robusta Blume**
Symbegonia arfakensis Gibbs
≡ **Begonia arfakensis (Gibbs) L.L. Forrest
& Hollingsw.**
Symbegonia beccarii Irmsch.
≡ **Begonia symbeccarii L.L. Forrest &
Hollingsw.**
Symbegonia bracteosa Warb.
≡ **Begonia symbracteosa L.L. Forrest &
Hollingsw.**
Symbegonia fulvo-villosa (Warb.) Warb.
≡ **Begonia fulvo-villosa Warb.**
Symbegonia geraniifolia Ridl.
≡ **Begonia symgeraniifolia L.L. Forrest &
Hollingsw.**
Symbegonia hirta Ridl.
≡ **Begonia symhirta L.L. Forrest &
Hollingsw.**
Symbegonia mooreana Irmsch.
≡ **Begonia mooreana (Irmsch.) L.L.
Forrest & Hollingsw.**
Symbegonia papuana Merr. & L.M. Perry
≡ **Begonia sympapuana L.L. Forrest &
Hollingsw.**
Symbegonia parvifolia Gibbs
≡ **Begonia symparvifolia L.L. Forrest &
Hollingsw.**
Symbegonia pulchra Ridl.
≡ **Begonia pulchra (Ridl.) L.L. Forrest &
Hollingsw.**
Symbegonia sanguinea Warb.
≡ **Begonia symsanguinea L.L. Forrest &
Hollingsw.**
Symbegonia strigosa Warb.
≡ **Begonia strigosa (Warb.) L.L. Forrest &
Hollingsw.**